Fundamentals of Green Hydrogen Supply Chain and Power-to-X

This book offers an engaging introduction to green hydrogen and its role in changing the global energy landscape. It explains theoretical and practical aspects of the green hydrogen supply chain, including production, storage, transportation, and utilization, and explores the techno-economic aspects relevant to each stage. Through hands-on examples, readers understand how green hydrogen can be implemented across various applications. It explains Power-to-X technologies, which enable the conversion of green hydrogen into eco-friendly fuels and chemical feedstocks. Designed for students and researchers, this book simplifies complex concepts and provides a comprehensive understanding of green hydrogen's benefits.

Key Features:

- Covers end-to-end the green hydrogen and Power-to-X technologies (PtX) supply chain.
- Provides a holistic and integrated approach of theoretical and practical knowledge utilizing examples and various case studies from around the world.
- Demonstrates the global potential of green hydrogen with sector-specific applications and case studies.
- Includes cutting-edge and recent developments in the green hydrogen landscape with clear visuals and technical explanations.
- Discusses forward-looking evaluations of Power-to-X technologies, hydrogen safety, and green hydrogen techno-economic aspects.
- Written in simple yet detailed language for a wide spectrum of readers.
- Tests readers' comprehension of the topics covered with self-assessment questions at the end of each chapter.

This textbook is an excellent foundational text for university students and researchers in renewable energy engineering, energy systems engineering, chemical, environmental, and mechanical engineering, and sustainable development. It is also valuable to professionals in energy, heavy industries, and the chemical and pharmaceutical sectors, and policymakers with a focus on innovative clean, renewable energy.

Fundamentals of Green Hydrogen Supply Chain and Power-to-X

Hani Muhsen, Adib Allaham,
Mohammed Al-Mahmodi, Rashed Tarawneh,
Asma Alkhraibat, and Ala'aldeen Al-Halhouli

CRC Press
Taylor & Francis Group
Boca Raton London New York

CRC Press is an imprint of the
Taylor & Francis Group, an **informa** business

Designed cover image: Shutterstock

First edition published 2026
by CRC Press
2385 NW Executive Center Drive, Suite 320, Boca Raton FL 33431

and by CRC Press
4 Park Square, Milton Park, Abingdon, Oxon, OX14 4RN

CRC Press is an imprint of Taylor & Francis Group, LLC

ISBN: 978-1-041-07176-1 (hbk)
ISBN: 978-1-041-07177-8 (pbk)
ISBN: 978-1-003-63919-0 (ebk)

DOI: 10.1201/9781003639190

Typeset in Times
by codeMantra

Access the Support Material: www.routledge.com/9781041071761

Contents

Preface

Green hydrogen and Power-to-X technologies are emerging solutions that aim to provide clean and versatile energy carriers and chemical feedstocks, thus minimizing the environmental impacts of numerous sectors worldwide including heavy industries, transportation, energy generation, and many more. In this book, the green hydrogen supply chain will be thoroughly examined to equip university students, researchers, industrialists, and policymakers with up-to-date knowledge relevant to this promising theme.

This book is primarily intended as a textbook for university students and as a reference for researchers, industrialists, and policymakers. It aims to cover the fundamental principles of the green hydrogen supply chain, foster an intuitive understanding by emphasizing the scientific theories behind each stage, and present a diverse set of technical examples to enhance the practical skills of readers.

Through the simplified explanations of concepts as well as the use of practical examples and figures, we hope that this book helps students sharpen their necessary skills to actively engage in the green hydrogen and Power-to-X theme.

About the Authors

Hani Muhsen is an associate professor in the Mechatronics and AI Department at the German Jordanian University. He earned his Ph.D. in electrical engineering from the Technical University of Chemnitz, Germany, and has immense experience in electrical engineering and renewable energy systems. He has published more than 25 papers in international journals and has worked on several projects related to the digital twin, peer-to-peer trading, microgrid modeling and optimization, and hydrogen technology.

Adib Allaham is an associate professor in the Mathematics, Physics, and Electrical Engineering Department at Northumbria University, Newcastle upon Tyne, UK. He earned his Ph.D. in control engineering from the University of Joseph Fourier, Grenoble, France, and has over 15 years of research experience. He has published over 30 papers in international journals on renewable energy solutions.

Mohammed Al-Mahmodi is a former research assistant at the Smart Grid Lab at the German Jordanian University in Amman, Jordan. He earned his M.Sc. degree in mechanical engineering/renewable energy from the University of Jordan. He worked in research for about 5 years at Yarmouk University and the German Jordanian University/Smart Grid Lab. He has published ten journal articles on green energy topics.

Asma Alkhraibat is a research assistant at the Smart Grid Lab at the German Jordanian University in Amman, Jordan. She earned her M.Sc. degree in computer engineering from the German Jordanian University. She is interested in microgrid applications, renewable energy resources, real-time implementation, energy management, and fault restoration. She has published nine journal and conference articles on green energy topics.

Rashed Tarawneh is a research assistant at the Smart Grid Lab at the German Jordanian University (GJU) in Amman, Jordan. He earned his M.Sc. degree in environmental and renewable energy engineering from the GJU. He has published four journal articles on green energy topics.

Ala'aldeen Al-Halhouli is the president of the German Jordanian University, and he is a professor of mechanical engineering at the Mechatronics Engineering Department. He earned his Ph.D. degree in mechanical engineering from the University of Jordan and a habilitation in microfluidics from the Technische Universität Braunschweig. He has published more than 70 papers in international journals and has received several awards, including the GJU Distinguished Researcher Award, Jordan (2014); Best Poster Award at Georgetown University, USA (2009); and the Cray Award for Microtechnology, Germany (2007).

1 Introduction to Green Hydrogen and Power-to-X

1.1 INTRODUCTION

Greenhouse gases (GHGs) are Earth's primary source of heat retention, causing global warming and climate change. Carbon dioxide (CO_2), methane (CH_4), nitrous oxide (N_2O), and fluorinated gases are considered GHGs. There are many primary sources of GHG emissions. One of these sources is the energy production and utilization sector, which includes burning coal, oil, and natural gas for electricity generation, heat, and transportation. It is the most significant contributor to global GHG emissions, accounting for about 72%. Industrial processes such as cement production and chemical manufacturing also release GHGs. These sectors are responsible for about 12% of global emissions. Also, agricultural activities, such as livestock farming and rice cultivation, contribute about 11% of global emissions due to methane and nitrous oxide produced by these activities. Burning fossil fuels for transportation (cars, trucks, ships, and airplanes) is considered a significant source of CO_2 emissions, and the transportation sector contributes 15% of the total CO_2 emissions. See Figure 1.1 [1].

Global warming is undeniably impacting the environment, with predictions of more severe consequences in the future. For instance, global warming can cause rising sea levels due to melting glaciers and ice sheets, threaten coastal communities, and displace millions. Also, it can cause extreme weather events, such as heat waves, droughts, floods, and storms. Global warming can disrupt the ecosystems by causing severe wildfires. Global warming has accelerated since 2010 by more than 50% over 1970–2010, with a warming rate of 0.18°C/decade [2]. Therefore, it was necessary to take action to reduce global warming and to collaborate to reduce carbon dioxide emissions.

1.2 THE PARIS AGREEMENT: A FRAMEWORK FOR GLOBAL CLIMATE ACTION

The 2015 Paris Agreement, a key component of the United Nations Framework Convention on Climate Change (UNFCCC), is pivotal in international efforts to address climate change. It creates a structure for global collaboration to fight climate change, with the ambitious goal of keeping global temperature rise below 2°C—ideally below 1.5°C—compared to preindustrial times [3]. This goal highlights the urgent need to lessen the severe effects of climate change, such as rising sea levels, more extreme weather, loss of biodiversity, and dangers to human health and well-being.

DOI: 10.1201/9781003639190-1

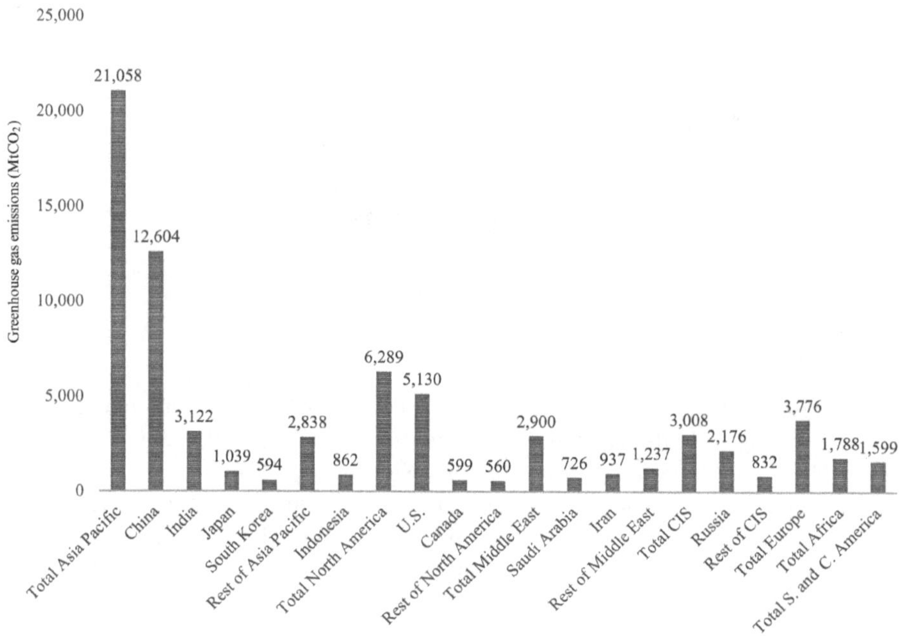

FIGURE 1.1 Carbon emissions from energy production and utilization in 2023. (Adapted from Energy Institute [1].)

The Paris Agreement signifies a crucial turning point toward a more collaborative and ambitious global approach to combating climate change. Its significance is rooted in several key aspects. Firstly, it achieves near-universal participation, uniting countries in a shared framework and fostering a sense of collective responsibility. Secondly, it establishes a clear long-term vision with a roadmap for achieving a low-carbon future, guiding policy and investment decisions. Thirdly, based on nationally determined contributions (NDCs), its dynamic framework allows flexibility while encouraging continuous improvement and increased ambition. However, the agreement faces significant challenges. A substantial "ambition gap" exists between current NDCs and the necessary emission reductions to meet temperature targets. Furthermore, translating national commitments into concrete actions requires overcoming political, economic, and technological hurdles. Finally, ensuring a just and equitable transition that addresses the needs of vulnerable countries and communities is essential for the agreement's ultimate success.

1.3 RENEWABLE ENERGIES: OPPORTUNITIES AND CHALLENGES

Meeting the sustainable global power demand is now considered the primary concern in different sectors. Renewable energy is produced by natural and sustainable resources and converted to usable power using specific technologies consumed by end-user levels. Solar, wind, geothermal, and biomass energy are leading examples of widespread renewable energy systems (RESs). Directly or indirectly, most RESs rely on sunlight. In such a solar panel case example, solar radiation and temperature are the main factors affecting energy production operations. Also, wind speed is the primary factor in wind farms, where the sun's heat drives the air and produces the required wind movement. Also, sunlight and wind are caused by the water evaporation operation, which, as a result, turns to rain and snow that flow into rivers, resulting in hydroelectric power production.

Renewable energy deployments involve many advantages that affect their prominent role in our society; these advantages include overall cost reduction, fewer maintenance requirements, toxic gas emission reductions, and assisting remote regions without grid access by providing electricity supplies to increase opportunities for commercial and economic growth. Hence, and in general, renewable energy projects distinctly meet the overall environmental and economic benefits that encourage all stakeholders (utilities, regulators, and customers) to collaborate in optimizing and spreading RES applications such as green hydrogen production.

Radiation and heat parameters of the sun mainly contribute to the production process of solar energy, which involves providing electricity through photovoltaic (PV) systems or hot water through thermal systems or using the innovative concept represented by concentrated solar power (CSP). CSP systems are mainly based on the use of mirrors or lenses in order to concentrate the sunlight and thereafter convert it into heat intensively. Furthermore, one of the valuable features of CSP technology is that CSP systems can reserve the heat in molten salts, which gives the CSP systems a preference in providing continuous electricity generation regardless of the time during the day. Ivanpah Solar Power Facility and solar energy generating systems (SEGS) are the main CSP projects in the United States, with a capacity of 395 and 361 MW, respectively. Another example of a solar thermal power project is the Ouarzazate Solar Power Station in Morocco, which has a capacity of 360 MW [4,5].

The advent of RESs as distributed energy resources (DERs) has facilitated and enhanced the spread of several applications in many regions worldwide. For instance, social, economic, and environmental impacts primarily represent the positive impacts of RES integration in the utility grid. Both economic and environmental impacts are generally linked to GHG emission reduction, operational cost saving, and decreasing the expensive investments of power networks and distribution systems. While the social impacts of RES integration are linked to the increase of acceptance and awareness of RES deployments by identifying the direct benefits of this integration in our daily life, another example of the social impacts is the ability to create job and education opportunities.

Therefore, renewable energy resources have significant potential to power the future with clean and affordable energy. According to the International Energy Agency (IEA), 2023 global renewable energy capacity additions are anticipated to increase to over 440 GW. This increase is attributed to the increasing competitiveness of renewable sources compared with traditional fossil fuel alternatives. Solar PV capacity is a major contributor, constituting two-thirds of the global increase in renewable capacity.

In 2024, as solar panel prices decrease, more solar power is expected to increase worldwide. Large markets such as China, Europe, the United States, and India will invest in more solar power [6]. However, under the current policies, investing in offshore wind power might slow down by 5%. The cumulative global renewable capacity is expected to exceed 500 GW by 2024; in an accelerated scenario, it could reach 550 GW. Figure 1.2 gives more insights into the world's renewable energy status by technology [7].

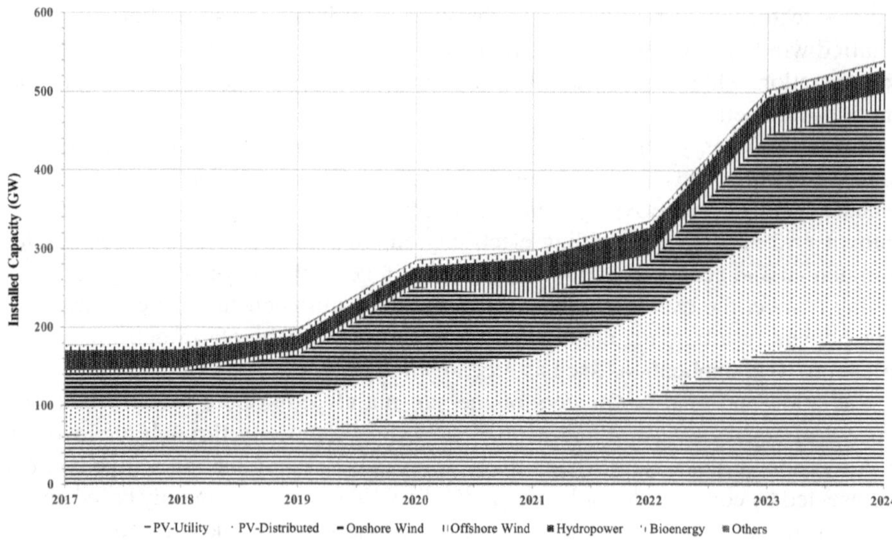

FIGURE 1.2 Net renewable electricity capacity additions by technology, 2017–2024. (Adapted from International Energy Agency [7].)

However, due to their intermittent nature, these resources cannot be considered completely reliable for achieving net-zero emissions in the energy sector. Thus, complementary technology is necessary to achieve independence of these resources rather than dependency on fossil fuels. However, replacing fossil fuels in all sectors will remain a big concern. Effective energy storage solutions are crucial to overcoming intermittency. Batteries, pumped hydro storage, and other technologies can store excess energy generated during peak production periods and release it when demand exceeds generation. However, these storage solutions have challenges, including cost, efficiency, and environmental impact. Emerging technologies such as green hydrogen and Power-to-X (PtX), which will be addressed in the next section, offer promising solutions to address the limitations of renewable energy. Green hydrogen is produced using renewable energy sources to split water into hydrogen and oxygen, providing a clean and versatile energy carrier. PtX technologies convert renewable electricity into other forms of energy, such as synthetic fuels or chemicals, which can be stored and transported more easily. Green hydrogen and PtX technologies offer significant advantages in addressing the challenges associated with renewable energy sources.

Firstly, they enable the long-term storage of large amounts of renewable energy, overcoming the intermittency issue posed by sources such as solar and wind power. This stored energy can be dispatched when needed, ensuring a consistent and reliable supply. Secondly, green hydrogen and PtX have versatile applications across various sectors. Green hydrogen can be used directly in transportation, industry, and power generation, while PtX technologies can create synthetic fuels suitable for aviation and other sectors that are difficult to decarbonize. Finally, these technologies can contribute to grid balancing by providing flexible demand and supply, ensuring stability even with fluctuating renewable generation. This flexibility enhances the reliability and resilience of energy systems that incorporate many renewable sources.

To this end, clean energy carriers and green fuels are needed to replace conventional ones. Green hydrogen is a significant candidate, as it could be used in different sectors and converted to synthetic fuel. Hydrogen is essential because it is an energy carrier and a green alternative to fossil fuels. It can be used directly in fuel cells and internal combustion engines or converted to other forms of energy carriers such as ammonia, methane, and other fuels through net-zero emission processes, making it a valuable advantage in transitioning to cleaner energy solutions.

Hydrogen (H_2) is a chemical element that is typically found in compounds such as water (H_2O) and hydrocarbons (C_nH_m); it is crucial to extract pure hydrogen through a separation process known as "hydrogen production," which requires an energy source.

Green hydrogen is a term for hydrogen gas produced using renewable energy sources, such as wind, solar, or hydroelectric power, as an alternative to hydrogen produced from fossil fuels, such as natural gas. The "green term" indicates environmentally friendly and sustainable production methods utilizing renewable energy resources.

Renewable energy generation is intermittent, dependent on weather conditions, and often does not cover the energy demand. Green hydrogen offers a valuable and efficient means of energy storage or acts as an energy carrier. It enables excess renewable electricity to be converted into hydrogen through electrolysis. This hydrogen can then be stored for later use.

No GHG emissions characterize green hydrogen, which is produced using renewable energy and water. In contrast, conventional hydrogen production methods, such as steam methane reforming or coal gasification, release significant carbon emissions. Thus, hydrogen has the potential to electrify different industries, such as heavy industry (e.g., steel and cement production) and transportation, which require energy-dense fuels. Green hydrogen can directly replace fossil fuels in these applications and is expected to reduce carbon emissions substantially.

1.4 POWER-TO-X CONCEPT

PtX is a technology that converts electricity from renewable sources, such as wind, solar, hydro, and geothermal power plants, into valuable energy carriers and chemicals, generically referred to as "X" [8]. This conversion can help capture renewable energy and enable decarbonization in multiple domains. The PtX concept mainly relies on utilizing green hydrogen along with captured carbon from industrial applications and nitrogen produced from organic processes to produce a wide range of fuels and chemical feedstocks.

The production of ammonia (NH_3) results from combining hydrogen (H_2) and nitrogen (N_2), where nitrogen is extracted from the air. Ammonia is an essential feedstock in the fertilizer industry and is also useful in the mining, pharmaceutical, and marine shipping industries.

In addition, renewable synthetic hydrocarbons (C_nH_m) are created using carbon (C) from renewable sources. The term renewable carbon is used when carbon is generated from nonfossil renewable materials such as biomass or reclaimed from industrial emissions through carbon capture and utilization (CCU) technology.

1.5 THE ROLE OF GREEN HYDROGEN AND PtX

Hydrogen has wide applications and can participate in decarbonizing several sectors. Electricity generation is one sector in which hydrogen is converted to electrical energy using devices known as fuel cells; more details will be introduced in Chapter 7. Hydrogen can also be used as energy storage, such as batteries, with several benefits over batteries. This topic will be discussed in Chapter 5. Further, hydrogen is equally important in transportation methods, which rely on hydrogen combustion. On the other hand, hydrogen-based fuel cells will also contribute to electrifying part of the transportation sector by generating electricity onboard vehicles, allowing vehicles to operate using hydrogen without combustion. In the chemical industry, hydrogen can be used in synthetic fuel production, oil refinery, steel manufacturing, baking, and cooking. Green hydrogen's application pathways extend far beyond this, one of its important derivatives being green ammonia, which is used for innovative, eco-friendly fertilizers and enabling energy storage. Green hydrogen's application pathways are shown in Figure 1.3.

Electric fuels, also called electro-fuels, are produced by combining green hydrogen and captured carbon alongside carbon capture and storage (CCS) technology. These fuels are comparable to conventional ones in their constituents, except for the difference in the hydrogen and carbon sources. The hydrogen of electro-fuels must be derived from water electrolysis, which requires renewable energy to break water into hydrogen and oxygen. They include methane, methanol, synthetic gas, and familiar fuels such as diesel and gasoline. For instance, electro-methane is manufactured using an existing technique called the Sabatier process. This process generates methane and water by reacting hydrogen with carbon dioxide at high temperatures and pressure. Syngas is another significant hydrogen derivative that can create numerous fuels. Moreover, syngas can be transformed into methanol and serve as a precursor

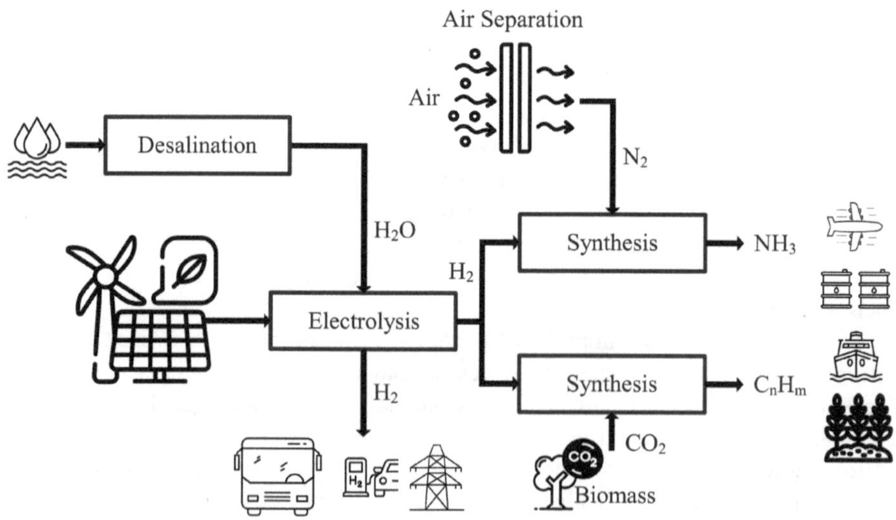

FIGURE 1.3 Green hydrogen and PtX uses.

for producing long-chain hydrocarbons and waxes using the Fischer–Tropsch process [9]. These hydrocarbons may undergo additional refinements to produce lighter, high-quality goods such as diesel, gasoline, and kerosene.

1.6 GREEN HYDROGEN POTENTIAL IN THE WORLD

The situation of low-emission hydrogen production is evolving rapidly. Although promising projects are on the rise, cost challenges might delay the deployment.

According to the IEA, the announced projects have substantial growth potential for low-emission hydrogen production by 2030. If all these projects were established, annual low-emission hydrogen production could reach 49 million tons (Mt) by 2030 [10]. However, it's clear that cost challenges are delaying widespread deployment, and 17 Mt is still in the early development stages [11]. This highlights the need for collaboration and the importance of each stakeholder's role in overcoming these challenges.

The current potential production from announced projects is 50% larger than indicated in the IEA's Global Hydrogen Review 2022. Notably, only 4% of this potential has reached a final investment decision (FID), doubling from the previous year to nearly 2 Mt. Electrolysis and low-emission electricity contribute to 27 Mt. At the same time, 10 Mt is derived from fossil fuels with carbon capture, utilization, and storage (CCUS). This emphasizes the significant growth in low-emission hydrogen initiatives. However, substantial obstacles, particularly in cost, must be addressed for widespread implementation.

A diverse landscape of green hydrogen projects has been announced worldwide, each at varying stages of development. In Australia, the Western Green Energy Hub, developed by InterContinental Energy, CWP Global, and Mirning Green Energy Ltd, is even more ambitious, boasting a planned 50 GW capacity and a targeted 3.5 Mt of annual green hydrogen production [12].

In Ostend declaration, Germany, Denmark, the Netherlands, and the United Kingdom have jointly set the target of achieving an electrolysis capacity of around 30 GW (onshore and offshore) by 2030. They want to expand their production further by 2050 [13].

Europe also showcases significant initiatives, such as the BrintØ Green Hydrogen Island in Denmark and Aqua Ventus in Germany. With planned capacities of 10 GW and expected production of 1 Mt of green hydrogen per year, both projects are currently in the planning phase [13,14]. Copenhagen Infrastructure Partners (CIPs) is leading the development of the BrintØ Green Hydrogen Island, while RWE Renewables and Aqua Ventus Förderverein are behind Aqua Ventus. While these large-scale projects are poised to impact the green hydrogen landscape significantly, smaller-scale projects are already operational, primarily in China [15].

China has launched several mega projects to produce green hydrogen. For instance, the Ningxia Baofeng Energy Group operates a 200 MW plant with a 27,000-ton annual green hydrogen production capacity. Also, Sinopec runs the Xinjiang Green Hydrogen Plant with a 260 MW capacity and 20,000-ton annual production [15].

In northwestern Mauritania, Project Nour, developed by Chariot Green Hydrogen and TE H2, aims for a 1.6 GW capacity and 150,000 tons of yearly green hydrogen production [16]. This project is currently in the planning stage.

In December 2023, Indonesia unveiled its National Hydrogen Strategy. The strategy bolsters renewable energy development, achieves decarbonization goals, and establishes hydrogen for export. Indonesia's renewable energy potential of nearly 3.7 TW makes it a global leader in green hydrogen production, with applications spanning industrial, transportation, and power generation for domestic and export sectors [17].

Chile, a country rich in renewable energy, is poised for significant growth. With a potential capacity of more than 1800 GW, over 70 times the current installed capacity, Chile's national green hydrogen strategy is set to develop 5 GW of electrolysis capacity by 2025, produce the most cost-efficient green hydrogen by 2030, and aim to be among the top three exporters by 2040 [18].

Argentina plans to produce 5 Mt of low-carbon hydrogen by 2050, 1 Mt for domestic use, and 4 Mt for export outside the country. This will require installing 55 GW of renewable energy sources and electrolysis with a more than 30 GW capacity [19].

Brazil sets ambitious goals, aiming to establish low-carbon hydrogen pilot plants across all regions of Brazil by 2025. Looking ahead to 2030, the objective is to consolidate Brazil as the most competitive global producer of hydrogen, followed by establishing low-carbon hydrogen hubs by 2035. The Ministry estimates the country's technical potential to produce 1.8 Gt of hydrogen annually by 2050, representing more than 14 times the world demand for hydrogen in 2018 [20].

Columbia plans to develop between 1 and 3 GW of electrolysis, which implies installing at least 1.5–4 GW of RES throughout the country. Low-carbon hydrogen production aims to be used domestically in different sectors, such as transportation, refining, steel production, and other minor industrial uses [21].

Mexico announced its potential to produce 230 kt of green hydrogen for domestic use by 2030 and 2.7 Mt by 2050. Therefore, it is estimated that 79 GW of renewable capacity and 51 GW of electrolysis must be installed by 2050 [22].

Panama's plan targets substantial production of green hydrogen to produce 500 kt by 2030. The plan aims to scale up to 2 Mt by 2040 and potentially reach 4 Mt by 2050 under optimal conditions. Significant infrastructure development is crucial to achieving these production goals, focusing on building 4.2 GW of electrolysis capacity by 2040. The strategy intends to use green hydrogen production to decarbonize critical sectors, including heavy transport, maritime, and aviation [23].

Many other countries focus on accelerating the transition toward low-carbon hydrogen for domestic applications to reduce carbon emissions or for import purposes. This section is focused on the main countries with high renewable energy potential. Finally, with increasing collaboration between countries worldwide, green hydrogen technology is expected to play an essential role in achieving global climate goals and transitioning to a more sustainable energy future.

1.7 GREEN HYDROGEN POTENTIAL IN THE MIDDLE EAST

The global shift toward clean energy goes beyond just generating it; it is about efficiently storing, transporting, and supplying energy globally. The Middle East is a potential top candidate in the global green hydrogen market due to its abundant renewable resources, strategic geographic location, and substantial investments in logistical infrastructure. This unique position of the region enables to meet the rising demand for green hydrogen worldwide.

NEOM Green Hydrogen Project in Saudi Arabia is considered one of the world's largest planned green hydrogen initiatives. This enterprising project aims to produce green ammonia using renewable energy sources for export. This project will position Saudi Arabia as a key player in the global hydrogen market. In addition, the country's Vision 2030 framework underscores its commitment to becoming a leading hydrogen producer and exporter. Saudi Arabia plans to use 4 GW of renewable energy to produce green ammonia in Neom. The country also aims to allocate a significant part of the $110 billion Jafurah field for blue hydrogen production [24]. In January 2022, Saudi Arabia launched eight agreements for hydrogen plants. Saudi Arabia is deepening cooperation with China on blue hydrogen [25], while South Korean companies are involved in various green hydrogen projects in the region [26].

Green hydrogen production in the United Arab Emirates (UAE) is advancing significantly. The UAE is working on creating a strong hydrogen economy, as presented by its National Hydrogen Strategy, which sets out precise production goals and plans for creating hydrogen hubs.

Several green hydrogen projects are underway in Abu Dhabi and Dubai, focusing on domestic use and export through strategic partnerships with international companies. The country's first green hydrogen plant is now operational, and plans are in place for additional hydrogen derivative facilities [27].

With an emphasis on green and blue hydrogen production, Oman is another significant actor in the hydrogen landscape of the area [28]. A collaborative platform called the Hy-Fly Alliance unites stakeholders to create hydrogen initiatives, primarily focusing on green hydrogen and the possibility of blue hydrogen [29]. Oman is strengthening its position in the regional hydrogen market by utilizing its plentiful solar and wind resources to create several large-scale green hydrogen projects. In order to feed Europe and Asia, Oman intends to build a large wind-and-solar plant with 25 GW of power and an electrolyzer [30].

Egypt is actively working to use its potential in renewable energy resources to create green ammonia manufacturing for export purposes. The country is also investigating how green hydrogen can be used in steel production to decarbonize the steel sector [31]. These initiatives demonstrate Egypt's dedication to using green hydrogen as an energy source and a way to lower emissions in heavy industries.

Morocco has comprehensively developed a green energy ecosystem that includes green hydrogen production for export and domestic use [32]. Morocco has successfully produced green steel using renewable energy, marking a significant milestone in decarbonizing the steel industry. This achievement highlights the potential of green hydrogen to transform various industrial processes and contribute to a more sustainable future [33].

Jordan is known for its high potential for renewable energy production. However, it suffers from water scarcity and is still in the early stages of involvement in green hydrogen and PtX projects. Several memoranda of understanding (MOU) have recently been signed with different investors to initiate hydrogen projects in Jordan, including green ammonia production. However, the issue of water shortage shall be considered when selecting the location and determining the capacities [34].

Other countries in the MENA region, such as Algeria and Tunisia, are exploring green and blue hydrogen production in collaboration with European partners [35]. These efforts demonstrate a growing regional interest in hydrogen as a clean energy solution and highlight the potential for international cooperation to accelerate the development of a global hydrogen economy.

1.8 DEVELOPMENT ASPECTS OF GREEN HYDROGEN

Different aspects must be considered when investing in green hydrogen technology. The technical aspect of green hydrogen is rapidly developing, as demonstrated by advancements in electrolysis technologies such as proton exchange membranes (PEM) and alkaline electrolyzers.

PEM electrolyzers are gaining prominence due to their high efficiency, compact design, and ability to operate under fluctuating power inputs, making them well-suited for integration with intermittent renewable energy sources. Recent research has focused on developing novel membrane and catalyst layer materials to enhance durability, reduce costs, and improve overall performance. Efforts are also directed toward reducing or replacing iridium with less expensive and more abundant materials while maintaining performance. In contrast, alkaline electrolyzers are a more mature technology known for their lower capital costs and longer lifespans. These systems typically use liquid alkaline electrolytes, such as potassium hydroxide (KOH).

Advancements in alkaline electrolysis are focused on improving the efficiency of the electrodes and the development of novel diaphragm materials that can withstand higher operating temperatures and pressures, leading to increased hydrogen production rates. Solid oxide electrolyzer (SOE) is a promising but less mature technology. SOEs operate at very high temperatures (700°C–1000°C) and offer the highest electrical efficiencies for hydrogen production. The high operating temperatures allow waste heat from industrial processes to be utilized, further improving overall system efficiency [36]. The steam is then electrolyzed to produce hydrogen. These innovations are essential for enhancing the efficiency and cost-effectiveness of green hydrogen generation.

In addition, efficient and safe storage and transportation of hydrogen are critical for its widespread adoption. Compressed gas storage is the most common method involving storing hydrogen at high pressures (up to 700 bar) in specialized tanks. The second storage option is storing hydrogen in liquid form at cryogenic temperatures (–253°C), which offers higher energy density but requires significant energy for liquefaction and specialized insulated tanks [37]. Hydrogen's most promising storage method is materials-based storage, such as metal hydrides, which offers a promising avenue for compact and safe hydrogen storage. However, challenges remain in achieving high storage capacities, fast kinetics, and suitable operating temperatures.

Integrating renewable energy sources, such as solar and wind, into the electrolysis process is another critical technical development, ensuring green hydrogen's sustainability and environmental benefits. Additionally, ongoing research focuses on improving storage methods and transportation technologies to facilitate the efficient handling and distribution of green hydrogen. Further, economic considerations play a vital role in developing green hydrogen. Cost-reduction strategies are essential to enhancing the technology's attractiveness compared with traditional sources.

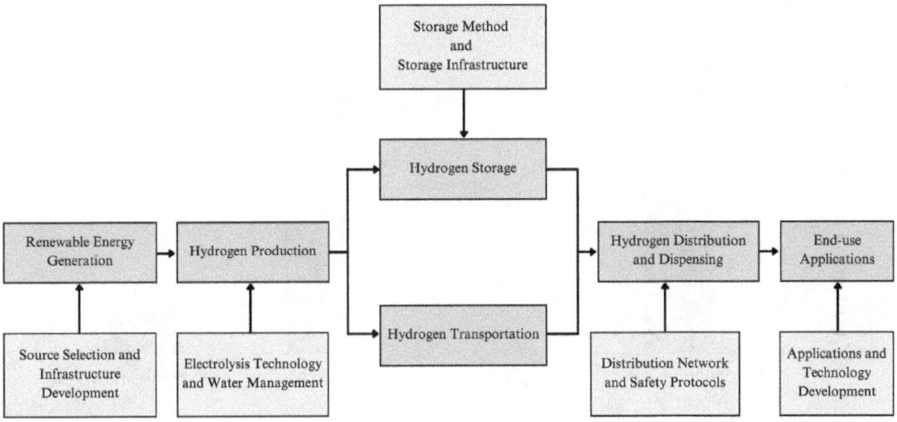

FIGURE 1.4 The green hydrogen supply chain.

The availability of investments, government incentives, and private financing mechanisms is crucial for scaling up green hydrogen projects until this technology becomes competitive. The lack of existing hydrogen transportation and distribution infrastructure is a significant economic and logistical challenge. Building new pipelines, repurposing existing natural gas infrastructure, and developing refueling stations for fuel cell vehicles require substantial investment and careful planning.

The social aspect of green hydrogen is also important. It involves various sides, such as job creation, and people benefit from such investments. Moreover, community awareness programs are essential to ensuring social acceptance and understanding of green hydrogen technologies and their role in energy transition [38]. Another point to consider: in regions with water scarcity, concerns about water usage for electrolysis may arise. It is important to emphasize that the amount of water required for electrolysis is relatively small compared with other industrial uses and agricultural practices. Moreover, advancements in water purification and desalination technologies, potentially powered by renewable energy, can further mitigate concerns about water stress.

The development of green hydrogen requires a holistic approach that addresses the technical, economic, and social dimensions of this emerging energy carrier. Continued innovation in electrolysis technologies, coupled with supportive policies, private sector investment, and community engagement, will be critical to unlocking the full potential of green hydrogen as a key enabler of a sustainable energy future. See Figure 1.4.

1.9 POTENTIAL CO_2 EMISSION REDUCTIONS

The expected reduction of CO_2 emissions due to green hydrogen and PtX technologies is substantial. Still, it is crucial to understand that it depends on various factors, including the scale of deployment, the specific PtX pathways considered, and the sectors to which they are applied. For instance, the Hydrogen Council, in collaboration with McKinsey, projects that hydrogen (including both green and blue) could abate up to 6 Gt of CO_2 emissions annually by 2050, which represents approximately 20%

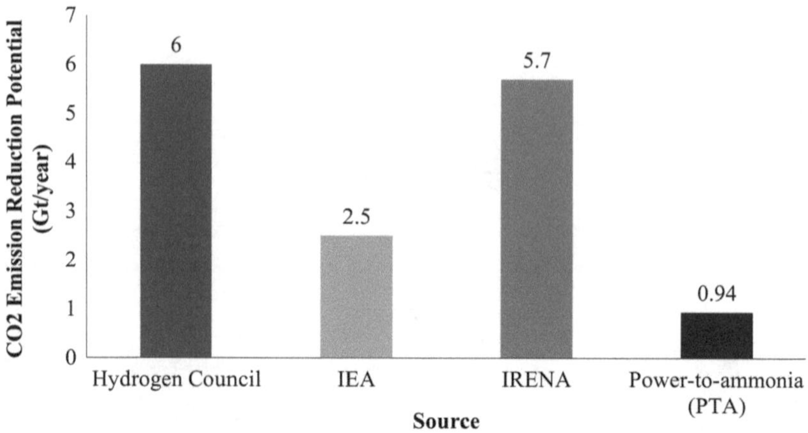

FIGURE 1.5 CO_2 emission reduction potential (Gt/year) by source.

of the reductions needed to align with global climate goals [39]. Similarly, the IEA suggests in its "Net Zero by 2050" roadmap that low-carbon hydrogen could reduce around 2.5 Gt of CO_2 per year by the same timeframe.

The International Renewable Energy Agency (IRENA) offers a more conservative estimate, indicating that green hydrogen and its derivatives could reduce global CO_2 emissions by up to 10% by 2050. Power-to-ammonia is another sector where the production of 620 Mt of green ammonia could help reduce emissions by up to 940 Mt CO_2-eq by 2050. See Figure 1.5.

The most substantial CO_2 reductions from green hydrogen and PtX are anticipated in sectors currently heavily reliant on fossil fuels and difficult to electrify directly. These "hard-to-abate" sectors include heavy industries such as steel, cement, and chemicals, where green hydrogen can replace fossil fuels as a source of high-temperature heat or as chemical feedstock [38]. In the transportation sector, long-haul trucking, shipping, and aviation can benefit significantly from green hydrogen-derived fuels (e-fuels), offering a lower-carbon alternative to conventional fuels where battery-electric solutions are not yet feasible due to weight, range, and refueling time constraints. Green hydrogen can also contribute to the decarbonization of the power sector by storing excess renewable energy over extended periods, helping to balance grid fluctuations, and replacing fossil-fuel-based backup power.

Green hydrogen offers a sustainable alternative in heavy industries such as steel and cement; green hydrogen presents a sustainable alternative for heavy industries such as steel, cement, and chemical production, which traditionally rely on fossil fuels for high-temperature heat and chemical feedstocks. The long-haul transportation sector—including trucking, shipping, and aviation—also faces challenges in adopting battery-electric solutions due to range, weight, and refueling issues. Green hydrogen can effectively address these challenges, especially for heavy-duty vehicles and long-distance travel.

In addition to its direct applications, green hydrogen plays a crucial role in energy storage and grid balancing by converting excess renewable electricity generated during peak production times into hydrogen. The stored hydrogen can be converted to electrical energy through fuel cells during periods of low renewable generation or high demand. This feature can enhance the reliability and stability of power systems as they increasingly integrate renewable energy sources.

While large-scale decarbonization efforts typically focus on heavy industry and transportation, green hydrogen and its derivatives, such as PtX technologies, also hold potential in other areas [40].

These include residential heating, distributed power generation, niche industrial applications, and fertilizer production. Although these sectors may not be the primary drivers of decarbonization, their adoption of green hydrogen technologies can still contribute to overall emission reduction goals.

1.10 DECARBONIZING HARD-TO-ABATE SECTORS

Several sectors are hard to decarbonize and are known as hard-to-abate sectors. These include long-haul transportation—such as aviation, heavy-duty trucking, and shipping—as well as heavy industries such as chemicals, cement, and steel production. Both transportation and heavy industries are expected to have low residual emissions by 2050 [41]. The most significant way to decarbonize the power industry is to redevelop plants to use zero- and low-carbon fuels or to capture and store carbon using CCUS advanced technologies. These sectors pose significant decarbonization challenges due to technological limitations, economic and geopolitical concerns, and substantial energy demands.

While the energy transition is making significant strides in sectors such as electricity supply and passenger road transport, specific industrial and transport subsectors are proving more difficult to decarbonize [42]. Despite the challenges posed by hard-to-abate sectors, combining technological solutions can facilitate their decarbonization. Therefore, the hard-to-abate industries should commit to CCS projects to keep pace with global decarbonization efforts [37].

Electrification is one option for decarbonizing systems or sectors. It depends on replacing fossil-fuel-based technologies with those powered directly by renewable electricity, which can significantly reduce emissions in various applications. For instance, electrification solutions include electric arc furnaces for steelmaking [43]. Heavy-duty battery-electric trucks (BETs) can electrify the sector and reduce CO_2 emissions [44]. Hence, electrification will be the best option for any application or sector that can be electrified, even if the electrification covers part of the system according to the system constraints [45].

Another option in the decarbonization efforts is using bioenergy and synthetic fuels. This option requires scaling up sustainable bioenergy production and can benefit decarbonizing sectors such as shipping and aviation. These fuels can also provide chemical feedstock and serve as a carbon source for synthetic fuels [46–48].

Finally, indirect electrification is an option that depends on hydrogen production via electrolysis. This can enable deep emission cuts in hard-to-abate sectors. For instance, green hydrogen can be used as a reductant in steel production [49], fuel for shipping and aviation [50,51], and feedstock for the chemical industry [52].

1.11 GREEN HYDROGEN AND PtX CHALLENGES

Although green hydrogen and PtX present opportunities for many industries to mitigate energy use and carbon emissions and meet renewable energy goals, these innovations face major techno-financial barriers. These include financial, sociotechnical, and policy challenges [32,53–55]. One such barrier is the relatively high production cost that green hydrogen incurs compared with fossil-fuel-based hydrogen. Green hydrogen costs substantially more than hydrogen produced from fossil fuels. Estimates suggest that green hydrogen costs between $3 and $10/kg, depending on the location and the quality of available renewable energy sources. On the other hand, grey hydrogen production can drop to $1 with favorable natural gas pricing, while blue hydrogen produced from natural gas with carbon capture is roughly $2. The expensive capital investment required for losing electrolysis as a process is also a primary driver of the exorbitant cost of green hydrogen. While there is an expectation for electrolyzer costs to decrease significantly by 2030, the subsidization of green hydrogen still creates strong limitations due to the project's capital intensiveness.

This is even more complicated by the sensitivity of renewable electric power, an input for green hydrogen production, to capital expenditures and financing terms [56–58].

Moreover, there are additional exigencies for the storage and transportation of hydrogen. This means that new transportation and storage infrastructure has to be built. These expenses will be incorporated into the overall cost of green hydrogen, thereby incentivizing the substitution with competitively priced new storage and transport materials. Environmentally, green hydrogen technology potentially affects water and land, notably in water-stressed countries. Additionally, there are known gaps in environmental issues, such as how to move carbon-negative transport and dispose of unsustainable brine waste from seawater desalination in a sustainable manner.

While these issues should be addressed, policies, including carbon pricing, energy tax modifications, and revisions to fossil fuel subsidies, must also be implemented. These feed-in tariffs, along with tax incentives and local content requirements, prompt the use of low-carbon energy alternatives and the enhanced participation of domestic industries [59,60]. These practices would allow stakeholders to participate in the transition process actively. Moreover, the establishment of bilateral and multilateral trade agreements must be accompanied by universally accepted methodologies for hydrogen certification in order to enhance consumer trust, as well as trade, while guaranteeing the hydrogen's quality and provenance. Such actions and policies are impossible without intra-department collaboration, the participation of the relevant regulatory and standardization institutions, and coordination with other countries.

From a social perspective, the fear surrounding this new fuel source remains an issue that must be addressed. This may be accomplished with greater public education regarding the safety practices that accompany technology deployment. Conversely, workers in this area must be adequately knowledgeable concerning the scope of national and international standards and their practical aspects. In addition, there is a need for postgraduate and professional skills training to facilitate the move toward green hydrogen and PtX technologies. Hence, these technologies, integral to the energy transition, made it important to have clear safety, compatibility, and market provisions for PtX and hydrogen technologies. As such, standards must be set for the design and construction of the necessary systems such as pipelines, refueling

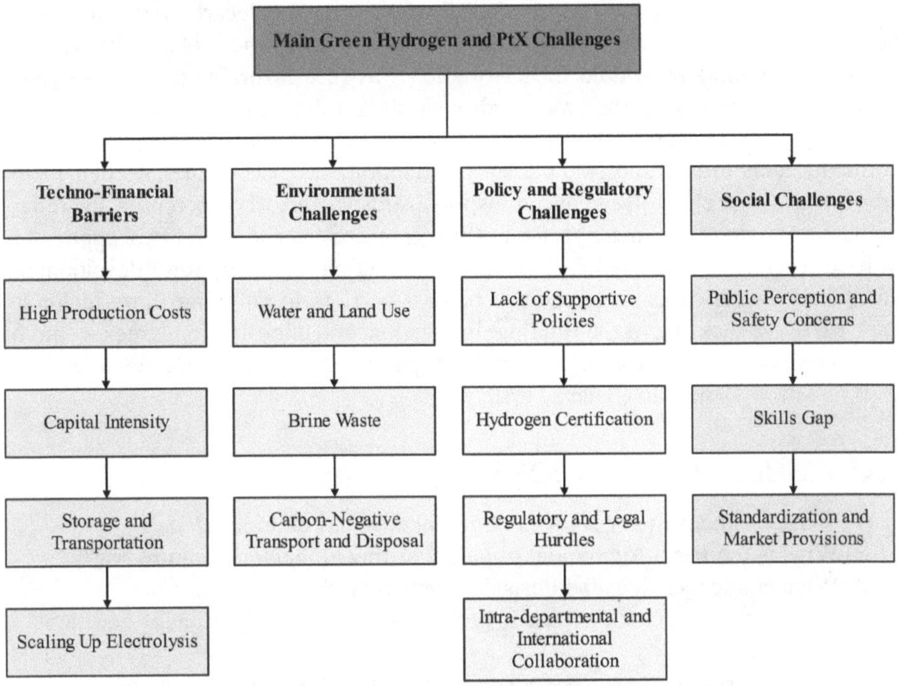

FIGURE 1.6 The main challenges for the widespread transition toward green hydrogen and PtX technologies.

stations, and transportation facilities. These have given rise to market changes, for example, the reorganization of electricity generation, transmission, and distribution (Figure 1.6) [61].

In order to implement green hydrogen technology, renewable energy generation must increase, which can be challenging and time-consuming legally and in terms of regulations in some countries. Scaling up electrolysis capacity is another difficult problem. The numerous strategies describe targets that will require more than 400 times the current capacities of electrolyzers, which makes meeting this target very difficult [53,62].

Ultimately, green hydrogen and PtX are anticipated to mark the shift toward decarbonized energy. To bolster these technologies, a wide range of considerable challenges must be overcome—not only technical ones related to production, storage, safety, and infrastructure but also environmental, social, and economic ones.

1.12 CHAPTER SUMMARY

This chapter explored the following topics: Developments in renewable technologies moved the world closer to mitigating emissions from fossil fuels and facilitated green hydrogen production together with PtX technologies.

Solar, wind, and geothermal energy sources were pivotal for decarbonization attempts. PtX processes, along with green hydrogen, had the potential to change the renewable energy industry for the better. Green hydrogen could be used in numerous sectors, including industry, transportation, and power generation. If synthetic fuels by PtX technologies were created, it would ease the decarbonization of aviation. The rising green hydrogen demand was clear from the global initiatives for hydrogen plants. The Middle Eastern region was in a favorable position to scope for expanding its role in the green hydrogen market due to the available renewable resources. The development of green hydrogen was not without obstacles, such as political, economical, and social issues. Attention was particularly needed in the areas of storage, electrolysis, and transport infrastructure. The potential for reducing CO_2 emissions became significant through green hydrogen and PtX application in hard-to-electrify sectors. High costs of production, infrastructural development, and policy frameworks remained the biggest barriers to adopting these technologies. Effective tackling of the funding, innovation, and coordination issues would be essential to capturing the complete value of green hydrogen and PtX methods in the shift to a more sustainable energy system.

SELF-ASSESSMENT QUESTIONS

1. What are the primary sources of greenhouse gas emissions?
2. What is the Paris Agreement's goal regarding global temperature rise?
3. What challenges does the Paris Agreement face?
4. What is the expected increase in global renewable energy capacity additions in 2023?
5. What are the challenges associated with renewable energy sources?
6. What are green hydrogen and Power-to-X (PtX) technologies?
7. What are the advantages of green hydrogen and PtX in addressing renewable energy challenges?
8. Why is hydrogen considered a green alternative to fossil fuels?
9. How is green hydrogen produced?
10. What are the applications of green hydrogen in different industries?
11. What is the Power-to-X (PtX) concept?
12. What are some examples of PtX products?
13. What is the role of green hydrogen in transportation?
14. What are electro-fuels?
15. According to the IEA, what is the potential for low-emission hydrogen production by 2030?
16. What is the status of the NEOM Green Hydrogen Project in Saudi Arabia?
17. What is the key feature of the Western Green Energy Hub in Australia?
18. What is the status of the Ningxia Baofeng Energy Group project in China?
19. What is the focus of the Aqua Ventus project in Germany?
20. Why is the Middle East considered a potential leader in the green hydrogen market?
21. What is the status of Saudi Arabia's NEOM Green Hydrogen Project?
22. What is the UAE's approach to green hydrogen?
23. What is Oman's focus in the hydrogen sector?
24. What are some key technical developments in green hydrogen production?
25. What are the economic considerations for green hydrogen development?
26. What are the social aspects associated with green hydrogen?

27. According to the Hydrogen Council, what is the potential for CO_2 emission reductions from green hydrogen and PtX?
28. In which sectors are the most substantial CO_2 reductions from green hydrogen and PtX anticipated?
29. What are the key challenges facing green hydrogen and PtX technologies?
30. What are the key requirements for accelerating the adoption of green hydrogen and PtX technologies?

REFERENCES

1. Energy Institute. 2024. "Statistical Review of World Energy." London, United Kingdom. ISBN: 9781787254084. Available at: https://www.energyinst.org/__data/assets/pdf_file/0006/1542714/684_EI_Stat_Review_V16_DIGITAL.pdf.
2. Hansen, James E., Makiko Sato, Leon Simons, et al. 2023. "Global warming in the pipeline." *Oxford Open Climate Change* 3 (1). https://doi.org/10.1093/oxfclm/kgad008.
3. Savaresi, Annalisa. 2016. "The Paris agreement: A new beginning?" *Journal of Energy & Natural Resources Law* 34 (1): 16–26. https://doi.org/10.1080/02646811.2016.1133983.
4. Lewis, Nathan S. 2016. "Research opportunities to advance solar energy utilization." *Science* 351 (6271). https://doi.org/10.1126/science.aad1920.
5. Fernández-García, Aránzazu, Eduardo Zarza, Loreto Valenzuela, et al. 2010. "Parabolic-trough solar collectors and their applications." *Renewable and Sustainable Energy Reviews* 14 (7): 1695–1721. https://doi.org/10.1016/j.rser.2010.03.012.
6. Jäger-Waldau, Arnulf. 2023. "Snapshot of photovoltaics." *EPJ Photovoltaics* 14: 23. https://doi.org/10.1051/epjpv/2023016.
7. International Energy Agency. 2023. "Renewable Energy Market Update June 2023." Available at: https://iea.blob.core.windows.net/assets/63c14514-6833-4cd8-ac53-f9918c2e4cd9/RenewableEnergyMarketUpdate_June2023.pdf.
8. Buffo, Giulio, Paolo Marocco, Domenico Ferrero, et al. 2019 "Power-to-X and power-to-power routes." In *Solar Hydrogen Production*, pp. 529–557. Elsevier eBooks. https://doi.org/10.1016/b978-0-12-814853-2.00015-1.
9. Blay-Roger, Rubén, Muhammad Asif Nawaz, Francisco M. Baena-Moreno, et al. 2024. "Tandem catalytic approaches for CO_2 enriched Fischer-Tropsch synthesis." *Progress in Energy and Combustion Science* 103: 101159. https://doi.org/10.1016/j.pecs.2024.101159.
10. International Energy Agency. 2024. "Global Hydrogen Review 2024—Analysis—IEA," Accessed December 1, 2024. Available at: https://www.iea.org/reports/global-hydrogen-review-2024.
11. International Energy Agency. 2023. "Global Hydrogen Review 2023—Analysis—IEA," Accessed December 1, 2024. Available at: https://www.iea.org/reports/global-hydrogen-review-2023.
12. Austrade International. 2023. "Australian Green Economy Prospectus." Accessed March 23, 2025. Available at: https://international.austrade.gov.au/en/news-and-analysis/publications-and-reports/australian-green-economy-prospectus.
13. Lenivova, Veronika. 2022. "Potential Development of Renewable Hydrogen Imports to European Markets Until 2030." *OIES Paper.* Oxford Institute for Energy Studies. https://ora.ox.ac.uk/objects/uuid:db083901-0382-47d4-ae48-3e3f48bc9280.
14. Choe, Changgwon and Hankwon Lim. 2025. "Life cycle-based strategy and feasibility of surplus-to-X-to-electricity on domestic surplus utilization in the Republic of Korea." *Korean Journal of Chemical Engineering.* https://doi.org/10.1007/s11814-025-00417-0.
15. Switzerland Global Enterprise. 2024. "China Hydrogen Economy Report." Report. Accessed March 5, 2025. Available at: https://www.s-ge.com/sites/default/files/publication/free/2024_e_campaign_clean_website_gatedcontent_text_ct5_en_china-hyd.pdf.

16. Chariot Limited. 2024. "Completion of Feasibility Study on Green Hydrogen Project in Mauritania." London: Chariot Limited. Available at: https://chariotenergygroup.com/app/uploads/2024/03/Nour-Feasibilty-Study-Completion-11.03.24.pdf.
17. Saria, Lana. 2024. "Indonesia's National Hydrogen Strategy." Accessed January 15, 2024. Available at: https://henca-tokyo.metro.tokyo.lg.jp/data/Ministry_of_Energy_and_Mineral_Resources_of_the_Republic_of_Indonesia.pdf.
18. Ministry of Energy and Government of Chile. 2020. "National Green Hydrogen Strategy." Santiago, Chile. Accessed January 15, 2024. Available at: https://energia.gob.cl/sites/default/files/national_green_hydrogen_strategy_-_chile.pdf.
19. Marcó Del Pont, Mercedes and Flavia Royón. 2023. "National Strategy for the Development of the Hydrogen Economy." Available at: https://www.argentina.gob.ar/sites/default/files/2023/07/enh2_-_english.pdf.
20. Ministry of Mines and Energy, Secretariats of Planning and Energy Development, and Secretariats of Oil, Natural Gas and Biofuels. 2022. "2031 Ten-Year Energy Expansion Plan." Accessed January 15, 2024. Available at: https://www.gov.br/mme/pt-br/assuntos/secretarias/sntep/publicacoes/plano-decenal-de-expansao-de-energia/pde-2031/english-version/relatorio_pde2031_introducao_eus.pdf.
21. Inter-American Development Bank. 2021. "Colombia's Hydrogen Roadmap." Report. Colombia's Hydrogen Roadmap. Accessed January 15, 2024. Available at: https://www.minenergia.gov.co/documents/5862/Colombias_Hydrogen_Roadmap_2810.pdf.
22. Mexican Hydrogen Association. 2022. "Green Hydrogen: The Energy Vector for Decarbonizing the Mexican Economy." Accessed January 15, 2024. Available at: https://h2mex.org/wp-content/uploads/GreenHydrogen_Study_H2Mex.pdf.
23. Secretaría Nacional de Energía. 2016. "Plan Energético Nacional 2015–2050." Available at: https://www.senacyt.gob.pa/wp-content/uploads/2018/12/3.-Plan-Energetico-Nacional-2015-2050-1.pdf.
24. Matthias Schimmel, Diego Bietenholz, and Karoline Steinbacher. 2022. "Hydrogen Cooperation Potential Between Saudi Arabia and Germany." Federal Ministry for Economic Affairs and Climate Action. Available at: https://www.bmwk.de/Redaktion/EN/Downloads/J/joint-study-saudi-german-energy-dialogue.pdf?__blob=publicationFile&v=1.
25. Alturki, Fahad, Abdullah Alkhudhiri, Majed, Al-Suwailem, et al. 2024. "Saudi-China Collaboration Under the Circular Carbon Economy: Priorities and Opportunities in CCUS." King Abdullah Petroleum Studies and Research Center. Available at: https://www.kapsarc.org/research/publications/saudi-china-collaboration-under-the-circular-carbon-economy-priorities-and-opportunities-in-ccus/.
26. Ministry of Energy and Government of Chile. 2024. "Green Hydrogen Action Plan for Chile." Available at: https://energia.gob.cl/sites/default/files/documentos/green-hydrogen_action-plan.pdf.
27. Argus Media. 2022. "Argus White Paper: Progress on the Path to Middle East Hydrogen Exports." Available at: https://www.middleeast-energy.com/content/dam/markets/emea/middle-east-energy/en/docs/Argus%20White%20Paper%20-%20Progress%20on%20the%20path%20to%20Middle%20East%20H2%20exports.pdf.
28. Al-Khayari, Hammad, Mohamed E. Farrag, and Mohamed A. Elgenedy. 2024. "Charting Oman's path to green hydrogen: Embracing opportunities and overcoming challenges." In *Paper Presented at the SPE International Health, Safety, Environment and Sustainability Conference and Exhibition*, Abu Dhabi, UAE, September 2024. Society of Petroleum Engineers. https://doi.org/10.2118/220470-MS.
29. Hydrogen Oman (Hydrom). 2024. "Oman Green Hydrogen Strategy 2024." Available at: https://hydrom.om/Media/Pdf/Oman-Green-Hydrogen-Strategy-2024.pdf.
30. Ministry of Energy and Minerals. 2022. "Sultanate of Oman—The Future of Green Energy." Oman. Available at: https://mem.gov.om/Portals/0/WEALTH%20Newsletter%20Eng.pdf.

31. Sprenger, Tobias, Linus Kalvelage, Michael Moritz, et al. 2023. "Towards a Green Shipping Gateway." Deutsche Gesellschaft für Internationale Zusammenarbeit (GIZ) GmbH. Available at: https://h2diplo.de/wp-content/uploads/2023/11/Towards-a-green-shipping-gateway-Establishing-a-green-hydrogen-economy-in-Egypt.pdf.
32. Bayssi, Oussama, Nouhaila Nabil, Meryeme Azaroual, et al. 2024 "Green hydrogen landscape in North African countries: Strengths, challenges, and future prospects." *International Journal of Hydrogen Energy* 84 (2024): 822–839. https://doi.org/10.1016/j.ijhydene.2024.08.277.
33. United Nations Industrial Development Organization | UNIDO. 2024. "Industrial Decarbonization Roadmap: Framework for the Cement and Steel Industry in Morocco." Available at: https://downloads.unido.org/ot/33/26/33262565/UNIDO%20Morocco%20industrial%20decarbonization%20roadmap.pdf.
34. Ministry of Energy and Mineral Resources. 2023. "Draft National Green Hydrogen Strategy for Jordan". Available at: https://www.memr.gov.jo/EBV4.0/Root_Storage/AR/EB_Info_Page/GH2_Strategy.pdf.
35. Tanchum, Michael, Alexander P. Martin, Hussein Suleiman, et al. 2024. "Renewable Energy and Electricity Interconnection Megaprojects in North Africa: Prospects for Euro-North Africa Cooperation." EuroMeSCo. Available at: https://www.euromesco.net/wp-content/uploads/2024/07/Policy-Study34-Megaprojects.pdf.
36. Akyüz, Emin Serhat, Esra Telli, and Murat Farsak. 2024. "Hydrogen generation electrolyzers: Paving the way for sustainable energy." *International Journal of Hydrogen Energy* 81: 1338–1362. https://doi.org/10.1016/j.ijhydene.2024.07.175.
37. Zhang, Tongyao, Mengshu Zhang, Ling Jin, et al. 2024. "Advancing carbon capture in hard-to-abate industries: Technology, cost, and policy insights." *Clean Technologies and Environmental Policy* 26 (7): 2077–2094. https://doi.org/10.1007/s10098-024-02810-5.
38. Almaraz, Sofía De-León, Tamás Kocsis, Catherine Azzaro-Pantel, et al. 2024. "Identifying social aspects related to the hydrogen economy: Review, synthesis, and research perspectives." *International Journal of Hydrogen Energy* 49: 601–618. https://doi.org/10.1016/j.ijhydene.2023.10.043.
39. Hydrogen Council. 2021. "Hydrogen for Net Zero—A Critical Cost-Competitive Energy Vector." Available at: https://publications.ieaghg.org/insightpapers/2021-IP28%20Hydrogen%20for%20net%20zero.pdf.
40. Simon Araya, Samuel, Cui, Xiaoti, Li, Na, et al. 2022. "Power-to-X: Technology Overview, Possibilities and Challenges." Available at: https://vbn.aau.dk/files/514146100/PtX_Report.pdf.
41. Anika, Ogemdi Chinwendu, Somtochukwu Godfrey Nnabuife, Abdulrauf Bello, et al. 2022. "Prospects of low and zero-carbon renewable fuels in 1.5-degree net zero emission actualisation by 2050: A critical review." *Carbon Capture Science & Technology* 5: 100072. https://doi.org/10.1016/j.ccst.2022.100072.
42. Zhang, Runsen and Tatsuya Hanaoka. 2022. "Cross-cutting scenarios and strategies for designing decarbonization pathways in the transport sector toward carbon neutrality." *Nature Communications* 13 (1): 3629. https://doi.org/10.5281/zenodo.6617098.
43. Wei, Max, Colin A. McMillan, and Stephane de la Rue Du Can. 2019. "Electrification of industry: Potential, challenges and outlook." *Current Sustainable/Renewable Energy Reports* 6: 140–148. https://doi.org/10.1007/s40518-019-00136-1.
44. Link, Steffen and Patrick Plötz. 2022. "Technical feasibility of heavy-duty battery-electric trucks for urban and regional delivery in germany—a real-world case study." *World Electric Vehicle Journal* 13 (9): 161. https://doi.org/10.3390/wevj13090161.
45. Von Wald, Gregory, Kaarthik Sundar, Evan Sherwin, et al. 2022. "Optimal gas-electric energy system decarbonization planning." *Advances in Applied Energy* 6: 100086. https://doi.org/10.1016/j.adapen.2022.100086.
46. Fragkos, Panagiotis. 2022. "Decarbonizing the international shipping and aviation sectors." *Energies* 15 (24): 9650. https://doi.org/10.3390/en15249650.

47. Samet, Mehdi Jahangir, Heikki Liimatainen, Mikko Pihlatie, et al. 2024. "Levelized cost of driving for medium and heavy-duty battery electric trucks." *Applied Energy* 361: 122976. https://doi.org/10.1016/j.apenergy.2024.122976.
48. The Royal Society. 2019. "Sustainable Synthetic Carbon Based Fuels for Transport: Policy Briefing." Available at: https://research.brighton.ac.uk/files/6852944/Sustainable_synthetic.pdf.
49. Choi, Wonjae and Sanggyu Kang. 2023. "Greenhouse gas reduction and economic cost of technologies using green hydrogen in the steel industry." *Journal of Environmental Management* 335: 117569. https://doi.org/10.1016/j.jenvman.2023.117569.
50. Atilhan, Selma, Sunhwa Park, Mahmoud M. El-Halwagi, et al. 2021. "Green hydrogen as an alternative fuel for the shipping industry." *Current Opinion in Chemical Engineering* 31: 100668. https://doi.org/10.1016/j.coche.2020.100668.
51. Maciorowski, Damian, Aleksandra Ludwiczak, and Adam Kozakiewicz. 2024. "Hydrogen, the future of aviation." *Combustion Engines* 197 (2): 126–131. https://doi.org/10.19206/CE-178375.
52. Rambhujun, Nigel, Muhammad Saad Salman, Ting Wang, et al. 2020. "Renewable hydrogen for the chemical industry." *MRS Energy & Sustainability* 7: E33. https://doi.org/10.1557/mre.2020.33.
53. Bade, Shree Om, Olusegun Stanley Tomomewo, Ajan Meenakshisundaram, et al. 2024. "Economic, social, and regulatory challenges of green hydrogen production and utilization in the US: A review." *International Journal of Hydrogen Energy* 49: 314–335. https://doi.org/10.1016/j.ijhydene.2023.08.157.
54. Jayachandran, M., Ranjith Kumar Gatla, Aymen Flah, et al. 2024. "Challenges and opportunities in green hydrogen adoption for decarbonizing hard-to-abate industries: A comprehensive review." *IEEE Access* 12: 23363–23388. https://doi.org/10.1109/ACCESS.2024.3363869.
55. Yue, Meiling, Hugo Lambert, Elodie Pahon, et al. 2021. "Hydrogen energy systems: A critical review of technologies, applications, trends and challenges." *Renewable and Sustainable Energy Reviews* 146: 111180. https://doi.org/10.1016/j.rser.2021.111180.
56. Taibi, Emanuele, Raul Miranda, Marcelo Carmo, et al. 2020. "Green Hydrogen Cost Reduction." Available at: https://www.h2knowledgecentre.com/content/researchpaper1613.
57. Lee, Moongyung and Deger Saygin. 2023. "Financing cost impacts on cost competitiveness of green hydrogen in emerging and developing economies." *OECD Environment Working Papers*. https://doi.org/10.1787/15b16fc3-en.
58. Deutsche Gesellschaft für Internationale Zusammenarbeit (GIZ) GmbH. 2023. "Green Hydrogen for the Commercial and Industrial (C&I) Sector." Berlin: German Energy Solutions Initiative of the German Federal Ministry for Economic Affairs and Climate Action (BMWK). Available at: https://www.giz.de/en/downloads/giz2023-en-h2pep-sector-analysis-jordan.pdf.
59. Yang, Lei, Shuning Wang, Zhihu Zhang, et al. 2023. "Current development status, policy support and promotion path of China's green hydrogen industries under the target of carbon emission peaking and carbon neutrality." *Sustainability* 15 (13): 10118. https://doi.org/10.3390/su151310118.
60. Côté, Elizabeth and Sarah Salm. 2022. "Risk-adjusted preferences of utility companies and institutional investors for battery storage and green hydrogen investment." *Energy Policy* 163: 112821. https://doi.org/10.1016/j.enpol.2022.112821.
61. Stamm, Andreas, Tilman Altenburg, Rita Strohmaier, et al. 2023. "Green hydrogen: Implications for international cooperation: With special reference to South Africa". No. 9/2023. *IDOS Discussion Paper*. https://doi.org/10.23661/idp9.2023.
62. Younas, Muhammad, Sumeer Shafique, Ainy Hafeez, et al. 2022. "An overview of hydrogen production: Current status, potential, and challenges." *Fuel* 316: 123317. https://doi.org/10.1016/j.fuel.2022.123317.

2 Green Hydrogen Production

2.1 INTRODUCTION

Hydrogen, with a molecular weight of 2.016 g/mol H_2, is the first element in the periodic table with the atomic number 1. It ranks among the most abundant elements in the universe, constituting roughly 75% of its total mass. Hydrogen, in its pure gaseous form, is seldom found in nature; instead, it is mostly present as a constituent in various compounds such as water (H_2O) and hydrocarbons (C_nH_m). Pure hydrogen can, however, be extracted through a separation process. This process, which entails breaking down compounds into their constituent parts to isolate hydrogen, is known as "hydrogen production." Such processes require energy inputs to facilitate the endothermic chemical reactions necessary to obtain hydrogen gas in a pure state [1–5]. Hydrogen gas can serve as an energy carrier, a fossil fuel alternative, or a surrogate fuel when internally combusted or used in fuel cells. Furthermore, it can be transformed into ammonia or methane fuels through non-emissive zero-net processes.

Hydrogen can be produced in different ways. When derived from conventional energy sources, it can be generated through steam reforming, gasification, and partial oxidation (POX). Nonetheless, these resources are not perpetually available, and their consumption leads to the emission of greenhouse gases. Consequently, such resources need to be replaced with appropriate and eco-friendly options. In this regard, green hydrogen stands out as an alternative, which is derived from the electrolysis of water using renewable energy systems. The hydrogen produced from each energy source is assigned a distinct color and so is how the potential emissions are handled. This chapter presents the various colors of hydrogen and examines the methods of producing hydrogen from both traditional and renewable sources. It also includes technologies to produce green hydrogen by electrolysis and the principal calculations related to electrolysis.

2.2 HYDROGEN COLORS

The 2015 Paris Agreement, a key component of the United Nations Framework Convention on Climate Change (UNFCCC), is pivotal in international efforts to address climate change. It creates a structure for global collaboration to fight climate change, with the ambitious goal of keeping global temperature rise below 2°C—ideally below 1.5°C—compared with preindustrial times [2]. This goal highlights the urgent need to lessen the severe effects of climate change, such as rising sea levels, more extreme weather, loss of biodiversity, and dangers to human health and well-being.

DOI: 10.1201/9781003639190-2

Hydrogen can be manufactured from various processes, which are classified according to the specific production techniques and environmental effects into various hydrogen colors. Given the dependency of steam methane reforming (SMR) or coal gasification on fossil fuels such as natural gas or coal, gray hydrogen is considered the most common today. It has no means of capturing and storing carbon emissions (CCS); thus, it is the most environmentally harmful option. Another fossil-fuel-based hydrogen production method is brown hydrogen. Unlike gray, it incorporates CO_2 capturing and storage. However, the efficiency of the captured and stored carbon, as well as the long-term viability of the stored carbon, remains an important issue.

Like brown hydrogen, blue hydrogen focuses on CCS and fossil fuel utilization for hydrogen production. Blue hydrogen seeks to reduce carbon emissions more than gray or brown hydrogen.

On the other hand, green hydrogen is produced through the electrolysis of water using renewable electricity from solar, wind, or hydropower. This method is the most environmentally friendly as there are no direct emissions of greenhouse gases.

Turquoise hydrogen refers to hydrogen produced from methane pyrolysis, which involves breaking down methane into solid carbon and hydrogen. When powered by renewable energy and with solid carbon sequestration, this process can be regarded as a low-carbon hydrogen fuel option. The solid carbon can be used instead of releasing CO_2 into the atmosphere for industrial purposes, such as construction materials.

Yellow hydrogen is produced via electrolysis with grid electricity, which typically contains a blend of renewable and fossil fuel sources. The grid electricity sources determine the impact on the environment.

Nuclear power is used for the electrolysis of pink hydrogen, offering a low-carbon alternative. However, it raises concerns regarding nuclear waste and safety. Purple hydrogen, or violet hydrogen, is produced by renewably powered electrolysis, referred to as advanced hydrogen, green hydrogen, or simply hydrogen emphasizing its production methods, advanced electrolysis technologies. High-temperature thermochemical water splitting defines red hydrogen, which can be powered by concentrated solar or nuclear, nuclear energy. The efficiency and scalability of this method are still under development.

The global decarbonization goals emphasize change in cleaner hydrogen production methods. Pink, purple, red, and green hydrogen showcase a pathway toward a sustainable hydrogen economy mark, while gray, brown, and blue represent a transitional solution with varying levels of environmental impact. The negative environmental impact of low-carbon hydrogen greatly relies on developing and deploying strong CCS technologies and increasing the capacity of renewable energies.

As noted earlier, various sources of hydrogen are distinguished by colors, which depict the source of energy and the treatment of emissions, if any, associated with them. Figure 2.1 shows the source and by-product of each method of hydrogen production referenced in color coding.

The advancements in geological science and detection technologies in the past few years have led to the exploration of new sources of hydrogen across the globe. White hydrogen was accidentally discovered at the oil and gas production sites. White hydrogen is generated naturally through continuous geochemical reactions in subsurface rock formations. This type of hydrogen is considered a potential low-carbon energy

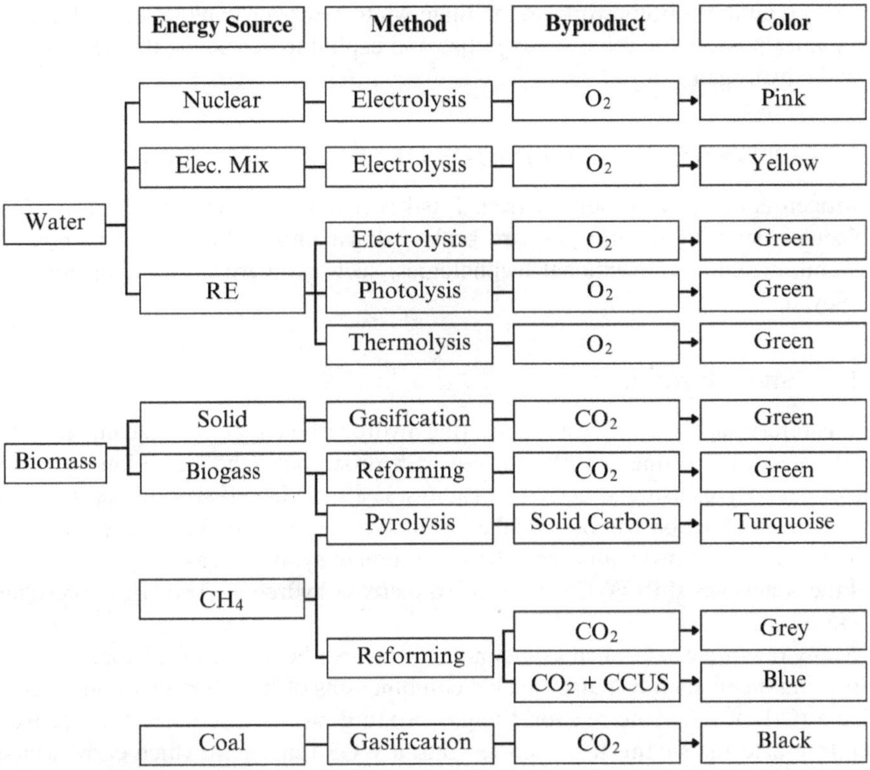

FIGURE 2.1 Hydrogen primary colors classification based on source and production method.

source. White hydrogen is characterized by its small, light molecular structure, facilitating potential escape from cap rocks. The white hydrogen is in the research phase to understand its formation and accumulation dynamics fully. Recently, investigations have focused on establishing key components of white hydrogen exploration by collecting field data and gaining practical experience. The global pressing primarily drives the promising interest in white hydrogen to decarbonize several sectors.

On the other hand, green hydrogen is expected to fulfill a significant portion of this demand. Still, green hydrogen faces high production costs from reliance on renewable energy and electrolysis, which render white hydrogen a potentially more economically viable alternative. The potential of white hydrogen comes from its production at significantly lower costs, particularly from reservoirs proximate to end-user markets.

Despite white hydrogen potential, it is not projected to become a dominant energy source in the near future. This view comes from current estimates that it will constitute a minor fraction of the future low-carbon hydrogen supply. However, successful pilot projects demonstrate technical and commercial feasibility coupled with supportive policy frameworks. Privately funded entities lead the growing white hydrogen technology with limited operational projects. Recently, some countries have adopted regulatory frameworks to facilitate white hydrogen exploration. Oil and gas companies possess the requisite expertise and capital to accelerate the development of white hydrogen.

2.3 CONVENTIONAL METHODS OF HYDROGEN PRODUCTION

Hydrogen is currently produced from fossil fuel resources, in which hydrocarbon molecules, such as natural gas, are broken down into hydrogen and carbon. To this end, different conventional technologies, such as hydrocarbon reforming, are employed.

2.3.1 Steam Reforming

Steam reforming is a process of reacting hydrocarbon fuel with steam. In other words, steam reforming is a way to convert hydrocarbons and steam into hydrogen and carbon oxides using a catalyst. A catalyst is a substance that accelerates the rate of a chemical reaction without being consumed in the process. The steam reforming process consists of two main steps: the generation of synthesis gas (syngas) via SMR, and the water–gas shift (WGS) reaction to increase hydrogen yield [1], as presented in Figure 2.2.

Many raw materials, such as methane (CH_4), can be used as feedstock in steam reforming. Feedstock can also involve combinations of light hydrocarbons such as ethane (C_2H_6). After the reactor, the gas mixture passes through a heat recovery step to create the steam. It is then fed into a WGS reactor in which carbon monoxide is reacted with water, resulting in additional hydrogen production and CO_2 formation. The chemical reactions in the steam reformer are expressed by the following chemical equations: reformer reaction and WGS in equations 2.1 and 2.2, respectively:

$$C_nH_m + nH_2O \rightarrow nCO + \left(n + \frac{1}{2}m\right)H_2 \tag{2.1}$$

$$CO + H_2O \rightarrow CO_2 + H_2 \tag{2.2}$$

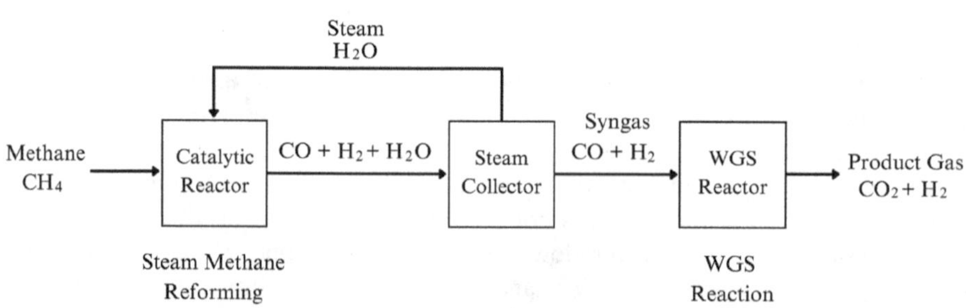

FIGURE 2.2 Steam methane reforming method for hydrogen production.

Applying $n=1$ and $m=4$ to the equations will represent the steam reforming of methane. SMR is the most common and mature method for large-scale hydrogen production.

At high temperatures above 800°C, methane (CH_4) reacts with steam (H_2O) to produce hydrogen (H_2) and carbon monoxide (CO). The presence of a nickel catalyst helps this process react more quickly and to retain more hydrogen gas. Generally, carbon can accumulate on the catalyst surface, leading to catalyst deactivation. However, nickel can react with carbon (forming nickel carbides) and prevent carbon buildup, thus prolonging the catalyst's lifespan and ensuring continuous hydrogen production. The hydrogen–carbon monoxide (the so-called synthetic gas) results from this stage. Carbon monoxide is not a useful reaction product and is harmful if released into the environment. Therefore, carbon monoxide from the reforming reaction enters the WGS reactor to produce more hydrogen.

This reactor is filled with water and an iron–chrome-based catalyst, which causes steam (H_2O) to break down into oxygen (O_2) and hydrogen (H_2). Hydrogen is then captured while oxygen reacts with carbon monoxide (CO) from the reforming reaction to produce carbon dioxide (CO_2). Carbon dioxide is less harmful to the environment than carbon monoxide. It can be captured using a carbon capturing and storage (CCS) system. CCS is a technology that captures carbon dioxide emissions from industrial processes and power plants, preventing them from entering the atmosphere and storing them underground [6]. The use of CCS changes hydrogen color from "gray" (high carbon emissions) to "blue" (lower carbon emissions due to carbon capture and storage implementation), indicating a more environmentally friendly production method [1].

The gas mixture leaving the WGS reactor is not pure and needs to be purified. The most used method for hydrogen purification is pressure swing absorption (PSA). The impure gas mixture is sent into a purifier, which encounters absorptive materials that trap contaminants under high pressure. The purified hydrogen is then extracted from the vessel. The trapped contaminants are released when the ship is depressurized. PSA effectively removes carbon dioxide, methane, carbon monoxide, and water from hydrogen. SMR is about 74% efficient, so a percentage of methane remains unreacted after completion. Carbon dioxide is often captured and used for enhanced oil recovery in the refining industry [1].

2.3.2 Partial Oxidation

POX is an exothermic chemical process for hydrogen production where feedstock reacts with a small amount of air and releases heat. The exothermic chemical process is a chemical reaction or a physical change in which energy is released to the surroundings in the form of heat. During exothermic reactions, the reactants have a higher energy level than the products, releasing excess energy in the form of heat. Raw materials such as methane, heavy oil, and coal can be used as feedstock in POX. POX reactions occur when a mixture of hydrocarbon and air is burned incompletely in a reformer, producing a gas called synthetic gas or syngas, which consists of hydrogen (H_2) and carbon monoxide (CO) [2].

In POX, sulfur should be removed from the feedstock before oxidation occurs. Once the sulfur is removed, pure oxygen (O_2) partially oxidizes the hydrocarbon feedstock. The produced syngas is then treated similarly to the product gas in the steam reforming process, which involves going through the WGS process to produce hydrogen. The following equations describe the chemical reaction of this method; the POX reaction and WGS reactor are presented in equations 2.3 and 2.4, respectively.

$$C_nH_m + \frac{n+m}{4} O_2 \rightarrow nCO + \frac{m}{2} H_2O \tag{2.3}$$

$$CO + H_2O \rightarrow CO_2 + H_2 \tag{2.4}$$

Typically, the POX of hydrocarbon gases occurs in the medium temperature range from 300°C to 1200°C [3], which mainly depends on the presence of the catalyst in the process. POX of hydrocarbon to syngas ($CO+H_2$) can happen at very high temperatures without a catalyst [4]. There are two types of POX: thermal POX (TPOX), which does not involve a catalyst, and catalytic POX (CPOX). TPOX requires temperatures of at least 1200°C, while CPOX operates at lower temperatures of 800°C–900°C. CPOX utilizes a range of feedstock from methane to naphtha, while TPOX can handle hydrocarbons such as methane, heavy oil, and coal. The choice between the two techniques depends on the sulfur content of the fuel. CPOX is preferred when the sulfur content is low, as it avoids catalytic poisoning caused by sulfur. On the other hand, TPOX is more suitable if the sulfur content is high.

2.4 GREEN HYDROGEN PRODUCTION METHODS

As presented in Figure 2.1, green hydrogen is mainly produced from water using renewable energy. Splitting water into hydrogen and oxygen using renewable energy can be achieved using different methods such as water electrolysis using electricity, thermolysis using thermal energy, photolysis using photonic energy, and biophotolysis microorganisms.

2.4.1 Water Electrolysis

Electrolysis involves breaking down water into its components, hydrogen and oxygen, using electricity. The only waste product from electrolysis after capturing hydrogen is pure oxygen gas. An electrochemical device such as "Electrolyzer" facilitates the process. Generally, two cited electrodes, anode and cathode, are placed in a liquid electrolyte, which is a conducting solution. Most devices used today for assembling hydrogen are electrolyzers because water is the original material. Water is free and available, making the process eco-friendly. They enable the separation of water molecules into hydrogen and oxygen by passing an electrical current through their electrolyte. In most cases, the water electrolysis reaction can be split into two half-cell reactions. First is the hydrogen evolution reaction (HER), in which water in the cathode is reduced to produce H_2. The second step in the anode to produce O_2 is called the oxygen evolution reaction (OER). Both HER and OER reactions can be expressed with equations 2.5 and 2.6 [5].

$$H_2O \rightarrow \frac{1}{2}O_2 + 2e^- + 2H^+ \quad (2.5)$$

$$2H^+ + 2e^- \rightarrow H_2 \quad (2.6)$$

Water electrolysis is classified into four types based on the type of electrolyzer used: its electrolyte composition, operating conditions, and ionic agents (OH, H^+, O_2). The four types of electrolyzers are alkaline electrolyzer (ALE), solid oxide electrolyzer (SOE), microbial electrolyzer (ME), and proton exchange membrane (PEM). All four types of electrolyzers share the same operational principle but differ in their electrolyte composition, operating conditions, and the specific ionic agents involved [7].

The ALE is the most widespread and developed technology for water electrolysis. "Alkaline" is indicative of the electrolyte, which at times is a potassium hydroxide (KOH) or sodium hydroxide (NaOH) solution. It includes four primary components: two electrodes (anode and cathode), an electrolyte, a catalyst, and a diaphragm to separate the produced gases from each electrode (refer to Figures 2.3 and 2.4). This should fulfill the exact requirements of a solid electrolyte. The most common diaphragm in alkaline electrolysis cells has historically been made of porous white asbestos ($Mg_3Si_2O_5(OH)_4$). For electrodes, a predominately used electrode material

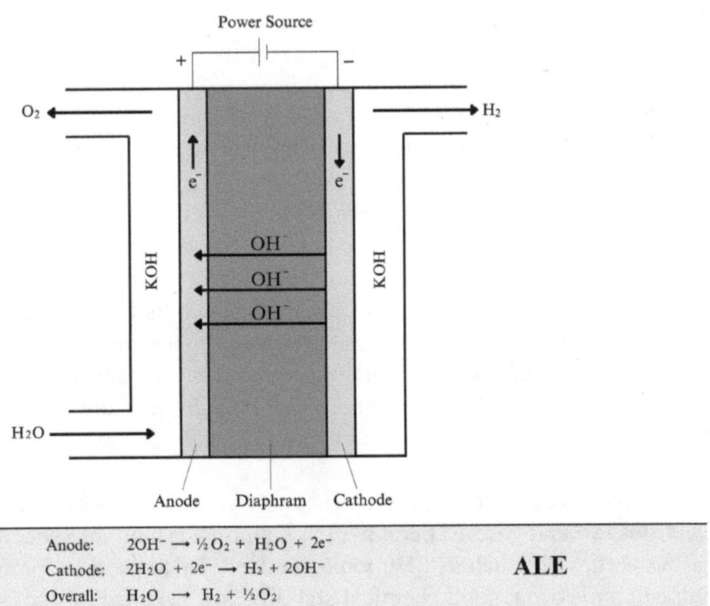

FIGURE 2.3 Alkaline electrolyzer components and operation.

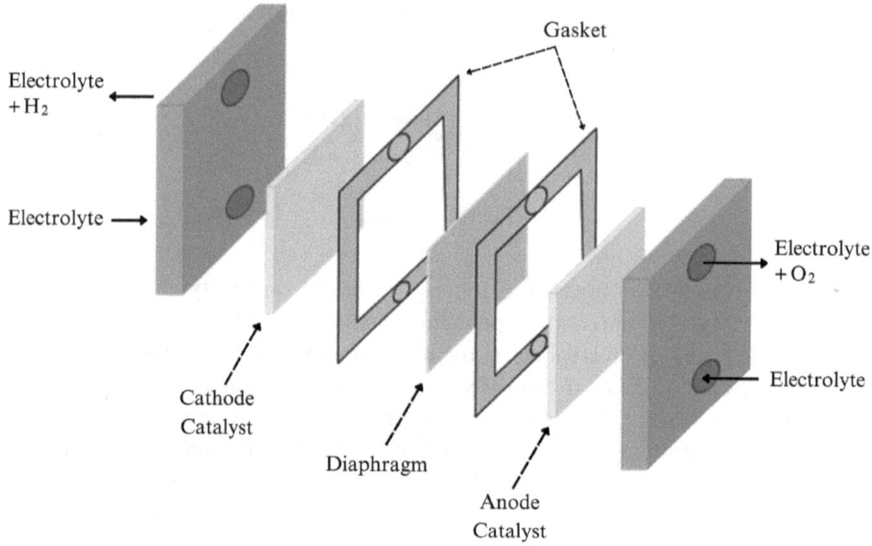

FIGURE 2.4 Construction of an alkaline electrolyzer.

is nickel because it is relatively inexpensive, possesses good conductivity, is resistant to corrosion, and is stable in a highly alkaline environment.

The principle of operation of an alkaline water electrolysis is quite simple. At the cathode, water (H_2O) is reduced to generate hydrogen gas (H_2) and hydroxide ions (OH^-), which then migrate through the diaphragm to the anode. At the anode, the hydroxide ions are oxidized to release oxygen gas (O_2) and regenerate water. Electrons flow through an external circuit from anode to cathode to complete the process.

In an ALE, the electrolyte is a concentrated aqueous solution of an alkali hydroxide, typically potassium hydroxide (KOH) or sodium hydroxide (NaOH). The catalysts used in ALEs are often nickel-based materials or other nonprecious metal compounds. These catalysts improve reaction kinetics, reduce overpotential—thus lowering energy requirements—, enhance product selectivity, and contribute to electrode durability, leading to more efficient and effective water electrolysis [8]. Figure 2.4 shows the construction of an ALE with its various components.

PEM electrolysis is the process of water separation using an electrolyzer with a solid polymer conductive membrane that can conduct the protons and separate the produced gases. PEM electrolyzer cell includes five key components: membrane, electrolyte, electrodes (anode and cathode), catalyst, and an external power source (see Figure 2.5).

The membrane allows protons (H^+) to pass through while blocking the flow of electrons (e^-) and gases. Therefore, the membrane shall have specific physicochemical properties [9], such as high ionic conductivity to promote proton migration and reduce ohmic drop, good chemical stability, low permeability to oxygen and hydrogen, good thermal stability up to 80°C–100°C, and good mechanical and dimensional stability, especially resistance to pressure.

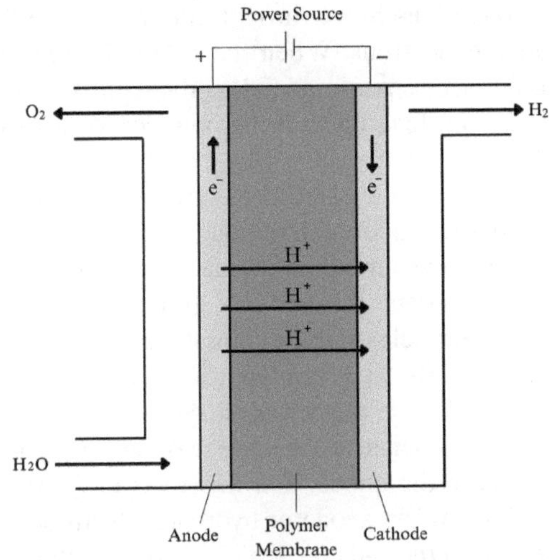

Anode:	$H_2O \rightarrow 2H^+ + ½O_2 + 2e^-$	**PEM**
Cathode:	$2H^+ + 2e^- \rightarrow H_2$	
Overall:	$H_2O \rightarrow H_2 + ½O_2$	

FIGURE 2.5 Proton exchange membrane electrolyzer components and operation.

The second component is the electrodes (anode and cathode). The anode is the positive electrode where the oxidation reaction takes place. At the anode, water decomposes into protons and molecular oxygen. This process is known as the OER. The anode material used in PEM electrolyzers is typically made of a precious metal called platinum (Pt) or platinum alloy. Platinum is favored for its excellent electro-catalytic properties, high conductivity, and corrosion resistance. The cathode is the negative electrode where the reduction reaction occurs.

At the cathode, hydrogen gas (H_2) is formed. This process is called the HER. The cathode material in PEM electrolyzers is often made of a mixture of platinum and other transition metals, such as nickel (Ni) or cobalt (Co), to improve efficiency and reduce the cost of the electrolysis process [7]. The catalyst in the PEM electrolyzer enhances the reaction rates at the anode and cathode. Typically, platinum or other precious metals are used as catalysts on each electrode to facilitate electrochemical reactions.

Another important component in a PEM electrolyzer is the external power source, which consists of a direct current (DC) power supply attached to the cell. This supply provides the electrical energy needed for electrolysis. The system moves electrons from the anode to the cathode via an external circuit.

PEM also includes PTL, a porous transport layer of a cell that enables the flow of the reactant water to the electrodes while eliminating oxygen gas and having high electrical conductivity.

The operational working of water electrolysis using a PEM is straightforward. Water flows into the cell at the anode, where it splits into hydrogen ions (H^+) and oxygen gas in the anode catalyst layer. The oxygen gas dissolves at the anode outlet.

Concurrently, the hydrogen ions move through the membrane to the cathode catalyst layer and strip off the electrons. When the external circuit is connected to the cathode, hydrogen ions' electrons are added to them in the cathode catalyst layer due to the inflow of current, leading to the production of hydrogen gas as per the aforementioned reactions. Thus, the cell yields oxygen at the anode and hydrogen at the cathode. The working principle of PEM electrolysis is shown in Figure 2.10 [8]. The creation of oxygen in the evaporation region is aided by a ruthenium and iridium catalyst layer on the membrane surface in the anode region [7]. The construction of a typical PEM electrolyzer is demonstrated in Figure 2.6.

A solid oxide electrolyzer cell (SOEC) consists of three main components: two porous electrodes and a solid oxide electrolyte (SOE), as seen in Figure 2.7.

In an SOE, at the cathode, the oxygen electrode, oxygen molecules (O_2) from a gas stream or water vapor dissociate at the surface of an electrode. O_2 ions (O^{-2}) then migrate through the SOE to the anode, and this reaction is referred to as "oxygen reduction reaction (ORR)." At the anode or hydrogen electrode, a reaction will take place: oxygen ions (O^{-2}) from the cathode, which yield from the reaction with the current, combined with electrons (e) supplied from outside. This electrochemical reaction is yielding hydrogen gas (H_2), hence the name "hydrogen evolution reaction." The SOE enables the flow of oxide ions (O^{-2}) from the cathode to the anode [7,9,10].

Common SOE materials include yttria-stabilized zirconia and doped ceria. These SOEs are requisite to high temperatures of 500°C–1000°C. SOEs may also be used

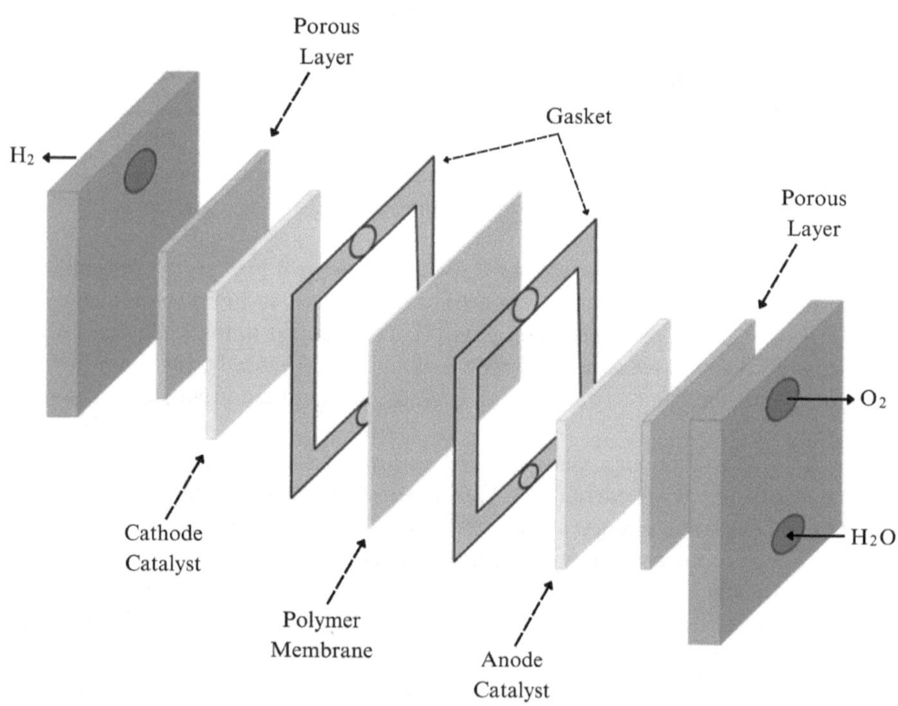

FIGURE 2.6 Proton exchange membrane electrolyzer construction.

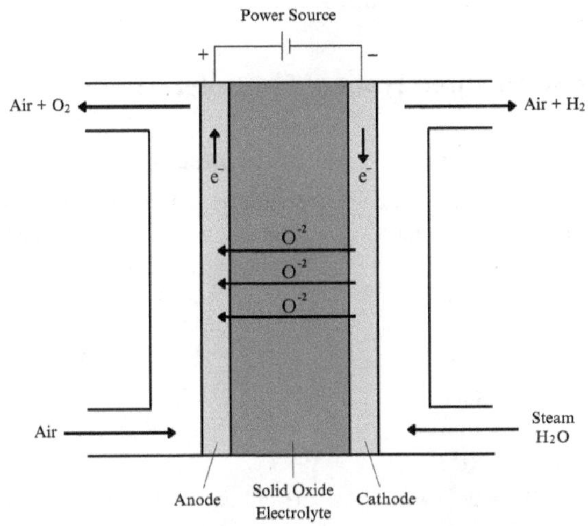

Anode:	$O^{2-} \rightarrow \frac{1}{2}O_2 + 2e^-$	
Cathode:	$H_2O + 2e^- \rightarrow H_2 + O^{2-}$	**SOE**
Overall:	$H_2O \rightarrow H_2 + \frac{1}{2}O_2$	

FIGURE 2.7 Solid oxide electrolyzer components and operation.

to produce syngas (a CO and H_2 mixture) from steam (H_2O) and CO_2 coelectrolysis and for water electrolysis (H_2O) to produce hydrogen. This will be covered in detail in Chapter 4. The three types of electrolyzers are compared in Table 2.1 [11].

Microbial electrolysis cells (MECs) are systems tailored to utilize a variety of synthetic nonthermophilic organic inputs, with a particular emphasis on waste streams. Organic-rich wastewater, nondomestic and agricultural sewage, and industrial effluents provide a key input source. Food processing brewery wastewater and agricultural runoff serve as primary exemplifiers because of their high organic loads. In addition to wastewater, MECs can use various solid organic waste streams well. These include food and agricultural waste, which comprises leftover crops and livestock manure, as well as the organic portion of the municipal solid waste. In more specialized research settings, specific organics such as sodium acetate, glucose, and other simple sugars or organic acids may be used to control the system's input more finely. Alongside the organic substrate, it is important to mention that the microorganisms colonized within the MEC have essential nutrient requirements such as nitrogen, phosphorus, and trace elements if effective functioning is expected. The substrate's biodegradability needs to be met along with the organic matter. The characteristics of the input stream will define the performance of the MEC, so, for example, elevated concentrations of some constituents may inhibit microbial activity. Other processes designed to enhance cellular methods for removing inhibitors or optimally pretreating the substrate bacteria aim to degrade may be needed.

The functioning of the microbial community is also strongly associated with the pH and temperature range provided within the input. In other words, MECs are

TABLE 2.1
Comparison between Three Types of Electrolyzers

Electrolyzer Type	PEM	Alkaline	SOE
Electrolyte/ Separator	Polymer (solid) perfluoro sulfonic acid polymers (PFSA).	Electrolyte KOH or NaOH (liquid). Diaphragm: asbestos	Ceramic (solid): Y_2O_3–ZrO_2
Operating Temperature (°C)	20–100	40–90	650–1000
Pressure Range (bar)	<200	<30	<20
Charge Carrier	H^+	OH^-	O^{2-}
Cathode Reaction	$2H^+ + 2e^- \rightarrow H_2$	$2H_2O + 2e^- \rightarrow H_2 + 2OH^-$	$H_2O + 2e^- \rightarrow H_2 + O^{2-}$
Anode Reaction	$H_2O \rightarrow \frac{1}{2}O_2 + 2e^- + 2H^+$	$2OH^- \rightarrow \frac{1}{2}O_2 + 2e^-$	$O^{2-} \rightarrow \frac{1}{2}O_2 + 2e^-$
Cell Voltage (V)	1.8–2.2	1.8–2.2	0.7–1.5
Current Density (A/cm^2)	0.2–0.6	0.2–0.4	0.3–1
Efficiency (%)	80	60–80	>90
Hydrogen Purity (%)	99.999	>99.8	99.9
Advantages	• Compact design • High current density • High purity of hydrogen • Quick response/start time • Simple design	• Low capital cost • High stability • Mature technology • Longer lifetime • Without precious catalyst	• Low capital cost. • Without precious catalyst. • High current density. • Lower energy need. • High efficiency.
Disadvantages	• High cost of the noble material membrane • Lower durability • Acidic medium • Precious catalyst	• Lower current density • Corrosivity of electrolyte • Gas permeation • Slow stop/go and load cycling	• Instability of electrodes • Safety problems • Improper sealing • Delamination of electrodes

designed with optimization in mind, allowing them to exploit practically any organic waste stream. This fact alone makes them much more useful for sustainable waste management and energy generation.

Therefore, MECs are considered one of the more modern methods of treating wastewater, in which microorganisms are utilized to convert organic waste into hydrogen gas, thus providing a more sustainable energy source. Each MEC contains an anode chamber and a cathode chamber divided by a membrane, constituting its core. Electrogenic bacteria in the anode chamber oxidize organic matter by forming

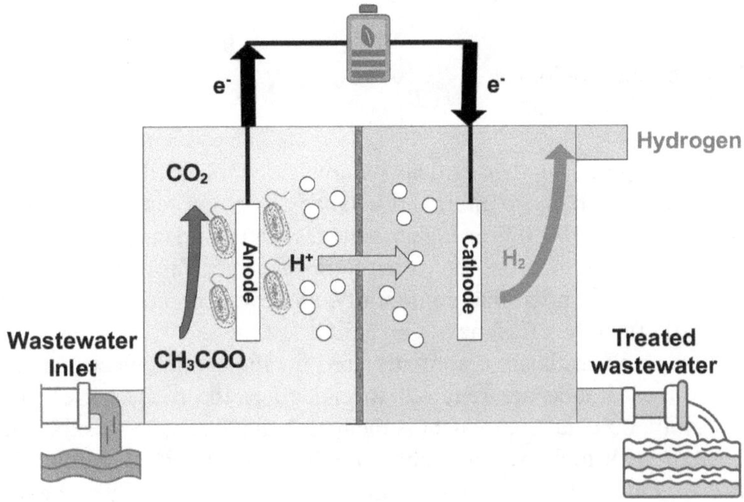

FIGURE 2.8 The structure of microbial electrolysis cells.

biofilms on electrodes (see Figure 2.8). Electrons and protons are released during this process. A prime example is the oxidation of acetate, which the following chemical formula can represent:

$$CH_3COO^- + 4H_2O \rightarrow 2HCO_3^- + 9H^+ + 8e^- \tag{2.7}$$

Electrons are released, which flow to the cathode through an external circuit while protons migrate through the membrane. Reduction of protons to hydrogen gas occurs in the cathode chamber when an external voltage is supplied, and the following equation is valid: $8H^+ + 8e^- \rightarrow 4H_2$. These processes are interlinked. Therefore, overall reactions will result in:

$$CH_3COO^- + 4H_2O \rightarrow 2HCO_3^- + 4H_2 + H^+ \tag{2.8}$$

Different values can be measured to calculate the efficiency of an MEC. Coulombic efficiency (CE) measures the proportion of electrons recovered as hydrogen, which is defined as

$$CE = \frac{\eta_{CE}}{\eta_{th}} \tag{2.9}$$

Where:

- η_{CE} is the moles of hydrogen produced from the measured current.
- η_{th} is the theoretical moles of hydrogen.

$$\eta_{CE} = \frac{CP}{2F} \tag{2.10}$$

Where:

- CP is the total Coulombs.
- F is Faraday's constant.

The applied voltage is another critical parameter, with the theoretical minimum around 0.14 V. However, due to internal resistance and overpotentials, practical applications often require 0.3–1.0 V or higher voltages. Current density is also crucial in hydrogen production to measure the electron transfer rate. Various electrode materials, the membrane, the microbial community, and even conditions such as pH and temperature all affect the efficiency of an MEC.

Carbon-based materials are commonly used for the anode's electrode, and other materials for the cathode are required to encourage the reduction of protons to hydrogen. Another crucial component is the membrane that separates the anode and cathode chambers. It must be permeable to protons, so the gases produced in both chambers do not mix. The microbial community within the anode chamber is crucial to the system since it catalyzes the oxidation of organic materials.

2.4.2 Electrolyzer-Associated Calculations

In chemistry, we employ the term Gibbs free energy (ΔG) to check if a reaction happens spontaneously (on its own). This value tells us how energy is transformed into a chemical reaction. Gibbs free energy (ΔG) is a thermodynamic potential defined as the maximum work obtainable from a thermodynamic system at constant temperature and pressure.

In the production of green hydrogen using an electrolyzer, for instance, reversible cell potential $E_{0,rev}$ has a pivotal role in electrochemical phenomena. It is defined as the lowest amount of electrical energy that can be used to dissociate H_2O into H_2 and O_2. Following is how it is calculated [12]:

$$E_{0,rev} = \frac{\Delta G}{(n \times F)} \tag{2.11}$$

Where:

- Gibbs energy (= 236483 J/mol).
- Faraday constant (= 96,485 C/mol).

Therefore,

$$E_{0,rev} = \frac{236.483 \text{ kJ/mol}}{(2 \times 96{,}485 \text{ C/mol})} = 1.23 \text{ V}$$

However, a thermal external source is not always available, so another concept is the so-called thermoneutral voltage at standard state. $E_{0,th}$ represents the voltage required for the reaction to occur at standard conditions (1 atm pressure and 298.15 K temperature) without any external heat source. It is calculated using the reaction's enthalpy change (ΔH), which is, for water splitting, approximately 285,830 J/mol. This value is the summation of Gibbs energy 236,483 J/mol, and the remaining is consumed for heating purposes [12], so:

$$E_{0,th} = \frac{\Delta H}{(n \times F)} \tag{2.12}$$

Hence,

$$E_{0,th} = \frac{285.83 \text{ kJ/mol}}{(2 \times 96{,}485 \text{ C/mol})} = 1.48 \text{ V}$$

The current flow through the electrolysis cell during the electrolysis or "water-splitting process" requires a specific voltage, referred to as the reversible cell potential. The actual voltage needed, however, is always higher. This additional voltage requirement primarily stems from losses within the cell that can never be recuperated. Such losses can be subdivided into two major classes: energy loss through Faraday's law, known as activation energy losses, or non-Faradaic losses, which are ohmic and mass transfer/concentration losses. Activation losses stem from the electrochemical reactions at the anode and cathode of the cells. This type of loss is paramount in explaining the low efficiency of the electrolysis cycle at low currents and high voltages. This observation is in accordance with Faraday's law that the amount of gas, in this case, hydrogen, produced in an electrochemical reaction is directly proportional to the current supply to the cell [12]. This can be expressed as

$$Q = n \times F \tag{2.13}$$

Where:

- Q represents the charge in Coulombs (C).
- n is the number of moles of electrons transferred during the reaction.

Any chemical reaction needs or performs work and is calculated using Equation 2.14 [13]. When it performs a work, it equals to

$$W = n \times F \times E_{cell} \tag{2.14}$$

However, water electrolysis requires work and energy, so the energy is required for hydrogen production.

$$E = -n \times F \times E_{cell} \tag{2.15}$$

Where:

- E is the energy in Joules.
- n is the number of electrons transferred (2 mol for water electrolysis).
- F is Faraday's constant (96,485 C/mol).

- E_{cell} is the cell potential (in volts).
- Negative signs because the reaction requires energy.
- The standard cell potential for water electrolysis is $E_{0,rev} = 1.23\ V$ if an external heat source is available, and it equals to $E_{0,th} = 1.48$, assuming no external heat source. In an ideal situation, 1.23 V of external potential would initiate water splitting in an electrochemical cell.

Example 2.1: Energy Required for Water Electrolysis

What is the energy required for water electrolysis, and how to calculate the energy required for this process, assuming there is an external heat source?

- The standard cell potential for this electrolysis setup is 1.23 V, with an external heat source.
- The Faraday constant (F) is approximately 96,485 C/mol (coulombs per mole).

Solution:

Water electrolysis involves the following balanced chemical reaction:

$$2H_2O(l) \rightarrow 2H_2(g) + O_2(g)$$

In this reaction, 2 mol of electrons (n) are transferred. The value of n=2 in water electrolysis comes from the above-balanced chemical equation for the reaction. Using (equation 2.15):

$$E = -n \times F \times E_{cell}$$

$$E = -2\ \text{mol} \times 96{,}485\ \text{C/mol} \times 1.23\ \text{V}$$

$$E = -238{,}509\ \text{J/mol}$$

The energy required to split one mole of water into hydrogen and oxygen gas without an external heat source is approximately 238,509 J.

To calculate the energy required to produce 1 kg of hydrogen. It is necessary to consider hydrogen molar mass and density. The molar mass of hydrogen is approximately 2.016 g/mol; in other words, 1 mol of hydrogen gas (H_2) has a mass of 2.016 g.

Example 2.2: Energy Required for Hydrogen Production

Calculate the energy (in kWh) required to produce 1 kg of hydrogen gas (H_2) through water electrolysis. Given that the standard cell potential for the electrolysis setup is 1.23 V.

Solution:

As shown in Example 2.1, the energy required to split 1 mol of water into hydrogen and oxygen gas with an external heat source is approximately 238,509 J. The first step in finding the required energy to produce 1 kg of hydrogen is to calculate the number of moles per kilogram as follows:

$$\text{Number of moles} = \frac{\text{Mass (kg)}}{\text{Molar mass (g/mol)}} = \frac{1000\ \text{g}}{2.016\ \text{g/mol}} = 49{,}603\ \text{mol}$$

The next step is to calculate the total electrical charge (Q), which is calculated by multiplying the number of moles of electrons by Faraday's constant (F). In this case, hydrogen H_2 has two electrons. Therefore, to find the total charge, the number of electrons in H_2 is multiplied by Faraday's constant, which is 96,485 C/mol.

$$Q = 496.03\ \text{mol} \times 2 \times 96{,}485\ \text{C/mol} = 95{,}718{,}909.1\ \text{C}$$

Hence, the required energy can be calculated as follows:

$$E = Q \times V$$

Where V is the cell potential. So,

$$E = 95{,}718{,}909.1\ \text{C} \times 1.23\ \text{V} = 117{,}734{,}258.193\ \text{J}$$

Finally, convert the joules to kWh: 1 kWh = 3,600,000 J. Therefore,

$$E = \frac{117{,}734{,}258.193\ \text{J}}{3{,}600{,}000\ \text{J/kWh}} = 32.7\ \text{kWh}$$

Considering that no external heat source is available, approximately 33 kWh of energy is needed to produce 1 kg of hydrogen through water electrolysis.

2.4.3 Thermolysis

Thermolysis, also known as thermal water splitting, is a technique that utilizes an external heat source to obtain hydrogen from water. It is possible to directly decompose water in a single step without relying on chemical reactions, as illustrated in equation 2.16 [14]. However, the process still requires 4000°C to completely break water's strong bonds through a water decomposition reaction that exists in the strongest form [15].

$$H_2O + \text{Heat} \rightarrow \frac{1}{2}O_2 + H_2 \tag{2.16}$$

Excessively high operating temperatures pose one of the main challenges for water thermolysis. The process must be started at least 1700°C, and even higher temperatures, up to 4000°C, are needed for complete decomposition.

Alternatively, water can be decomposed in multiple steps, termed "thermochemical cycles," because it operates at a lower temperature than the direct decomposition method. Thermochemical cycles are substitutes for one-step thermolysis, as they use lower temperatures, creating a safer, more controlled environment, unlike single-step thermolysis. In these cycles, heat can be provided by renewable energy sources such as concentrated solar power [16].

2.4.4 Photo-Electrolysis

Photo-electrolysis is the cleavage of water into hydrogen and oxygen using light energy. It involves the use of a photoelectric chemical cell (PEC) or a photo-electrolyzer, which typically repairs into five principal parts. A photoelectrode (also known as the light electrode), the primary part to collect light, for both oxidation and reduction electrodes, a cathode and anode, which serve as electrolytes, and lastly, a transparent conductive substrate (CS) [14].

As noted earlier, the photoelectrode is the part constituting the semiconductor absorbing the semiconductor. TiO_2 and tungsten trioxide (WO_3), alongside other metal oxides, serve as semiconductors with appropriate bandgap, which can be accomplished through sulfonation, thus increasing suitability for mercerization. The platinum electrodes serve as the nucleus where electro-reduction processes occur. An anode is made from a conductive material such as platinum or other metals. This allows high efficiency during enzymatic reactions where protons (H^+) and electrons (e^-) are converted into hydrogen gas in the presence of a catalyst alongside an electrolyte (E) component [14].

The electrolyte provides the supporting environment through which current ions can circulate within the photo-electrolyzer. Providing an appropriate environment, the substrates, which are made of indium tin oxide or fluorine-doped tin oxide, enable protons H^+ to pass during electrolysis, enabling the flow of ions during the process of photoelectric conversion while also being transparent and hence allowing light to reach the photoelectrode.

In a photo-electrolyzer, an external connection between the photoelectrode and the cathode is required to enable the flow of electrons generated during the photoexcitation process and the subsequent electrochemical reactions, allowing for the separation and collection of hydrogen and oxygen gases. Also, a light source is required to provide the photons necessary for the photo-electrolysis process, typically sunlight or simulated sunlight [17]. Figures 2.9 and 2.10 represent schematics of research topics and experiments of photoelectrolysis in the literature. You can explore these references for more insights into this technology [6,18].

2.5 CHAPTER SUMMARY

This chapter presented several methodologies for hydrogen production utilizing both fossil fuels and renewable resources. It discussed hydrogen's categorization by color and its associated characteristics, focusing on the source, feedstock, and production processes employed. Special attention was given to green hydrogen, defined as hydrogen produced from renewable resources.

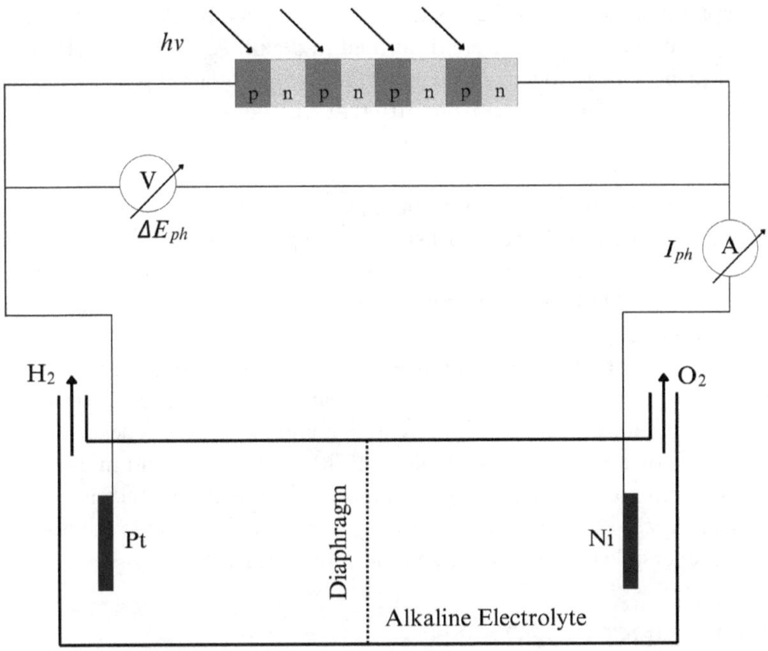

FIGURE 2.9 Case study on water photo-electrolysis on TiO_2 surface. (Adapted from Juodkazis et al. [18].)

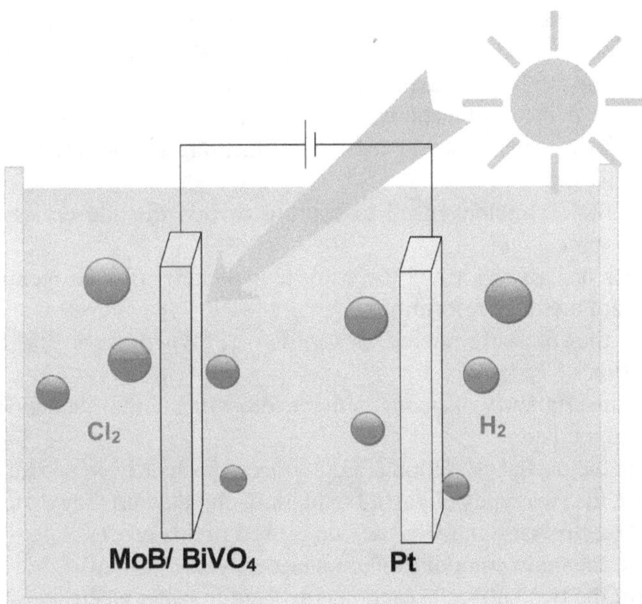

FIGURE 2.10 Seawater photo-electrolysis. (Adapted from Gao et al. [6].)

The chapter also described the electrolyzer, the device used to perform electrolysis. Four types of electrolyzers were described in detail: ALE, SOE, ME, and PEM. In addition to green hydrogen, the chapter addressed other types, including gray, blue, and pink hydrogen, derived from different feedstocks.

Furthermore, the chapter explored conventional methods of hydrogen production, such as steam reforming and POX, which utilize fossil fuel resources. The steam reforming process, involving the reaction of hydrocarbon fuel with steam in the presence of a catalyst, was explained, focusing on the production of synthesis gases and the water–gas shift reaction. The chapter also detailed the POX process, an exothermic reaction where feedstock reacts with a small amount of air, releasing heat and producing synthetic gas.

Moreover, the chapter provided an overview of green hydrogen production methods, emphasizing water electrolysis using renewable energy. The electrolysis process, involving the breakdown of water into hydrogen and oxygen, was described, and the four main types of electrolyzers—alkaline, PEM, solid oxide, and microbial—were compared in terms of their electrolyte composition, operating conditions, and ionic agents. Electrolyzer-associated calculations were also presented, including Gibbs free energy, reversible cell potential, and thermoneutral voltage.

SELF-ASSESSMENT QUESTIONS

1. How is hydrogen classified into different colors?
2. What are the three main methods for green hydrogen production from water?
3. What is the primary goal of the 2015 Paris Agreement about climate change?
4. What does the term "green hydrogen" specifically refer to?
5. What is the most common method for large-scale hydrogen production?
6. Describe the process of steam methane reforming.
7. What is the purpose of a catalyst in steam reforming?
8. What happens during steam methane reforming in the Water–Gas Shift (WGS) reactor?
9. What is the technology used to capture carbon dioxide emissions from industrial processes?
10. How does using CCS affect the color classification of hydrogen produced from steam methane reforming?
11. Explain the pressure swing absorption (PSA) process for hydrogen purification.
12. What is the efficiency of steam methane reforming, and what happens to the unreacted methane?
13. Describe the partial oxidation (POX) process for hydrogen production.
14. What are the two types of partial oxidation, and how do they differ?
15. What is electrolysis, and what device is used in this process?
16. What are the main components of an electrolyzer?
17. What are the two half-cell reactions involved in water electrolysis?
18. Name the four types of electrolyzers based on their electrolyte composition and operating conditions.

19. What is the most common type of water electrolysis technology?
20. Describe the principle of operation of an alkaline electrolyzer.
21. What is the key characteristic of a PEM electrolyzer?
22. What materials are typically used for the anode and cathode in a PEM electrolyzer?
23. What is the role of the external power source in a PEM electrolyzer?
24. Explain the operating principle of a PEM electrolyzer.
25. What are the main components of a solid oxide electrolyzer cell (SOEC)?
26. At what temperatures do SOECs typically operate?
27. What is Gibbs free energy, and why is it important in understanding chemical reactions?
28. What does reversible cell potential ($E_{0,rev}$) represent in electrochemical processes?
29. What is the thermoneutral voltage at the standard state ($E_{0,th}$)?
30. Why is the voltage required for water electrolysis higher than the reversible cell potential?
31. What are the two main types of losses in an electrolysis cell?
32. What does Faraday's law state in the context of electrolysis?
33. How do you calculate the energy required for hydrogen production through water electrolysis?
34. What is thermolysis or thermal water splitting?
35. What is the main challenge of water thermolysis?
36. What are thermochemical cycles, and why are they used as an alternative to one-step thermolysis?
37. Describe the process of photo-electrolysis for hydrogen production.
38. What are the key components of a photo-electrolyzer?
39. What is the role of the photoelectrode in a photo-electrolyzer?

REFERENCES

1. Anthony Velazquez Abad and Paul E. Dodds. 2017. "Production of hydrogen." In *Encyclopedia of Sustainable Technologies*, pp. 293–304. Elsevier. https://doi.org/10.1016/B978-0-12-409548-9.10117-4.
2. Kumar, Shashi, Surendra Kumar, and Jitendra K. Prajapati. 2009. "Hydrogen production by partial oxidation of methane: Modeling and simulation." *International Journal of Hydrogen Energy* 34 (16): 6655–6668. https://doi.org/10.1016/j.ijhydene.2009.06.043.
3. Makaryan, Iren A., Eugene A. Salgansky, Vladimir S. Arutyunov, and Igor V. Sedov. 2023. "Non-catalytic partial oxidation of hydrocarbon gases to syngas and hydrogen: A systematic review." *Energies* 16 (6): 2916. https://doi.org/10.3390/en16062916.
4. Speight, James G. 2017. "Gasification processes for syngas and hydrogen production." In *Gasification for Syngas and Hydrogen Production*, edited by J. G. Speight, pp. 119–146. Elsevier. https://doi.org/10.1016/B978-0-85709-802-3.00006-0.
5. Chi, Jun and Hongmei Yu. 2018. "Water electrolysis based on renewable energy for hydrogen production." *Chinese Journal of Catalysis* 39 (3): 390–394. https://doi.org/10.1016/S1872-2067(17)62949-8.
6. Gao, Rui-Ting, Xiaotian Guo, Shujie Liu, et al. 2022. "Ultrastable and high-performance seawater-based photoelectrolysis system for solar hydrogen generation." *Applied Catalysis B: Environmental* 304: 120883. https://doi.org/10.1016/j.apcatb.2021.120883.

7. Shiva Kumar Sampangi and Vurimindi Himabindu. 2019. "Hydrogen production by PEM water electrolysis—A review." *Materials Science for Energy Technologies* 2 (3): 442–454. https://doi.org/10.1016/j.mset.2019.03.002.
8. Keçebaş, Ali, Muhammet Kayfeci, and Mutlucan Bayat. 2020. "Electrochemical hydrogen generation." In *Hydrogen Energy and Fuel Cells*, edited by Francesco Calise, Massimo Dentice D'Accadia, Massimo Santarelli, Andrea Lanzini, and Domenico Ferrero, pp. 299–317. Elsevier. https://doi.org/10.1016/B978-0-12-814853-2.00009-6.
9. Naimi, Youssef and Amal Antar. 2018. "Hydrogen generation by water electrolysis." In *Advances in Hydrogen Generation Technologies*, edited by Murat Eyvaz. Rijeka: IntechOpen. https://doi.org/10.5772/intechopen.76814.
10. Hauch, Anne, Rainer Küngas, Peter Blennow, et al. 2020. "Recent advances in solid oxide cell technology for electrolysis." *Science* 370 (6513): eaba6118. https://doi.org/10.1126/science.aba6118.
11. Burton, Nathan A., Ricardo Vazques Padilla, Andrew L. Rose, et al. 2021. "Increasing the efficiency of hydrogen production from solar powered water electrolysis." *Renewable and Sustainable Energy Reviews* 135: 110255. https://doi.org/10.1016/j.rser.2020.110255.
12. Daniela Sofia Falcão and Alexandra Ferreira Rodrigues Pinto. 2020. "A review on PEM electrolyzer modelling: Guidelines for beginners." *Journal of Cleaner Production* 261: 121184. https://doi.org/10.1016/j.jclepro.2020.121184.
13. Jadhav, Divya and Mohan Thakare. 2022. "Mathematical modeling and simulation of a fuel cell in MATLAB." In *Proceedings of the Third International Conference on Contents, Computing & Communication (ICCCC-2022)*, pp. 1–10. SSRN. https://dx.doi.org/10.2139/ssrn.4056056.
14. Dincer, Ibrahim and Canan Acar. 2015. "Review and evaluation of hydrogen production methods for better sustainability." *International Journal of Hydrogen Energy* 40 (34): 11094–11111. https://doi.org/10.1016/j.ijhydene.2014.12.035.
15. Ji, Mengdi and Jianlong Wang. 2021. "Review and comparison of various hydrogen production methods based on costs and life cycle impact assessment indicators." *International Journal of Hydrogen Energy* 46 (78): 38612–38635. https://doi.org/10.1016/j.ijhydene.2021.09.142.
16. Mehrpooya, Mehdi and Roghayeh Habibi. 2020. "A review on hydrogen production thermochemical water-splitting cycles." *Journal of Cleaner Production* 275: 123836. https://doi.org/10.1016/j.jclepro.2020.123836.
17. Nikolaidis, Pavlos and Andreas Poullikkas. 2017. "A comparative overview of hydrogen production processes." *Renewable and Sustainable Energy Reviews* 67: 597–611. https://doi.org/10.1016/j.rser.2016.09.044.
18. Juodkazis, Kęstutis, Jurga Juodkazytė, Edgaras Jelmakas, et al. 2010. "Photoelectrolysis of water: Solar hydrogen–achievements and perspectives." *Optics Express* 18 (102): A147–A160. https://doi.org/10.1364/OE.18.00A147.

3 Techno-Economic Aspects of Green Hydrogen

3.1 INTRODUCTION

The production of green hydrogen could prove beneficial to the decarbonization of an economy's transport and industrial sectors. It may be produced by the electrolysis of water, which requires renewable energy resources such as solar, wind, concentrated solar power, hydropower, or any renewable energy source, resulting in complete zero-carbon emissions. Therefore, the technical system for green hydrogen production requires meticulous planning and design to guarantee sustainable production. The economic and financial viability facets shall also be integrated into the optimal system design.

An effective system design for PV or wind renewable energy hydrogen systems incorporates all relevant variables for maximum efficiency. Factors such as the uncertain and fluctuating nature of weather conditions and the associated renewable energy reliability must be addressed during the plant design process. In addition, the efficiency of the electrolyzer also impacts the amount of renewable energy needed to produce a specific amount of hydrogen.

3.2 TECHNICAL ASPECTS

A technical assessment of green hydrogen is necessary to ensure system reliability and sustainability. Thus, considering system parameters and resource availability, a precise design is required to start a renewable energy system for hydrogen production. The hydrogen plant component is illustrated in Figure 3.1.

The power produced by the PV system depends on many factors, including solar radiation and the size of the system; one of the crucial factors to consider is the temperature, as any increase or decrease in the temperature will affect the output power, for instance, an increase or decrease in 1°C affects the production with 0.5% [1]. The hydrogen production from electricity can be calculated using the calorific ratio between the hydrogen and the electricity, which is almost 39.4 kWh/kg, i.e., 39.4 kWh to produce 1 kg of hydrogen. However, this number is for a 100% utilization of PV power to produce hydrogen and for a 100% efficient electrolyzer.

Therefore, more PV power should be produced for 1 kg of hydrogen production. For example, a 65% utilization and 60% electrolyzer efficiency mean that 101 kWh should be produced to produce 1 kg of hydrogen successfully. This will be discussed later in this chapter.

DOI: 10.1201/9781003639190-3

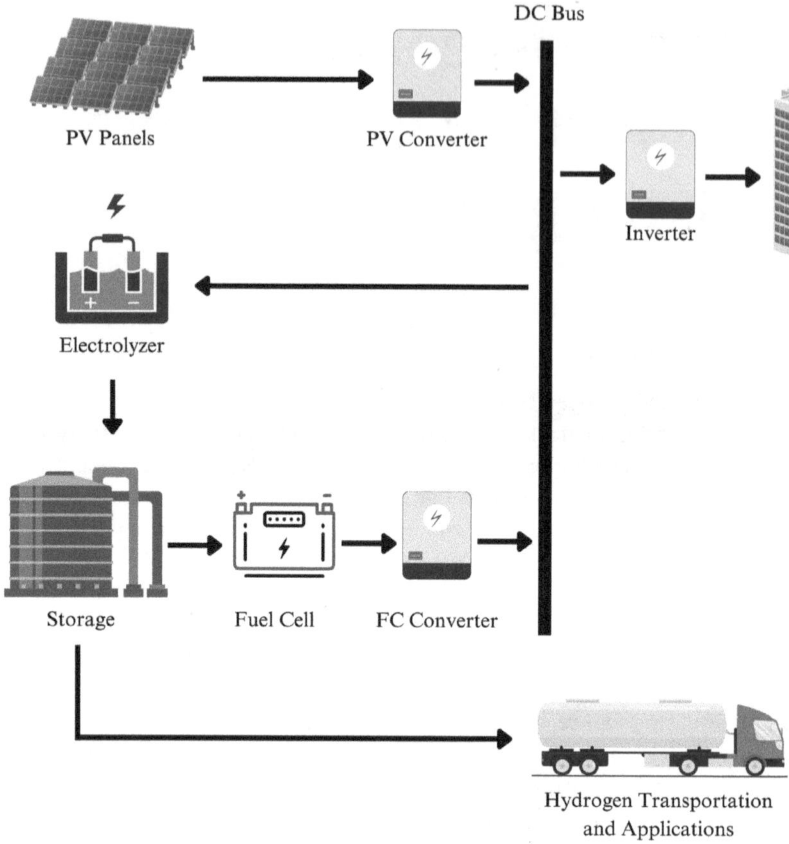

FIGURE 3.1 Green hydrogen production plant component.

After estimating the amount of hydrogen that can be produced, the number of electrolyzers is selected based on their capacities. The available electrolyzer efficiency reaches 75% (52.5 kWh to produce 1 kg) [2–4].

3.2.1 Energy Consumption for Hydrogen Production

In order to estimate daily and monthly hydrogen production from renewable energy, it is necessary to calculate the amount of renewable energy available at the desired location and the efficiency of the hydrogen production process. This can be done through a process as follows:

- Determining renewable energy resources. The amount of renewable energy available at the location depends on the specific resource and related weather conditions, such as solar radiation and wind speed. Estimating the amount of renewable energy available by analyzing historical data or using software that considers the location's weather, such as PVsyst, SAM, and PVgis.

- Calculating the energy available for hydrogen production. If renewable energy is available, estimating the amount of energy that can produce hydrogen is necessary. This calculation will depend on electrolysis efficiency, which ranges from 60% to 80%.
- To estimate the monthly hydrogen production, you multiply the daily production by the number of days in the month, and the same is true for yearly production.

This section discusses the energy required to produce hydrogen, specifically the power required by the electrolyzer.

P_e can be calculated using the following general equation:

$$P_e = V_e \times I_e \tag{3.1}$$

Where:

- V_e: input voltage
- I_e: input current

So, the mass of the produced hydrogen m_{H_2} from utilizing P_e is calculated using the following equation:

$$m_{H_2} = \frac{\eta_e \times \left(P_e \times HV_{P_e}\right)}{HV_{H_2}} \tag{3.2}$$

where

- η_e is the efficiency of the electrolyzer.
- HV_{P_e} is the heating value of the electrical power supplied to the electrolyzer.
- HV_{H_2} is the heating value of hydrogen.

Knowing that HV_{P_e} and HV_{H_2} equals 860 and 33,889 kcal/kWh, respectively [2]. Table 3.1 estimates the energy required to produce 1 kg of green hydrogen under different electrolyzer efficiencies.

TABLE 3.1
Required Input Power to Produce 1 kg H_2 Green Hydrogen under Different Efficiencies of the Electrolyzer

Electrolyzer Efficiency (%)	Required Input Power to Produce 1 kg H_2 (kWh)
100	39.4
90	43.8
80	49.2
70	56.3
65	60.6
60	65.0

The energy required to produce 1 kg of hydrogen can vary depending on the efficiency of the electrolysis process used. The minimum theoretical energy required to produce 1 kg of hydrogen using water electrolysis is approximately 39.4 kWh/kg, assuming 100% efficiency.

However, in practice, the efficiency of water electrolysis is typically around 70%–80%. Therefore, the energy required to produce 1 kg of hydrogen would be higher than the theoretical minimum. If we assume an electrolysis efficiency of 80%, the energy required to produce 1 kg of hydrogen would be approximately:

$$\frac{39.4\ \text{kWh/kg}}{0.8} = 49.25\ \text{kWh/kg}$$

The amount of electrical energy required to produce 1 kg of hydrogen depends on whether it is produced from liquid water or steam. If it is produced from liquid water, the theoretical amount of electrical energy needed is about 39 kWh/kg, at 100% electrolyzer efficiency, the higher heating value of hydrogen. However, if it is produced from steam, it requires less energy, about 33 kWh/kg, and at 100% electrolyzer efficiency, which is hydrogen's lower heating value [5].

3.2.2 Design and Sizing Steps

In Section 3.2, hydrogen production is discussed in terms of the electricity needed for electrolysis, considering the efficiency of the electrolyzer. This section will introduce the design steps and sizing procedure of the renewable energy system and electrolyzer. As shown in Figure 3.2, the designers of green hydrogen systems need to have more insights into the design steps of the whole system, starting from designing renewable energy systems for operating the plant, producing hydrogen, and treating water, and ending with the electrolyzer and compressor sizing. Therefore, these steps should be thoroughly considered in system design.

3.2.3 System Requirements

The system design requirements for a green hydrogen system include several components that are very important. First, there's selecting the type of renewable energy system, which could be a photovoltaic (PV) or wind system that will act as the primary energy provider to the system. All systems must have a certain total capacity that needs to be defined considering hydrogen production and the electrolysis energy demand.

Providing system performance figures during different times of the year requires adequate and precise input data such as solar irradiance and wind speed. Another very important parameter to consider is the electrolyzer efficiency, particularly how the energy converted to hydrogen changes with the energy supplied.

Countries in the MENA region have an enormous potential for harnessing renewable energy resources such as solar energy. For instance, Jordan has high solar irradiation along with huge PV field projects around the country. The volume of solar energy required to be available to reach a specific hydrogen production target must

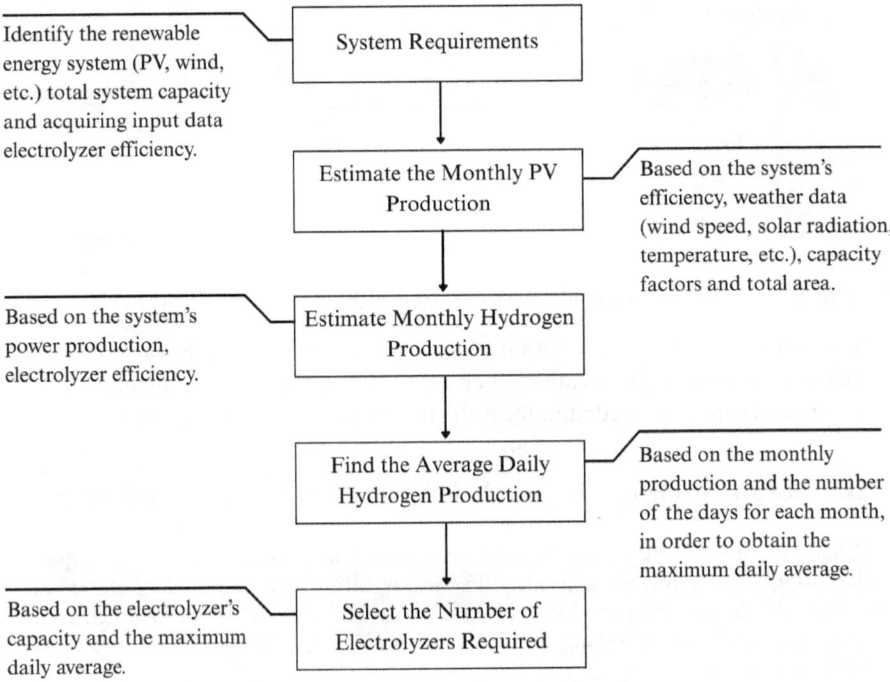

FIGURE 3.2 Green hydrogen production plant design and sizing steps.

be estimated to determine what size of solar energy system would be appropriate for hydrogen production.

Based on Table 3.1, the energy required to produce 1 kg of hydrogen is about 39.4 kWh, given that the electrolysis process's efficiency is 100%. Hence, to find the extent of the solar PV array required, we need to determine the amount of solar energy needed to generate 39.4 kWh of energy in a day while maintaining a constant production rate of hydrogen at 1 kg/day.

The solar energy potential from the location depends on the solar irradiance at the site for system installation, measured in peak sun hours. If we take 5 peak sun hours a day as the average, we may find the necessary area of the solar PV array using the equation below:

$$\text{PV capacity (kWp)} = \frac{\text{Energy (kWh/day)}}{\text{Peak sun hours}} \tag{3.3}$$

Using the above equation, the target is to produce only 1 kg of hydrogen daily. If the electrolysis efficiency is 80% (i.e., 39 kWh is needed to produce 1 kg of hydrogen), we can calculate the required size of the solar PV system: PV capacity (kWp) = (39 kWh)/5.

Thus, we would need a solar PV system with a capacity of approximately 7.8 kWp to produce 1 kg of hydrogen/day, assuming a 100% efficiency electrolyzer and an average of 5 peak sun hours/day.

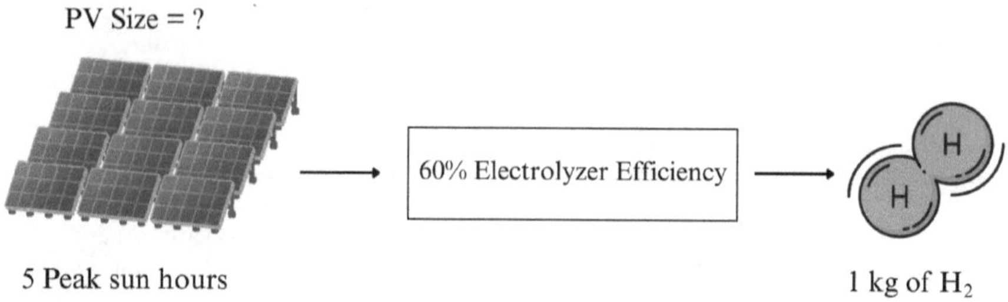

FIGURE 3.3 Determining the photovoltaic system size for hydrogen production.

In real implementation, the actual PV system size required may vary depending on different factors in the location, such as solar irradiance and temperature, so a higher capacity may be needed to meet the required energy; see Figure 3.3.

Example 3.1: Sizing of a Photovoltaic System for Hydrogen Production

What size of solar PV system would be required to produce 1 kg of hydrogen/day using an electrolyzer with an efficiency of 60%? If the amount of electrical energy needed to produce 1 kg of hydrogen is 65 kWh/kg and an average of 5 peak sun hours/day. Knowing that HV_{P_e} and HV_{H_2} equals860 and 33,889 kcal/kWh, respectively.

Solution:

The amount of electrical energy needed to produce 1 kg of hydrogen is calculated using equation 3.2:

$$1\ (\text{kg}) = \frac{0.60 \times \left(P_e\ (\text{kWh}) \times 860\ (\text{kcal/kWh})\right)}{33{,}889\ (\text{kcal/kWh})}$$

$$P_e = 65\ \text{kWh/kg}$$

The amount of electrical energy needed to produce 1 kg of hydrogen with an efficiency of 60% is 65 kWh/kg.

Using (equation 3.3), the capacity of PV system can be found:

$$\text{PV capacity}\ \ (\text{kWp}) = \frac{65\ \text{kWh}}{5} = 13\ \text{kWp}$$

In this case, we would need a solar PV system with a capacity of approximately 13 kWp to produce 1 kg of hydrogen/day, assuming an electrolyzer efficiency of 60% and an average of 5 peak sun hours/day.

The intermittent nature of wind energy introduces a significant challenge in sizing the wind energy system for continuous hydrogen production. The wind turbine capacity must be determined considering the wind resource's capacity factor to ensure a consistent hydrogen output. The capacity factor reflects the ratio of actual energy output to the maximum potential output. A low-capacity factor necessitates a higher installed wind turbine capacity to compensate for periods of low wind availability. The following equation 3.4 explains the parameters that affect the wind power density P:

$$P = \frac{1}{2}\rho A V^3 \tag{3.4}$$

where

- A is the swept area in m^2.
- ρ is the air density in kg/m^3.
- V is wind velocity in m/s.

Dividing both sides in (equation 3.4) by A yields

$$\frac{P}{A} = \frac{1}{2}\rho V^3 \tag{3.5}$$

However, wind speed is not always given at the desired height, so to estimate the wind speed value under a certain elevation, see Figure 3.4, the following equation is used [6]:

$$\frac{P}{A} = V(Z_R) = V(Z) \times \frac{\ln(Z_R/Z_0)}{\ln(Z/Z_0)} \times \frac{1}{2}\rho V^3 \tag{3.6}$$

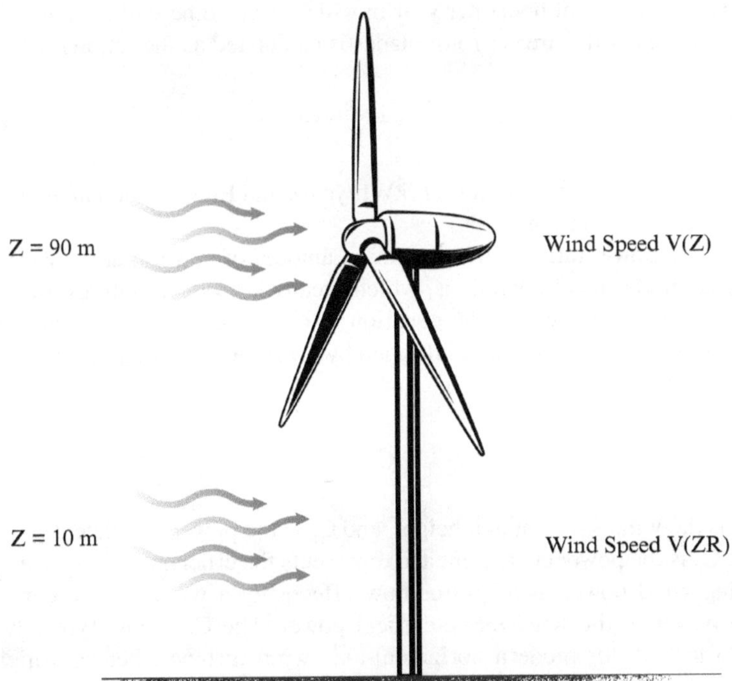

FIGURE 3.4 Illustration of the wind speed value under a specific elevation.

TABLE 3.2
Values of Z_0 at Different Terrain Descriptions [3]

Terrain Description	Z_0 (m)
Open, flat terrain, few isolated obstacles	0.05
Low crops, occasional large obstacles	0.1
Parkland, bushes, numerous obstacles	0.5
Regular large obstacle coverage (suburb, forest)	1.0
City center with high- and low-rise buildings	≥2.0

The wind speed at the hub height of the wind turbine is denoted by $V(Z_R)$, while V(Z) represents the wind speed measured at the meteorological station, which is generally located at a height of 10m; Z is the desired height. The roughness height, Z_0, depends on the terrain and can be found in Table 3.2 [6].

The actual useful energy density is the amount of kilowatt hour per year per square meter kWh/(m²/year) as shown in the following equation:

$$E_D = \frac{P}{A} \times n \tag{3.7}$$

where n is the number of hours per year in which the turbine will operate.

The rotor area or the area of rotor blades is calculated as the following:

$$A = \frac{E}{E_D} \tag{3.8}$$

where E is the actual energy required (kWh/year), and E_D is the actual annual useful energy density (kWh/m²/year).

The turbine power rating is the maximum amount of power that the wind turbine can generate under ideal conditions, which means it produces energy 100% of the time at optimal wind speeds. The equation to calculate the turbine's actual power rating is the actual power density multiplied by the rotor area of the turbine, as shown below:

$$P_T = \frac{1}{2} C_p \times \rho A V^3 \tag{3.9}$$

where V is the wind speed at hub height, and C_p is the power coefficient.

While C_p is the power coefficient and represents the efficiency of the wind turbine in utilizing wind power. It measures how effectively a wind turbine converts the available power in the wind into electrical power. The C_p value typically ranges from 0.25 to 0.45 for modern horizontal-axis wind turbines; hence, a higher C_p value indicates a more efficient turbine.

However, the wind turbine is not operating 24 hours at this rated power as the wind speed is not optimal throughout the year.

This means that the energy harvested from wind turbine/s is less than the rated energy produced in perfect conditions. This so-called capacity factor of the wind turbine or wind farm C_F is the percentage of the energy produced to the energy at rated power.

$$C_F = \frac{E_{actual}}{E_{rated}} \tag{3.10}$$

This means that the size of wind turbine/s in (kW) to produce specific energy equals the following:

$$P_{size} = \frac{P_T}{C_F} \tag{3.11}$$

Recent studies have shown that the capacity factor of Jordanian wind farms ranges from 23% to 50% [6].

Example 3.2: Sizing Wind Energy System for Hydrogen Production

Calculate the wind turbine size required to produce 100 kg of hydrogen/day using an electrolyzer with an efficiency of 60%. Hydrogen's heating value is 33,889 kcal/kWh, and the heating value of the supplied electrical power to the electrolyzer is 860 kcal/kWh.

Assume wind speed at hub height is 11 m/s. Air density $\rho = 1.23$ kg/m³, 24 operation hours every day, $C_p = 0.3$ and $C_F = 45\%$.

Solution:

The amount of electrical energy needed to produce 1 kg of hydrogen is calculated using equation 3.2. Hence, the amount of electrical energy needed to produce 1 kg of hydrogen with an efficiency of 60% is 65 kWh/kg. Consequently, the energy needed to produce 100 kg of hydrogen is 6500 kWh.

Thus, the rotor sweep area of the wind turbine is calculated as equation 3.8, but the energy density is calculated as equation 3.7.

Wind power is calculated as the following equation 3.6:

$$P = \frac{1}{2}\rho A V^3$$

So,

$$\frac{P}{A} = \frac{1}{2}\rho V^3 = \frac{1}{2}(1.23)(11)^3$$

$$\frac{P}{A} = 818.6 \text{ W/m}^2$$

$$E_D = \frac{P}{A} \times n = 818.6 \times 24 = 196.46 \text{ kWh}$$

The rotor sweep area of wind turbine/s is

$$A = \frac{E}{E_D} = \frac{6500}{196.46} = 33 \text{ m}^2$$

Then, we calculate the wind turbine/s capacity as the following:

$$P_T = \frac{1}{2} C_p \times \rho A V^3 = \frac{1}{2}(0.3)(1.23)(33)(11)^3 = 8103.8 \text{ W}$$

However, the energy harvested from wind turbine/s is less than the rated energy produced in perfect conditions by capacity factor. So, the size of the wind system to ensure the production of 100 kg of hydrogen is calculated as follows:

$$P_{size} = \frac{P_T}{C_F} = \frac{8103.8 \text{ W}}{0.45} = 18 \text{ kW}$$

3.2.4 Electrolyzer Sizing

Establishing the required number of electrolyzer units for a hydrogen production target implies accurate knowledge of the electrolyzer's production capacity and operating parameters for a given renewable energy source. The hydrogen production capacity is measured in kilograms per hour (kg/h) or normal cubic meters per hour (Nm^3/h), which becomes the most important value in the system's design. To compute the number of electrolyzer units to achieve the required hydrogen production quantities, the nominal production capacity of one electrolyzer stack must be determined. Several parameters can also affect the electrolyzer capacity, such as the technology used for an electrolyzer (alkaline, PEM, or SOEC), as well as an active electrode area and operating current density.

The variation in renewable energy sources such as solar PV and wind turbine arrays introduces severe uncertainty in the duty cycle of the electrolyzer. Thus, the capacity factor of the renewable energy installation, which is the ratio of energy produced to the possible maximum output, needs careful consideration. This parameter directly affects the average hydrogen production rate per unit of electrolyzer. The total installed electrolyzer capacity must be lowered to achieve the desired hydrogen production volume during the energy scarcity period. This reduction requires an increase in the size factor established from the statistical relationship between renewable energy resources and hydrogen output parameters.

As previously discussed, from the mass equation of the produced hydrogen m_{H_2} from utilizing P_e, the power supplied to the electrolyzer depends on the desired hydrogen quantity and the electrolyzer efficiency, as indicated in equation 3.2. This equation shall be used to identify the power needed for a certain mass of hydrogen derived from equation 3.2.

$$P_e = \frac{(m_{H_2} \times HV_{H_2})}{\eta_e \times HV_{P_e}} \tag{3.12}$$

TABLE 3.3
Specifications of the EX-425D System

Specification		Value
Input	Stack power consumption	Up to 1 MW
	Voltage and frequency	480 VAC, 60 Hz (USA)
	Stack power consumption	400 VAC, 50 Hz (EU)
	Water consumption	13 L/kg of H_2 produced
Output (hydrogen gas)	Volume	200 Nm^3/h
	Mass	425 kg/day
	Purity	Up to 99.999%
	Pressure	40 bars
Operational	Start up time	30 seconds warm/<5 minutes cold
	Average stack efficiency	49.9 kWh/kg (about 80%)
	Load following	Instantaneous
Physical environment	Installed footprint	29.3 m^2
	Ambient temperature	−20°C to +40°C (wider temperature range optional)

The per-hour mass of hydrogen production can be calculated as the following:

$$N_e = \frac{m_{H_2}}{C_e} \tag{3.13}$$

Before providing a numerical example for performing electrolyzer-related calculations, let us demonstrate a commercial electrolyzer to familiarize ourselves with the specifications of the available products.

For example, the Plug EX-425D electrolyzer was developed by the Plug company. This electrolyzer is designed to provide on-site hydrogen production, up to 425 kg/day. It adjusts the hydrogen production rate according to the available electric capacity, making such products a solution for integration with the grid or renewable energy resources. The specifications of the EX-425D system are in Table 3.3.

Example 3.3: Electrolyzer Sizing

If a company intends to produce 25,500 kg of hydrogen every month from a PV system in a location with an equivalent daily sunny hour of 5 hours, and the available electrolyzer capacity is 85 kg/h, how many electrolyzers are needed?

Solution:

Based on the above discussion, employing equation 3.12:

$$m_{H_2} = \frac{25{,}500 \text{ kg/month}}{5 \text{ hours} \times 30 \text{ days}}$$

$$m_{H_2} = 170 \text{ kg/h}$$

The number of electrolyzers is calculated as follows:

$$N_e = \frac{m_{H_2}}{C_e}$$

$$N_e = \frac{170}{85} = 2$$

Two electrolyzers with an 85 kg/h capacity are needed.

3.2.5 Compressor Sizing

Hydrogen is produced by electrolysis in a low-pressure state. It must be stored at higher pressures for different applications requiring the hydrogen flow rate at a specific value and pressure needed along with the appropriate type of compressor [7]. The compressors can be classified into different units such as mechanical (reciprocating, diaphragm, and screw) or thermal and electrochemical, where each has a specific set of functional and pressure ranges and a target application [8]. Selecting the right material is inherently difficult, particularly due to the increasing energy costs associated with the embrittlement of hydrogen, making it critical to optimize efficiency to meet high energy demand for compression. Specific end applications dictate the value of hydrogen's flow rate and pressure. For fuel cell vehicles, the storage pressure is about 700 bar [9].

Transportation and storage of hydrogen, especially in the case of pipelines, are essential whenever they are under specific pressures. Compressing hydrogen poses an investment and recurring expense when considering the overall cost, making it an important consideration in connection with economic viability. Dealing with high-pressure environments also requires dependable and safe operations. Thus, regular maintenance becomes vital. Material selection remains a challenge owing to the embrittlement of hydrogen. Advanced alloys and coatings can avoid compressor component failure [10]. Compression of hydrogen in the energy sector, which relies on pressure ratios and flow rates, is costly. Some techniques, such as intercoolers or variable speed drives, increase efficiency. It is important to effectively integrate the compressor within the electrolyzer system so that system efficiency increases and costs decrease. Novel compressor designs and materials are expected to transform the field of electrochemical compression. An effort was made to reduce the cost and increase the efficiency of the compression. The strategy of compressing centrally or decentrally is more strategic concerning the quantity of hydrogen and its purpose [11]. The realized price of hydrogen could be best optimized through electrochemical compression to cover the total expenses.

The amount of hydrogen produced by the electrolyzer must be compressed to find the compressor's total size. Hence, equation 3.14 can be used as follows:

$$N_c = \frac{m_{H_2}}{C_c} \tag{3.14}$$

TABLE 3.4
Specifications of the MHE 400–22Ex Compressor

Parameter	Value
Model name	MHE 400–22Ex
Year of construction	2017
Gas type	Hydrogen
Delivery capacity	173 Nm^3/h
Suction pressure	40–100 bar
Final pressure	350 bar
Drive power	22 kW

where C_c is the per-hour capacity of the compressor.

Storing hydrogen as a gas requires storage tanks with high pressure, which can carry hydrogen at pressures of 350–700 bars (5000–10,000 psi) or even higher. In this high-pressure tank, hydrogen is stored at an ambient temperature (around 15°C–25°C). To better understand compressor specifications, let us demonstrate a commercial example that could be used for hydrogen compression after electrolysis.

MHE 400–22Ex is an example of such a compressor. Its specifications are in Table 3.4.

Now, let us consider Example 3.3 again. Regarding the compressor's sizing, see Example 3.4 below.

Example 3.4: Sizing of Hydrogen Compressor

If we have two electrolyzers, each with a production rate of 200 Nm^3/h, and the available compressor delivery capacity is 173 Nm^3/h, how many compressors are needed?

Solution:

The two-electrolyzer production rate is 400 Nm^3/h, thus:

$$N_c = \frac{m_{H_2}}{C_c} = \frac{400}{173} = 2.3 \approx 3$$

The number of compressors needed is 3, which must be connected in parallel.

3.2.6 Water Use and Desalination

The increasing need for green hydrogen has made hydrogen production through water desalination one of its core pillars. The need for water input is apparent when considering that freshwater required to produce hydrogen through electrolysis is approximately 9 L/kg of hydrogen [2]. As is evident, relying on freshwater resources for hydrogen production poses a great challenge to regions struggling with water scarcity. However, the abundance of seawater makes it an attractive alternative, along with the need for effective desalination technologies.

Reverse osmosis is the primary method to convert seawater into the purified water required for electrolysis. While one of the most effective methods used to desalinate water, the RO technique also comes with a major drawback: the high energy cost of RO desalination presents a significant operational cost. For the process to be more economical, it must be integrated with renewable energy sources. Additionally, there is also a need for careful pretreatment and posttreatment procedures to ensure the electrolyzers work for a longer period without interruption.

In addition to energy usage, the implications for the environment regarding brine disposal, concentrated salt solution that comes as a by-product of RO, must be considered with caution. To lessen the negative effects of large-scale desalination [12], sustainable brine management strategies such as mineral extraction and industrial utilization are necessary. Also, the specific water quality required by various types of electrolyzers requires certain RO system designs, which further escalates the understanding of the relationship between the two technologies: desalination and electrolysis.

Evaluating the RO desalination plants for commercial use focusing on hydrogen production offers much regarding the systems' integration of technical and economic value. Preconditions such as the plant capacity, the energy quantity, in terms of megajoules, used for each cubic meter of the desalinated water, the membranes' period of operation before repair, and the cost amount of the desalination systems provide the needed base for hydrogen production project feasibility studies. In addition, knowing how to incorporate RO systems physically with the electrolyzer plants is important for enhanced system efficiency and reduced operational costs.

Powered by research and development, the advancements in RO technology are expected to yield significant improvements in energy efficiency and cost-effectiveness. It is also likely that hydrogen production costs can be lowered by introducing advanced membrane materials, energy recovery systems, and hybrid desalination processes, which can potentially reduce the environmental impacts of the processes [13]. As a clean, versatile energy carrier, the full potential of hydrogen can only be realized by integrating effective and sustainable water desalination technologies.

The 10-SWRO Watermaker is a commercial desalination system designed to produce freshwater from seawater. Table 3.5 shows its key specifications.

Let us consider the following example for calculating and sizing a desalination system dedicated to hydrogen production.

Example 3.5: Sizing of Desalination System

Producing 25,500 kg of hydrogen per month from seawater using reverse osmosis desalination to treat the seawater: The available desalination system production capacity is 10,000 L/day, and the energy consumption per 1 hour is 4.5 kWh. What is the size of the PV system to power this process?

Assume that the water consumption of the electrolyzer is 13 L/kg of hydrogen produced.

TABLE 3.5
Specifications of the 10-SWRO Watermaker

Specification		Value
Hydraulic data	Permeate production capacity	10,000 L/day
	Operating pressure	50–65 bar
	Total dissolved solids (TDS)	<550 mg/L
	Inlet flow rate	2.7 m^3/h
	Feed water temperature	5°C–40°C
	Salt rejection	up to 99.8%
	Membrane area	15 m^2
	Inlet flow rate	2.7 m^3/h
Electrical data	Nominal power	5 kW
	Energy consumption per 1 hour	4.5 kWh
	Shaft speed	1450 rpm
	Supply voltage	400 ± 10% VAC (3 ph)
	Supply frequency	50/60 Hz
	Consumption current	9.5 A

Solution:

Thirteen liters of freshwater are needed for every 1 kg of hydrogen produced.

$$V = \frac{25{,}500\ \text{kg/month}}{30\ \text{days}} \times 13\ \text{L} = 11{,}050\ \text{L/day}$$

So, producing 25,500 kg/month to desalinate (= 850 kg/day) of hydrogen/day, 11,050 L of freshwater/day are required.

The size of the system should cover the above-mentioned capacity, so based on the available system, the number of units needed to meet water quantity is

$$E = \frac{11{,}050\ \text{L/day}}{10{,}000\ \text{L/day}} = 1.15\ \text{units}$$

The energy required per day is

$$1.15\ \text{units} \times 4.5\ \text{kWh/h} \times 24\ \text{hours/day} = 124.2\ \text{kWh/day}$$

Assuming an average of 5 peak sun hours/day, we can calculate the required size of the solar PV array using the following equation:

$$\text{PV capacity (kWp)} = \frac{\text{Energy (kWh/day)}}{\text{Peak sun hours}}$$

The required size of the solar PV system:

$$\text{PV capacity}\left(\text{kWp}\right) = \frac{124.2\ \text{kWh}}{5} = 24\ \text{kWp}$$

Example 3.6: Water Desalination Energy Consumption

For the previous Example 3.5,

a. What is the required energy for desalinating 13 L of water to produce 1 kg of hydrogen?
b. What is the percentage of the energy required for desalination to the total energy required?

Note that the electrolyzer consumption is $49.9\ \text{kWh/kg}_{H_2}$.

Solution:

The desalination production capacity is 10,000 L/day, so, the per hour water volume is

$$V_{\text{Per Hour}} = \frac{10{,}000\ \text{L/day}}{24\ \text{hours}} = 416.67\ \text{L/h}$$

The energy required for desalination per kg of hydrogen.

$$E = \frac{4.5\ \text{kWh/h}}{416.67\ \text{L/h}} \times 13\ \text{L} = 0.14\ \text{kWh/kg}_{H_2}$$

The percentage of the energy required for desalination to the total energy required is

$$E = \frac{0.14\ \text{kWh/kg}_{H_2}}{0.14\ \text{kWh/kg}_{H_2} + 49.9\ \text{kWh/kg}_{H_2}} \times 100\% = 0.27\%$$

Thus, the percentage of the energy required for desalination is less than 1% of the total system consumption.

3.3 ECONOMIC ASPECTS

Analyzing the financial elements of the green hydrogen industry provides an economic perspective that is crucial in attracting stakeholders' attention to invest in this technology. Thus, a projection of the project's costs and benefits is a prerequisite in conducting an economic analysis. In the case of the project's feasibility, it can be said that the project is viable if the expected costs are minimized out of all alternatives and the benefits maximized.

Each aspect of cost–benefit analysis is integrated holistically into evaluating economic viability.

3.3.1 Cash Flow and NPV

As illustrated in Figure 3.5, the cash flow diagram is an important business analysis tool that graphically depicts cash inflows and outflows within a business entity over time. Cash flows are arranged chronologically on a horizontal matrix (each box representing an accounting period) and are denoted by arrows; upward arrows signify cash inflows and downward arrows mark outflows. It is vital to comprehend that cash

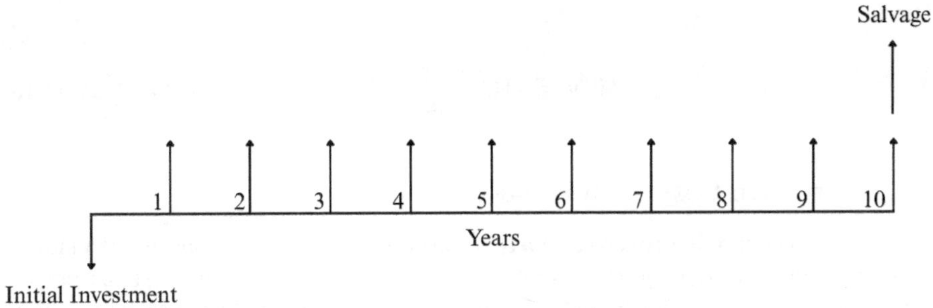

FIGURE 3.5 The cash flow illustration.

flows occurring in different periods cannot be directly compared with one another because of the time value of money. In cash flows that occur at different intervals, it is essential to use interest rates to factor their value against time for the value of money.

With this computed cash flow, the stockholder can determine the project's net present value (NPV). The NPV is defined as the present value of all expected cash inflows, together with the salvage value at the conclusion of the investment term, less the initial investment. A positive NPV suggests that the project is feasible and will earn a reasonable return on the investment; losing such value indicates that the project is unacceptable.

To calculate the NPV, the following equation is used [14]:

$$NPV = \sum_{i=0}^{n} \frac{CF_n}{(1+i)^n} \tag{3.15}$$

where i represents the interest rate, n is the year number of the cash flow, and CF represents the cash flow amount, which equals:

$$CF_n = TB_n - TC_n \tag{3.16}$$

where TB_n is the total benefits or revenue in year n, and TC_n is the total cost or expenses in year n. In other words, NPV equals the cash flow of the first year and the summation of the following cash flows, considering the interest rate. So, the NPV is calculated as

$$NPV = CF_0 + \sum_{i=1}^{n} \frac{CF_n}{(1+i)^n} \tag{3.17}$$

However, at year 0, there are no benefits yet, there is only the initial cost, so

$$CF_0 = TB_0 - TC_0$$

While

$$TB_0 = 0$$

Then

$$NPV = -TC_0 + \sum_{i=1}^{n} \frac{CF_n}{(1+i)^n} \tag{3.18}$$

3.3.2 Levelized Cost of Hydrogen

The levelized cost of hydrogen (LCOH) is a crucial metric for evaluating the viability of the project, considering its lifespan per unit. It considers all the costs associated with construction, operation, and maintenance, as well as the revenue from selling the hydrogen produced.

The LCOH is calculated by dividing the generator's total lifetime cost by its total lifetime production of hydrogen and is expressed in dollars per unit (per kg in the hydrogen case). The generator's lifetime production of hydrogen is the total amount of electricity produced by the generator over its expected lifetime [14].

$$LCOH = \frac{\sum_{i=0}^{n} \frac{TC_n}{(1+i)^n}}{\sum_{i=0}^{n} \frac{Q_n}{(1+i)^n}} \tag{3.19}$$

or:

$$LCOH = \frac{IC + \sum_{i=1}^{n} \frac{TC_n}{(1+i)^n}}{\sum_{i=1}^{n} \frac{Q_n}{(1+i)^n}} \tag{3.20}$$

where IC represents the initial investment cost at year 0, and Q_n represents the quantity of the hydrogen produced in kg.

Example 3.7: Levelized Cost of Hydrogen

Let us consider the following costs for a hydrogen system to produce 90,910 kg/year [14]:

Component	Initial Cost (IC) ($)	OMC/Year ($)
Electrolyzers	1,920,000	120,000
Compressors	90,000	1500
Storage tanks	60,000	1500
Power converters	151,260	6050
PV systems	3,000,012	50,000
Total	5,221,272	179,050

Suppose the interest rate is 8%. Calculate the levelized cost of hydrogen over a period of 25 years.

Solution:

The levelized cost of hydrogen is calculated as introduced in equation 3.20:

$$LCOH = \frac{IC + \sum_{i=1}^{n} \frac{TC_n}{(1+i)^n}}{\sum_{i=1}^{n} \frac{Q_n}{(1+i)^n}}$$

Consider the following costs for a hydrogen system [14].

IC = \$5,221,272
OC = \$179,050
Q = 90,910 kg/year

$$LCOH = \frac{5{,}221{,}272 + \sum_{i=1}^{25} \frac{179{,}050}{(1+0.08)^n}}{\sum_{i=1}^{25} \frac{90{,}910}{(1+0.08)^n}}$$

Use a calculator to find the answer.

Note: This example assumes that the lifetime is 25 for all components. However, this is not accurate as different components have different lifetimes. For example, compressors are 5 years old, converters are 10 years old, and electrolyzer operations take about 50,000 hours (it differs from one electrolyzer to another). So, to make more accurate calculations, the replacement cost and salvage value shall be considered.

The LCOH serves as a single measurement encompassing all expenses incurred over the life of a hydrogen production project. A significant portion of LCOH cost is capital expenditures (CAPEX) associated with the electrolyzer's technology maturity [15]. More modern solid oxide electrolysis cells (SOECs) or advanced anion exchange membrane electrolyzers have more significant CAPEX than currently established alkaline electrolyzers [16,17]. The reason for this cost premium is low production volumes resulting from the limited scale of manufacturing, the need to pay back research and development efforts, and the usage of new, often costly materials. Therefore, projects using novel technologies incur a so-called "first-of-a-kind" premium where most incurred costs stem from the vast uncertainties and risks of deploying such cutting-edge technologies.

In contrast, mature technologies such as alkaline electrolyzers have experienced a decrease in CAPEX due to economies of scale. Further cost reductions have been achieved as manufacturing processes and materials have improved over time [18]. It is important to capture the "learning curve" from newer technologies. With greater output and more efficient manufacturing processes, CAPEX for these technologies is likely to drop by significant volumes. These expected cost savings warrant inclusion in LCOH calculations if they are to be realistically assessed for long-term socioeconomic feasibility. Projects incorporating less mature technologies are more financially risky due to the unreliability of forecasting cost savings and performance capabilities. Nonetheless, these investments are critical for fostering creativity and steepening the learning curve, leading to a better and more competitive hydrogen economy. For instance, PEM electrolyzers have higher CAPEX than alkaline electrolyzers; however, their greater current density and improved dynamic response make them attractive to the industry, ensuring that costs will decrease in the future.

3.4 CHAPTER SUMMARY

This chapter provided a comprehensive analysis of the techno-economic factors influencing green hydrogen production systems. The significance of conducting thorough technical and economic evaluations to ascertain project viability and long-term sustainability was emphasized. Detailed insights into the design and sizing procedures for green hydrogen systems were presented, carefully considering renewable energy resources, electrolyzer efficiency, and the appropriate sizing of system components. The methodologies for determining the optimal size of PV arrays, wind energy systems, electrolyzers, and compressors to meet specific hydrogen production targets were illustrated with examples. For regions facing freshwater scarcity, the critical role of water desalination processes in ensuring sustainable hydrogen production was underscored.

Furthermore, the chapter provided a detailed examination of the economic factors relevant to green hydrogen projects. This included cash flow analysis, NPV calculations, and the LCOH assessment. These economic analyses are crucial for evaluating the financial viability of projects and attracting potential investors. By delivering a comprehensive overview of technical and economic considerations, this chapter equips the readers with valuable insights into the complexities of green hydrogen production.

SELF-ASSESSMENT QUESTIONS

1. What is the definition of green hydrogen?
2. What are the main components of a green hydrogen production plant?
3. How does temperature affect the power output of a PV system?
4. What is the calorific ratio between hydrogen and electricity?
5. What factors influence the amount of renewable energy required for a certain amount of hydrogen production?
6. What are the steps involved in designing and sizing a green hydrogen system?
7. What are the key requirements for designing a green hydrogen system?
8. How do you calculate the required size of a solar PV array for hydrogen production?
9. What factors can affect the actual size of a PV system needed for hydrogen production?
10. How do you estimate wind speed at a specific height for wind energy system sizing?
11. What is the capacity factor of a wind turbine or wind farm?
12. How does an electrolyzer work?

13. What factors determine the sizing of compressors in a hydrogen production system?
14. Why is water desalination important for hydrogen production?
15. What is the typical water consumption for producing 1 kg of hydrogen?
16. What is the purpose of a cash flow diagram in financial analysis?
17. How is the net present value (NPV) of a project calculated?
18. What does a positive NPV indicate for a project?
19. What is the levelized cost of hydrogen (LCOH)?
20. What factors are considered in the calculation of LCOH?
21. Calculate the required size of a solar PV system to produce 5 kg of hydrogen/day, assuming an average of 6 peak sun hours/day and an electrolyzer efficiency of 70%.
22. Estimate the wind turbine size required to produce 50 kg of hydrogen/day, assuming an electrolyzer efficiency of 65%, a wind speed at hub height of 10 m/s, air density of 1.2 kg/m^3, 20 operation hours/day, a power coefficient (C_p) of 0.35, and a capacity factor (C_F) of 40%.
23. Determine the number of electrolyzers needed to produce 100 kg of hydrogen per hour, if the available electrolyzer capacity is 25 kg/h.
24. Calculate the number of compressors required if each electrolyzer has a production rate of 150 Nm^3/h and the available compressor delivery capacity is 200 Nm^3/h.
25. Determine the size of a PV system needed to power a desalination process that consumes 5 kWh/h and produces 8000 L/day of freshwater, assuming an average of 5.5 peak sun hours/day.

REFERENCES

1. Alnasser, Tamadher M.A., Aedah M.J. Mahdy, Khaleel I. Abass, et al. 2020. "Impact of dust ingredient on photovoltaic performance: An experimental study." *Solar Energy* 195: 651–659. https://doi.org/10.1016/j.solener.2019.12.008.
2. Zghaibeh, Manaf, Paul C. Okonkwo, Ikram Ben Belgacem, et al. 2022. "Analytical model for a techno-economic assessment of green hydrogen production in photovoltaic power station case study Salalah City-Oman." *International Journal of Hydrogen Energy* 47 (31): 14171–14179. https://doi.org/10.1016/j.ijhydene.2022.02.180.
3. Okonkwo, Paul C., Ikram Ben Belgacem, Manaf Zghaibeh, et al. 2022. "Optimal sizing of photovoltaic systems based green hydrogen refueling stations case study Oman." *International Journal of Hydrogen Energy* 47 (75): 31964–31973. https://doi.org/10.1016/j.ijhydene.2022.07.140.
4. Esposito, Daniel V. 2017. "Membraneless electrolyzers for low-cost hydrogen production in a renewable energy future." *Joule* 1 (4): 651–658. https://doi.org/10.1016/j.enconman.2021.115156.
5. Shafiey Dehaj Mohammad and Hassan Hajabdollahi. 2021. "Multi-objective optimization of hybrid solar/wind/diesel/battery system for different climates of Iran." *Environment, Development and Sustainability* 23 (7): 10910–10936. https://doi.org/10.1007/s10668-020-01094-1.
6. Alquraan, Ayman, Mohammed Al-Mahmodi, Ashraf Radaideh, et al. 2020. "Comparative study between measured and estimated wind energy yield." *Turkish Journal of Electrical Engineering and Computer Sciences* 28 (5): 2926–2939. https://doi.org/10.3906/elk-2002-85.

7. Makridis, Sofoklis. 2017. "Hydrogen storage and compression." arXiv preprint arXiv:1702.06015. https://doi.org/10.48550/arXiv.1702.06015.
8. Marciuš, Doria. 2024. "The concept of electrochemical hydrogen compression and purification technology." In *Challenges and Solutions in the Hydrogen Value Chain: State of the Art Perspectives on Decarbonization and the Green Energy Transition*, pp. 103–144. Cham: Springer Nature Switzerland. https://doi.org/10.1007/978-3-031-69051-8_4.
9. Hua, Thanh Q., Hee-Seok Roh, and Rajesh K. Ahluwalia. 2017. "Performance assessment of 700-bar compressed hydrogen storage for light duty fuel cell vehicles." *International Journal of Hydrogen Energy* 42 (40): 25121–25129. https://doi.org/10.1016/j.ijhydene.2017.08.123.
10. Heshmat, Hooshang. 2014. "Oil-free centrifugal hydrogen compression technology demonstration." No. DOE-MiTi-18060-01. Mohawk Innovative Technology Inc., Albany, NY. https://doi.org/10.2172/1225266.
11. Trattner, Alexander, Martin Höglinger, Marie-Gabrielle Macherhammer, and Markus Sartory. 2021 "Renewable hydrogen: Modular concepts from production over storage to the consumer." *Chemie Ingenieur Technik* 93 (4): 706–716. https://doi.org/10.1002/cite.202000197.
12. Backer, Sumina Namboorimadathil, Ines Bouaziz, Nabeela Kallayi, et al. 2022. "Brine solution: Current status, future management and technology development." *Sustainability* 14 (11): 6752. https://doi.org/10.3390/su14116752.
13. Osman, Ahmed I., Zhonghao Chen, Ahmed M. Elgarahy, et al. 2024. "Membrane technology for energy saving: Principles, techniques, applications, challenges, and prospects." *Advanced Energy and Sustainability Research* 5 (5): 2400011. https://doi.org/10.1002/aesr.202400011.
14. Vartiainen, Eero, Christian Breyer, David Moser, et al. 2022. "True cost of solar hydrogen." *Solar RRL* 6 (5): 2100487. https://doi.org/10.1002/solr.202100487.
15. Cavaliere, Pasquale. 2023. "Electrolysis economy." In *Water Electrolysis for Hydrogen Production*, pp. 793–830. Cham: Springer International Publishing. https://doi.org/10.1007/978-3-031-37780-8_20.
16. Hauch, Anne, Rainer Küngas, Peter Blennow, et al. 2020. "Recent advances in solid oxide cell technology for electrolysis." *Science* 370 (6513): eaba6118. https://doi.org/10.1126/science.aba6118.
17. Krishnan, Subramani, Vinzenz Koning, Matheus Theodorus de Groot, et al. 2023. "Present and future cost of alkaline and PEM electrolyser stacks." *International Journal of Hydrogen Energy* 48 (83): 32313–32330. https://doi.org/10.1016/j.ijhydene.2023.05.031.
18. Arsad, Siti Rosilah, Akmal Zaini Arsad, Pin Jern Ker, et al. 2024. "Recent advancement in water electrolysis for hydrogen production: A comprehensive bibliometric analysis and technology updates." *International Journal of Hydrogen Energy* 60: 780–801. https://doi.org/10.1016/j.ijhydene.2024.02.184.

4 Hydrogen Derivatives

4.1 INTRODUCTION

Hydrogen is a valuable chemical element that has broad applications. For instance, hydrogen is used as an input in fuel cells and power plants to generate electricity [1]. Also, hydrogen can provide heating in boiler-based systems. Furthermore, hydrogen plays a role in many industries, including petroleum refining, steel reforming, material treatment, food processing, and fertilizer production [2]. Hydrogen can combine with other chemical components and form what is known as hydrogen derivatives. These derivatives include ammonia when it is combined with nitrogen. Electrofuels such as e-methane, e-methanol, synthetic gas, and synthetic gas fuels are formed when hydrogen is combined with carbon under different conditions [3]. Hence, comprehending the fundamentals of hydrogen derivatives remains vital in understanding their possible applications in the decarbonization effort of various sectors [4]. This chapter will examine hydrogen derivatives to illustrate their potential and comprehensive uses. This chapter examines hydrogen derivatives to illustrate their potential and comprehensive uses.

4.2 GREEN AMMONIA

Green ammonia has been tightly connected to green hydrogen due to its massive potential in the hydrogen storage, transportation, and utilization stages. Green ammonia has a volumetric energy density of (12.92–14.4) Megajoules per Liter (MJ/L) as stated in [5], which is significantly higher than that of pure hydrogen, which equals 0.0107 MJ/L [6]. This makes the transportation and storage of green ammonia easier and more feasible. It is worth mentioning that ammonia is not a newly discovered substance, as it has been used since the 1850s for refrigeration purposes [7]. Hence, its maturity has led to its prevalence worldwide. As a result, it has an adequate number of terminals worldwide, making its existing infrastructure suitable for the global market. Currently, ammonia is mainly used as feedstock for chemical applications, with 80% used in agricultural fertilizers. On the contrary, prospects and strategies worldwide indicate that ammonia will be blended more frequently in other sectors, especially in the shipping market. According to the International Energy Agency (IEA), one of the crucial requirements to reach net-zero emission (NZE) conditions worldwide by 2050 is to utilize green ammonia fuel to account for 45% of the global shipping energy demand by 2050 [8,9].

4.2.1 Concept of Green Ammonia Production

Green ammonia is produced by combining green hydrogen with nitrogen via the Haber–Bosch process. The Haber–Bosch process was invented by the German Chemist Fritz Haber in 1908 and then developed for industrial applications by the German chemist Carl Bosch in 1913.

DOI: 10.1201/9781003639190-4

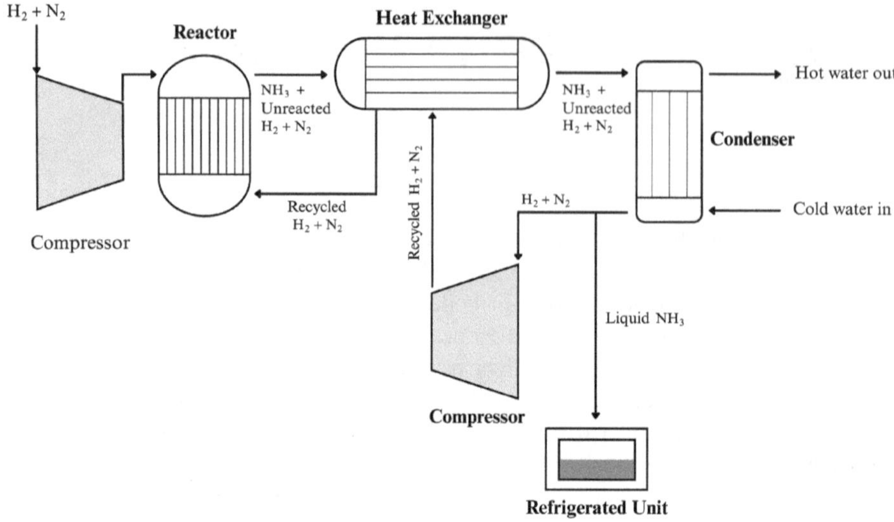

FIGURE 4.1 The components of the Haber–Bosch process.

The breakthrough led by the two Nobel Prize winners expanded the field of ammonia applications, most notably in fertilizer production, which enhanced global agricultural food production in a revolutionary manner. The Haber–Bosch process is a nitrogen fixation process that relies on converting atmospheric nitrogen (N_2) to ammonia (NH_3) by reaction with hydrogen (H_2) using a metal catalyst [10]. The schematic given in Figure 4.1 shows the typical Haber–Bosch process.

As shown in Figure 4.1, nitrogen and hydrogen are fed into the reactor by passing through a compressor. The compressor ensures that the gases are supplied at a pressure of between 250 and 350 bars to maintain sufficient surface coverage of the catalyst bed with nitrogen. The ratio of nitrogen to hydrogen is 1:3, meaning that to achieve conversion, each mole of nitrogen must be accompanied by 3 moles of hydrogen [11]. The reactor is a vessel that contains a heater and a catalyst bed; the flow of nitrogen and hydrogen within the reactor is shown in Figure 4.2. Nitrogen and hydrogen gases enter the reactor at high pressure after passing through the compressor.

The nitrogen–hydrogen stream then starts to heat up gradually due to mixing with the recycled nitrogen–hydrogen stream from the heat exchanger and the elevated temperature within the reactor enclosure. The gases then get heated via the heater to a temperature of 450°C–550°C before being sent to the catalyst bed. The catalyst bed is mainly composed of finely divided iron, which acts as a booster that speeds up the reaction rate. It is a requirement for the industrial production of ammonia since nitrogen is unreactive as its atoms are bonded with strong triple bonds. The ammonia production reaction is shown below [12]:

$$N_2 + 3H_2 \rightleftharpoons 2NH_3 \tag{4.1}$$

The reversible reaction (equation 4.1) governing ammonia production is subject to multiple parameters, with temperature and operating pressure being particularly

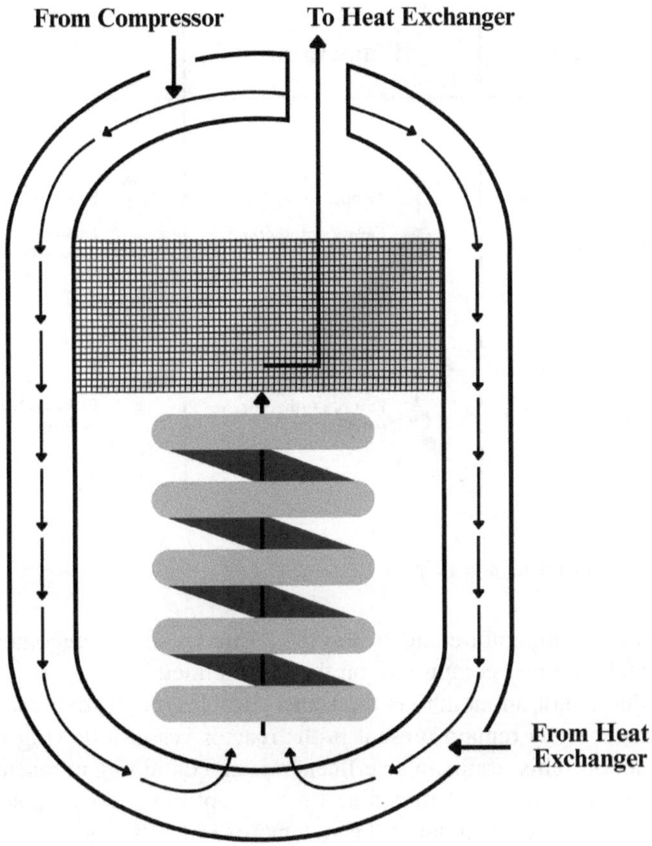

FIGURE 4.2 The Haber–Bosch reactor.

influential. At room temperature, the equilibrium is skewed in favor of ammonia, but the reaction rate is slow due to the reaction's high activation energy, which leads to the process's inefficiency. At a temperature of 150°C–200°C, the equilibrium constant is unity, indicating that reactants and products are being produced at the same rate. However, temperatures above 200°C and low pressures of around 1 atm lead to an unfavorable reaction for ammonia production. To enhance the effectiveness of the iron catalyst, a temperature of over 400°C is required, but this increase in temperature also negatively impacts the rate of ammonia production. Therefore, the Haber–Bosch process follows Chatelier's principle (Figure 4.3), which implies that increasing the pressure will favor the side of fewer moles, which is ammonia in our case [13].

In conclusion, ammonia synthesis requires high temperatures to fully capture the benefits of the iron catalyst, which boosts the reaction. In contrast, high pressure must be maintained during the process to ensure that the reaction favors ammonia production, according to Le Chaterlier's principle. This makes use of highly pressurized reactants at a pressure exceeding 150 atm, necessary to counter the negative

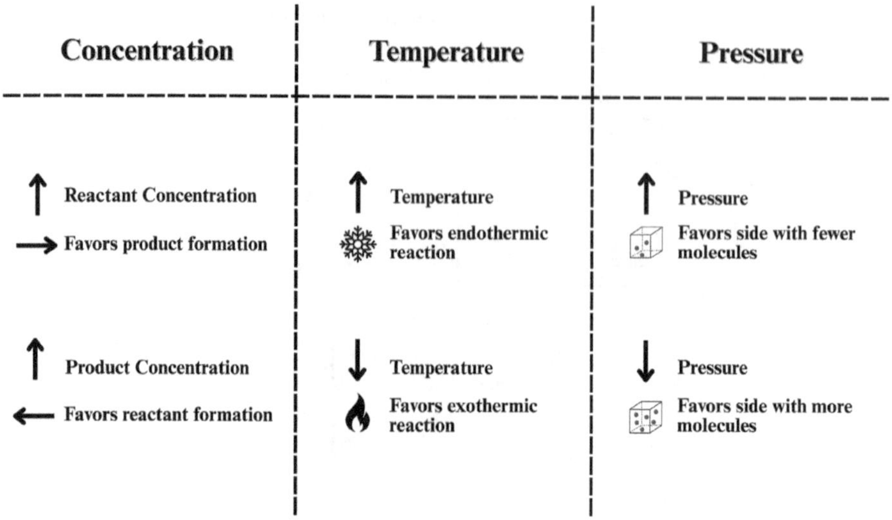

FIGURE 4.3 Le Chaterlier's principle.

effects of elevated temperature and ensure the desired reaction outcomes. This also explains the inclusion of a compressor on the system inlet.

In the production of ammonia via the Haber–Bosch process, the reactants, nitrogen and hydrogen, may remain present in the reactor vessel following the completion of the reaction. This results in an effluent stream containing unreacted nitrogen, hydrogen, and the produced ammonia. Energy-capturing devices, such as heat exchangers, are employed to enhance the system's overall efficiency. One such device is illustrated in Figure 4.1, which utilizes the heat from the effluent stream, leaving the reactor to preheat the recycled stream of nitrogen and hydrogen, thereby reducing the energy required to bring the stream to the reaction temperature of 400°C [14].

The effluent stream, composed of ammonia, nitrogen, and hydrogen, enters a condenser for ammonia separation. The boiling points of these gases are –33.3°C, –195.8°C, and –252.8°C, respectively (Table 4.1).

As a result, only the ammonia condenses and liquefies due to its relatively high boiling point, while the unreacted nitrogen and hydrogen are discharged in their gaseous phase. These gases are then compressed and passed through the heat exchanger before being fed back into the reactor. The produced liquid ammonia is then directed to a refrigeration unit for further cooling to remove any residual heat of vaporization. For large-scale industrial applications, the condenser is typically cooled by a water-cooled system, such as a cooling tower [15].

Example 4.1: Stoichiometry of the Haber–Bosch

Considering the general form of the ammonia production reaction (equation 4.1), balance the following reactions:

TABLE 4.1
Boiling Points for Common Gases at 1 atm

Gas	Boiling Point (°C)
Ammonia	−33.3
Helium	−268.9
Methane	−161.5
Nitrogen	−195.8
Propane	−42.1
Argon	−185.9
Hydrogen	−252.8
Carbon dioxide	−78.4
Oxygen	−183
Refrigerant-134a	−26.1

a. $1.5N_2 + xH_2 \rightleftharpoons yNH_3$
b. $xN_2 + 5H_2 \rightleftharpoons yNH_3$
c. $3.2N_2 + 9.6H_2 \rightleftharpoons xNH_3$
d. $xN_2 + yH_2 \rightleftharpoons 7NH_3$

Solution:

a. The stoichiometric coefficient y is determined by the number of nitrogen and hydrogen atoms that should be present in the products, which must equal the number of atoms on the reactant side. Hence, y can be considered the greatest common divisor for 3 and 9, which equals 3. Hence, $x=4.5$, $y=3$.

Following the same steps in (part a):

b. $x = 1.\overline{6}$, $y = 3.\overline{3}$
c. $x=6.4$
d. $x=3.5$, $y=10.5$

4.2.2 Important Evaluation Parameters of Ammonia Production

In practice, the measurement unit of ammonia production is commonly quantified in terms of mass flow rate, specifically Metric Tons per Day (MTPD), or volumetric units, specifically cubic meters per hour (m^3/h). Further, it is crucial to understand that hydrogen and nitrogen gases do not fully react in a single pass through the Haber–Bosch reactor, and only a small portion of them react to produce ammonia in operational conditions. The ratio between the number of reacted moles and the number of moles fed is known as the fractional conversion factor, which is represented by the following equation [16]:

$$f_x = \frac{\dot{n}_{reacted,x}}{\dot{n}_{fed,x}} \tag{4.2}$$

Where f_x is the fractional conversion factor for substance x, $\dot{n}_{reacted,x}$ is the number of reacted moles per second of substance x, and $\dot{n}_{fed,x}$ is the number of fed moles per second of substance x. The single pass conversion factor is typically within the range of (10%–15%) at nominal operating conditions (400°C–500°C and 100–300bar). Thus, recycling the unreacted hydrogen and nitrogen gases is necessary to achieve an overall fractional conversion factor from 97% to 99%. Another important parameter related to ammonia production is how much of the reactants are converted into a product in a chemical reaction. The following equation can be utilized to calculate the extent of the reaction [17]:

$$\xi = \frac{f_x \times \dot{n}_{fed,x}}{\alpha_x} \tag{4.3}$$

where ξ is the extent of the reaction, and α_x is the stoichiometry coefficient for substance x. The extent of reaction can be utilized to calculate the number of reacted moles of a substance in a chemical reaction and, subsequently, the number of produced moles through the following equations:

$$\dot{n}_{reacted,x} = \alpha_x \times \xi \tag{4.4}$$

$$\dot{n}_{produced,x} = \dot{n}_{fed,x} \pm \dot{n}_{reacted,x} \tag{4.5}$$

where $\dot{n}_{reacted,x}$ is the number of reacted moles per second of substance x, and $\dot{n}_{produced,x}$ is the number of unreacted or produced moles per second of substance x. The positive sign in (equation 4.5) is applicable when x is a product, while the negative sign is applicable when x is a reactant.

Further, the extent of the reaction can be applied to assess the concentration of a substance before and after a chemical reaction. In this case, the following equation can be utilized [11]:

$$\xi\% = \frac{C_{initial,x} - C_{final,x}}{C_{initial,x}} \tag{4.6a}$$

where $C_{final,x}$ is the final concentration of substance x, and $C_{initial,x}$ represents the initial concentration of substance x.

Example 4.2: Fractional Conversion and Extent of Reaction

An ammonia production plant consumes 200 moles of nitrogen gas per second to produce ammonia, with a single-pass fractional conversion factor of 10%. Knowing that the initial nitrogen concentration is 0.95 moles per liter (mol/L), calculate the final concentration of nitrogen and the output flow rates of hydrogen, nitrogen, and ammonia for a single-pass reaction if a single pass takes 1 second.

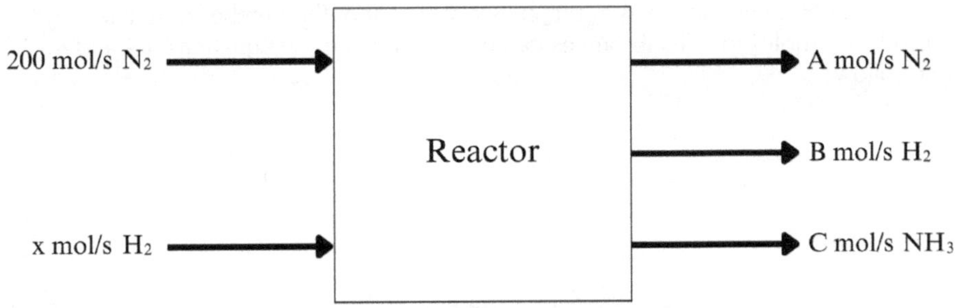

FIGURE 4.4 Haber–Bosch reactor example in moles per second.

Solution:

The diagram illustrated in Figure 4.4 represents the problem under study:

Considering the chemical reaction in equation 4.1, x, A, B, and C can be calculated as follows:

$$N_2 + 3H_2 \rightleftharpoons 2NH_3$$

In order to calculate x, the chemical reaction must be balanced in terms of atoms for reactants and products as in Example 4.1.

Therefore,

$$x = 600\,\text{mol/s}$$

By applying equation 4.2, considering a single-pass fractional conversion factor of 10%, the reacted moles of nitrogen can be calculated as follows:

$$f_{N_2} = \frac{\dot{n}_{\text{reacted},N_2}}{\dot{n}_{\text{fed},N_2}}$$

$$0.1 = \frac{\dot{n}_{\text{reacted},N_2}}{200}$$

$$\dot{n}_{\text{reacted},N_2} = 0.1 \times 200 = 20\text{ mol}$$

Further, by applying equation 4.3, the extent of the reaction can be calculated considering nitrogen gas as follows:

$$\xi = \frac{f_{N_2} \times \dot{n}_{\text{fed},N_2}}{\alpha_{N_2}}$$

$$\xi = \frac{0.1 \times 200}{1} = 20$$

Note that the extent of reaction is equal to the number of reacted moles of nitrogen in the case of the Haber–Bosch process, which can be justified by equation 4.4.

After obtaining the value of the extent of reaction, the number of reacted and produced moles for all substances can be calculated using equations 4.4 and 4.5 as follows:

Nitrogen:

$$\dot{n}_{produced,N_2} = \dot{n}_{fed,N_2} \pm \dot{n}_{reacted,N_2} = 200 - 20 = 180 \text{ mol/s} = A$$

Hydrogen:

$$\dot{n}_{reacted,H_2} = \alpha_{H_2} \times \xi = 3 \times 20 = 60 \text{ mol/s}$$

$$\dot{n}_{produced,H_2} = \dot{n}_{fed,H_2} \pm \dot{n}_{reacted,H_2} = 600 - 60 = 540 \text{ mol/s} = B$$

Ammonia:

$$\dot{n}_{reacted,NH_3} = \alpha_{NH_3} \times \xi = 2 \times 20 = 40 \text{ mol/s}$$

$$\dot{n}_{produced,NH_3} = \dot{n}_{fed,\ NH_3} \pm \dot{n}_{reacted,NH_3} = 0 + 40 = 40 \text{ mol/s} = C$$

The final concentration of nitrogen can be calculated using equation 4.6a as follows:

$$\xi\% = \frac{C_{initial,N_2} - C_{final,N_2}}{C_{initial,N_2}}$$

$$20\% = \frac{0.95 - C_{final,N_2}}{0.95}$$

$$C_{final,N_2} = 0.76 \text{ M}$$

As previously mentioned, the ammonia production rate is dealt with regarding MTPD. In order to calculate the production rate in MTPD for any substance, the following equations must be applied:

$$\dot{m}_{x,produced} = \dot{n}_{x,produced} \times MW_x \times 86.4 \times 10^{-3} \tag{4.6b}$$

$$\dot{m}_{x,reacted} = \dot{n}_{x,reacted} \times MW_x \times 86.4 \times 10^{-3} \tag{4.6c}$$

And,

$$\dot{m}_{NH_3,produced} = \dot{m}_{N_2,reacted} + \dot{m}_{H_2,reacted} \tag{4.7}$$

where $\dot{m}_{x,produced}$ is the production mass flow rate in MTPD, $\dot{n}_{x,produced}$ is the production molar flow rate in mol/s, MW_x is the molar weight in g/mol of substance x, $\dot{m}_{x,reacted}$ is the reacted mass flow rate in MTPD, and $\dot{n}_{x,reacted}$ is the reacted molar flow rate in mol/s. The conversion factor that can be utilized to obtain the flow rate in MTPD for a given flow rate in mol/s equals 86.4×10^{-3} (Table 4.2).

Example 4.3: Production Capacity of Ammonia

An ammonia production plant consumes 350 moles of nitrogen gas per second to produce ammonia. The molar weight values are 28.014, 2.016, and 17.031 grams per mole (g/mol) for nitrogen gas, hydrogen, and ammonia, respectively. Assuming an overall fractional conversion factor of 97%, calculate the production rates of nitrogen, hydrogen, and ammonia in MTPD.

Solution:

The schematic in Figure 4.5 represents the problem under study:

Considering the general form of the Haber–Bosch chemical reaction equation 4.1, we can calculate the molar flow rate for hydrogen and ammonia as follows:

$$N_2 + 3H_2 \rightleftharpoons 2NH_3$$

TABLE 4.2
Molar Weight of Common Gases

Gas	MW (g/mol)
Ammonia	17.031
Hydrogen	2.016
Nitrogen	28.014
Carbon dioxide	44.01
Methane	16.04
Oxygen	31.998

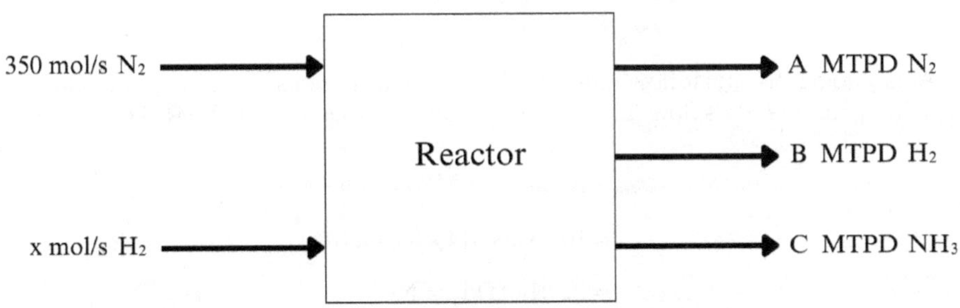

FIGURE 4.5 Haber–Bosch reactor example in MTPD.

We can use the reaction stoichiometry to calculate the molar flow rate of hydrogen by:

$$\text{x mol/s of } H_2 \times 1 \text{ mol/s of } N_2 = 3 \text{ mol/s of } H_2 \times 350 \text{ mol/s of } N_2$$

Therefore,

$$x = 1050 \text{ mol/s}$$

By applying equation 4.2 considering an overall fractional conversion factor of 97%, the reacted moles of nitrogen and the extent of reaction can be calculated as follows:

$$f_{N_2} = \frac{\dot{n}_{reacted,N_2}}{\dot{n}_{fed,N_2}}$$

$$0.97 = \frac{\dot{n}_{reacted,N_2}}{350}$$

Therefore,

$$\dot{n}_{reacted,N_2} = 0.97 \times 350 = 339.5 \text{ mol/s} = \xi$$

Similar to example 4.2, the number of reacting and produced moles of nitrogen, hydrogen, and ammonia are as follows:

$$\dot{n}_{produced,N_2} = 10.5 \text{ mol/s} = A$$

$$\dot{n}_{reacted,H_2} = 1018.5 \text{ mol/s}$$

$$\dot{n}_{produced,H_2} = 31.5 \text{ mol/s} = B$$

$$\dot{n}_{reacted,NH_3} = \dot{n}_{produced,NH_3} = 679 \text{ mol/s}$$

After finding the molar flow rates of both reactants in mol/s, we can use equation 4.6a to find the mass flow rates of nitrogen and hydrogen gases in MTPD:

$$\begin{aligned}\dot{m}_{N_2,produced} &= \dot{n}_{N_2,produced} \times MW_{N_2} \times 86.4 \times 10^{-3} \\ &= 10.5 \times 28.014 \times 86.4 \times 10^{-3} \\ &= 25.41 \text{ MTPD of } N_2\end{aligned}$$

And,

$$\begin{aligned}\dot{m}_{H_2,produced} &= \dot{n}_{H_2,produced} \times MW_{H_2} \times 86.4 \times 10^{-3} \\ &= 31.5 \times 2.016 \times 86.4 \times 10^{-3} \\ &= 5.48 \text{ MTPD of } H_2\end{aligned}$$

Finally,

$$\begin{aligned}\dot{m}_{NH_3,produced} &= \dot{n}_{NH_3,produced} \times MW_{NH_3} \times 86.4 \times 10^{-3} \\ &= 679 \times 17.031 \times 86.4 \times 10^{-3} \\ &= 999.13 \text{ MTPD of } NH_3\end{aligned}$$

Alternatively, we can obtain the mass flow rate of ammonia in MTPD by applying equation 4.7:

$$\begin{aligned}\dot{m}_{NH_3,produced} &= \dot{m}_{N_2,reacted} + \dot{m}_{H_2,reacted} \\ &= 821.73 + 177.4 \\ &= 999.13 \text{ MTPD of } NH_3\end{aligned}$$

The determination of the ammonia production rate requires detailed chemical analyses that are accurate and realistic. These analyses can be divided into two categories: experimental and theoretical. Experimental approaches use physical observations of the system, such as measuring the concentrations of the reactants and products, while theoretical approaches utilize mathematical models and simulations. The Peng–Robinson Equation of State (EOS) is an example of a theoretical method that can estimate the extent of reaction of ammonia production. Unlike simpler EOSs, it is more accurate because it accounts for the size and interactions of the molecules participating in the reaction [18].

Other specialized software packages can also assist in estimating the production rate of ammonia, including the computational fluid dynamics (CFD) tools. These packages, such as ANSYS Fluent and Autodesk CFD, enable the simulation and analysis of the flow of reactants and products through a reactor vessel [19,20].

This helps to gain a better understanding of the system's behavior and optimize production. In general, for estimating the ammonia production rate, sophisticated experimental and theoretical analyses are required.

Complex computational analyses and simulations as well as profound knowledge of chemical sciences, thermodynamics, software, and tools are needed to conduct these methods.

4.2.3 Ammonia Storage and Transfer

After ammonia is produced, it can be stored in a variety of ways, such as in compressed gas or liquefied gas cylinders, in ammonia refrigeration systems, or in underground storage tanks. Ammonia exists in a gaseous state, and it can be stored in carbon steel or stainless-steel tanks at room temperature and at a pressure of 10–15 bar. Just like gaseous ammonia, liquid ammonia is also stored in carbon steel or stainless steel tanks, but the temperature and pressure vary. Liquid ammonia can be stored either at high pressure, more than 8 bar at room temperature, or at 1 bar at −33°C. Ammonia can, however, be easily refrigerated, under temperatures of −10°C and −40°C, at a pressure range of 10–20 bar [21].

When used as an energy carrier, ammonia has better traits when compared with hydrogen in its pure form. These include but are not limited to, higher volumetric density, higher energetic density, stability in volatility, combustion, and many other characteristics.

For example, the energy density of ammonia, approximately 12.7 MJ/L, exceeds the value of liquid hydrogen (8.5 MJ/L). Having greater energy density than liquid hydrogen enables more effective energy transfer within a fixed volume. Moreover, ammonia is stored in a liquid state at −33°C, while hydrogen must be cooled to −253°C before it can be stored in a liquid state. Thus, the energy required to liquefy the energy carrier for storage and transfer becomes lower. Additionally, the combustion of ammonia is less controllable, which makes it safer than the reduction of hydrogen due to the risk of unanticipated combustion of hydrogen. On the other hand, the efficiency required in the storage and transportation of hydrogen in its gaseous state is low due to the fact that hydrogen gas under normal conditions has a density of 0.09 kg/m^3, making it 14 times lighter than air. As such, uncompressed hydrogen cannot be stored because it requires huge amounts of volume, making the transportation of hydrogen unfeasible in these conditions. In order to increase the density of uncompressed hydrogen from 0.09 to 42 kg/m^3, it must be compressed to a pressure of 700 bars, which requires a massive amount of energy [22].

4.2.4 Hydrogen Production from Ammonia

Hydrogen can be produced from ammonia using the Haber–Bosch process. Ammonia is an energy carrier due to its favorable physical and chemical characteristics, making the transportation of ammonia easier than that of hydrogen. Furthermore, ammonia as a hydrogen carrier will be utilized more efficiently because of the mature development of ammonia infrastructure and transportation in several countries. Ammonia can be split into hydrogen and nitrogen by electrolysis, as illustrated in Figure 4.6, and thermal decomposition, which is shown in Figure 4.7. This attribute makes ammonia one of the most convenient options for transporting hydrogen, especially in shipping delivery techniques.

Ammonia electrolysis is carried out by passing an electric current through an aqueous solution containing ammonia in order to produce hydrogen and nitrogen gases. Like conventional water electrolysis, an ammonia electrolyzer comprises the same components: a power source, a cathode, and an anode. The theoretical external voltage required for the electrolysis of ammonia is 0.06 V, which is significantly less than the external voltage required for water electrolysis, which equals 1.23 V. Hence, the electrolysis of ammonia is not energy-intensive. The efficiency of ammonia

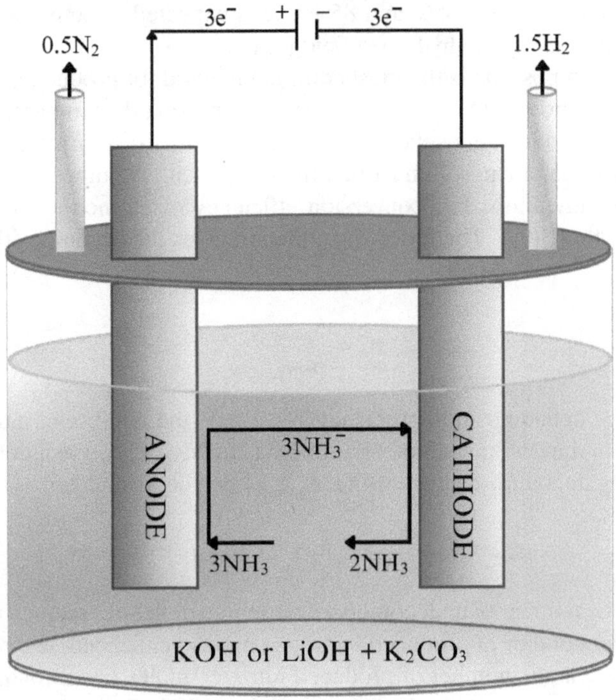

FIGURE 4.6 Ammonia electrolysis process.

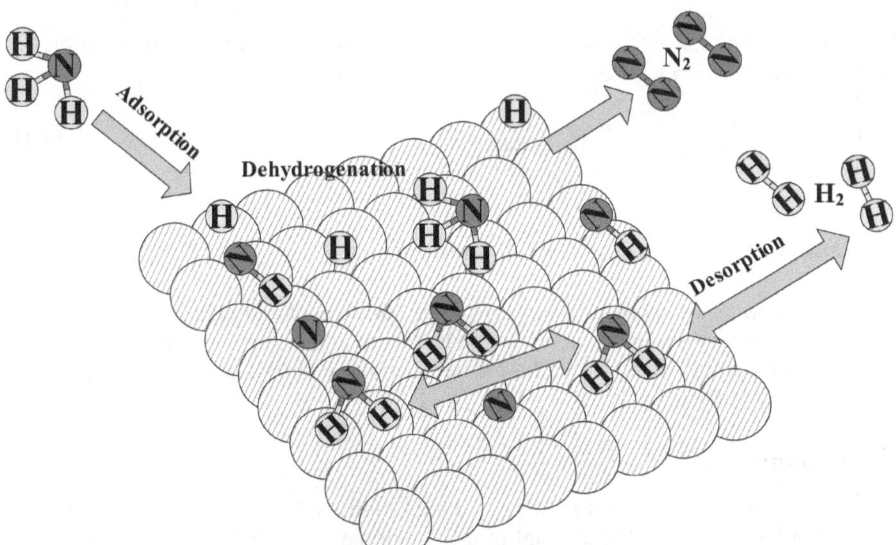

FIGURE 4.7 Ammonia thermal decomposition.

electrolysis ranges between 65% and 85% and is expected to be improved with further investment in the green hydrogen field [23].

Thermal decomposition is the most common method for producing hydrogen and nitrogen from ammonia. This process relies on increasing the temperature of ammonia up to 400°C under atmospheric pressure. In order to minimize the activation energy of ammonia, a catalyst must be utilized, which is composed of iron, nickel, platinum, and ruthenium. The conversion efficiency of ammonia decomposition is typically more than 90%. The following equation gives the decomposition efficiency [24,25]:

$$\eta_d = \frac{\dot{n}_{produced}}{\dot{n}_{NH_3,fed}} \tag{4.8}$$

Where η_d is the decomposition efficiency, $\dot{n}_{produced}$ is the number of moles produced per second from the decomposition of ammonia, and $\dot{n}_{NH_3,fed}$ is the number of ammonia moles being decomposed per second. $\dot{n}_{produced}$ can be formulated as the following:

$$\dot{n}_{produced} = \dot{n}_{undecomposed,NH_3} + \dot{n}_{produced,H_2} + \dot{n}_{produced,N_2} \tag{4.9}$$

where NH_3 is the number of undecomposed ammonia moles per second, n dot sub produced, H_2 is the number of produced hydrogen moles per second, and n dot sub produced, $\dot{n}_{produced,N_2}$ is the number of produced nitrogen moles per second. The amount of undecomposed ammonia can be calculated through the following equation:

$$\dot{n}_{NH_3,undecomposed} = \dot{n}_{NH_3,fed} \times (1 - \eta_d) \tag{4.10}$$

Also, recalling the stoichiometry of the ammonia production reaction equation 4.1, the following relation applies:

$$\dot{n}_{produced,H_2} = 3 \times \dot{n}_{produced,N_2} \tag{4.11}$$

Example 4.4: Decomposition of Ammonia

An ammonia decomposition plant consumes 800 TPD of ammonia with a decomposition efficiency of 94%. Calculate the hydrogen and nitrogen production rate in moles per second, considering the following molar weights: $MW_{H_2} = 2.016$ g/mol, $MW_{N_2} = 28.014$ g/mol, $MW_{NH_3} = 17.031$ g/mol.

Solution:

The schematic in Figure 4.8 represents the problem under study:

In order to convert the amount of fed ammonia from TPD to mol/s, equation 4.6a must be applied as follows:

$$\dot{m}_{NH_3,fed} = \dot{n}_{NH_3,fed} \times MW_{NH_3} \times 86.4 \times 10^{-3}$$

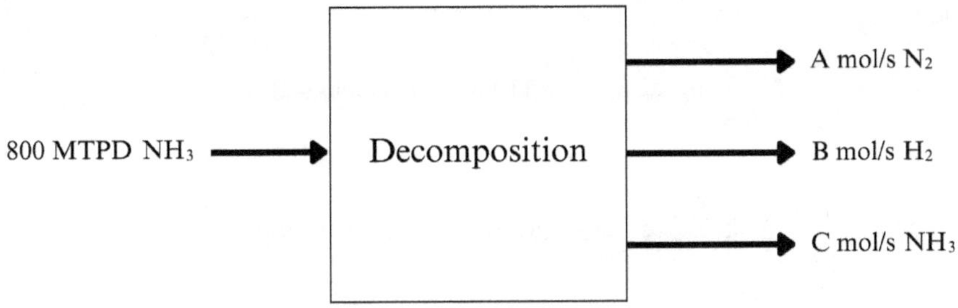

FIGURE 4.8 Ammonia decomposition example.

$$\dot{n}_{NH_3,fed} = \frac{800}{17.031 \times 86.4 \times 10^{-3}} = 543.67 \text{ mol/s}$$

Hence, by applying equation 4.8, the following can be obtained:

$$\eta_d = \frac{\dot{n}_{produced}}{\dot{n}_{NH_3,fed}}$$

$$\dot{n}_{produced} = 543.67 \times 0.94 = 511.05 \text{ mol/s}$$

In order to find and calculate the molar flow rate for the produced ammonia, hydrogen, and nitrogen, equation 4.9 must be utilized as follows:

$$\dot{n}_{produced} = \dot{n}_{undecomposed,NH_3} + \dot{n}_{produced,H_2} + \dot{n}_{produced,N_2}$$

Where:

$$\dot{n}_{undecomposed,NH_3} = \dot{n}_{NH_3,fed} \times (1 - \eta_d)$$

And,

$$\dot{n}_{produced,H_2} = 3 \times \dot{n}_{produced,N_2}$$

Therefore,

$$\dot{n}_{produced} = \dot{n}_{undecomposed,NH_3} + \dot{n}_{produced,H_2} + \dot{n}_{produced,N_2}$$

$$511.05 = \dot{n}_{NH_3,fed} \times (1 - \eta_d) + 3 \times \dot{n}_{produced,N_2} + \dot{n}_{produced,N_2}$$

$$511.05 = 543.67 \times (1 - 0.94) + 3 \times \dot{n}_{produced,N_2} + \dot{n}_{produced,N_2}$$

$$\dot{n}_{produced,N_2} = \frac{478.43}{4} = 119.6 \text{ mol/s} = A$$

Hence,

$$\dot{n}_{produced,H_2} = 3 \times 119.6 = 358.8 \text{ mol/s} = B$$

And,

$$\dot{n}_{undecomposed,NH_3} = 543.67 \times (1 - 0.94) = 32.6 \text{ mol/s} = C$$

4.2.5 Ammonia Applications

Ammonia can be used in internal combustion engines and fuel cells. According to the literature, when used in an internal combustion engine, the efficiency of the power conversion process ranges between 25% and 36%. The benefits of using ammonia as a green fuel for internal combustion engines include reducing transport-related emissions since ammonia is a carbon-free fuel that can be synthesized with minimal harmful emissions. Ammonia can also be used in various internal combustion engines, including spark ignition and compression ignition (Figure 4.9).

On the other hand, hydrogen constitutes 17.65% of the mass of ammonia, making ammonia a rich hydrogen provider for fuel cell applications for electricity generation. Furthermore, direct ammonia fuel cells can have an electrochemical efficiency of around 60%, making them an up-and-coming technology for utilizing green ammonia in the future [26].

Ammonia can also produce plastics, fibers, explosives, textiles, pesticides, dyes, and, most importantly, fertilizers. As previously mentioned, 80% of ammonia produced worldwide is utilized to produce fertilizers. For fertilizers, ammonia can be

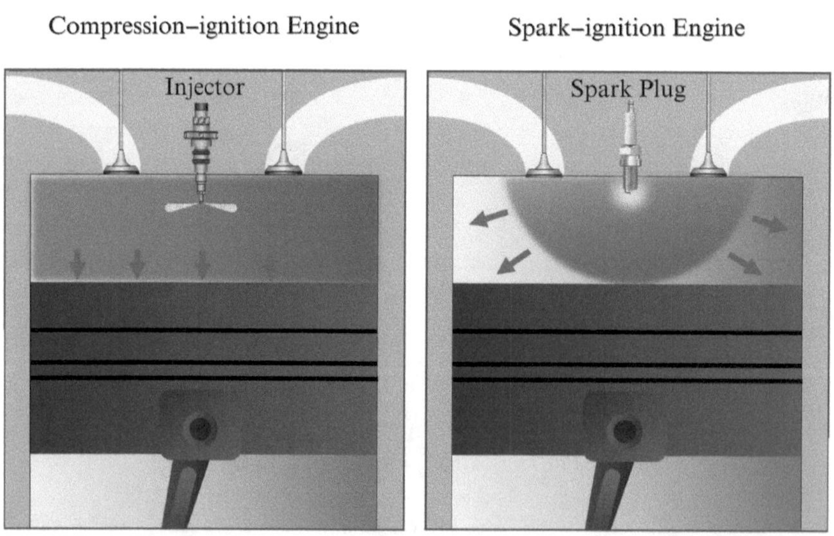

FIGURE 4.9 Spark–ignition and compression–ignition internal combustion engines.

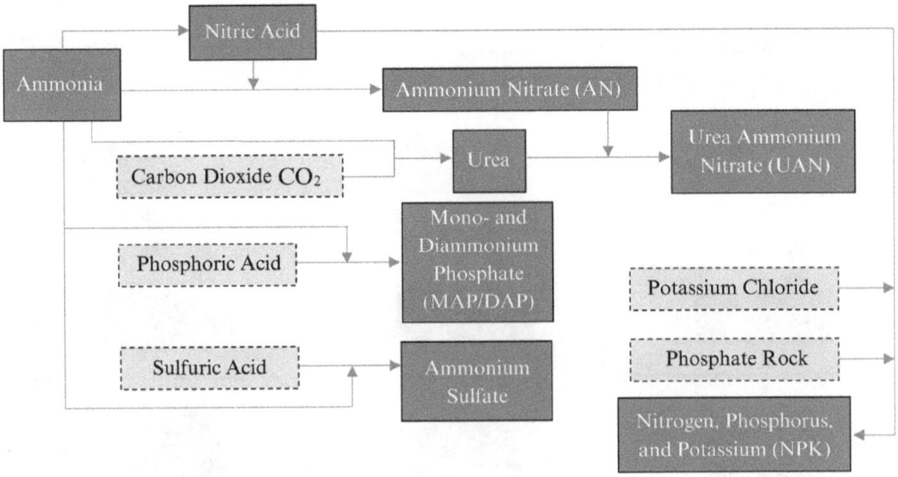

FIGURE 4.10 Ammonia-based fertilizers.

utilized in its pure form or in the form of nitric acid, ammonium nitrate, or urea. The major and most common fertilizers that rely on ammonia directly or indirectly are shown in Figure 4.10 [8,27].

In conclusion, understanding the involvement of ammonia within the hydrogen supply chain is one of the most crucial aspects of green hydrogen and power-to-X.

4.3 ELECTRO-FUELS

Electro-fuels are the second most important derivative of green hydrogen and are expected to become more significant. In general, electro-fuels are those synthesized by reacting captured carbon dioxide or carbon monoxide with green hydrogen produced from renewable energy [28]. Electro-fuels can easily replace conventional fuels since they are the same but produced in a more environmentally friendly manner. Conventional fossil fuels are extracted from the ground to be utilized, meaning that their carbon content is eventually emitted into the atmosphere instead of being enclosed within the ground, which makes using them in a carbon-positive process. Hence, the share of carbon emissions roaming the atmosphere is increasing. Capturing carbon from the atmosphere and storing it in different phases underground are the most environmental ways to handle carbon emissions. This can be implemented through carbon capture, use, and storage (CCUS) technologies. Unfortunately, this cannot be globally implemented soon since numerous applications still rely on carbon-based fuels, and their carbon withdrawal will probably be more challenging than anticipated. Electro-fuels are already synthesized from hydrogen, carbon monoxide, or dioxide in the atmosphere, thus creating a neutral carbon cycle [29,30]. Electro-fuels include electro-methane, synthetic gas, electro-methanol, and synthetic petroleum fuels [31]. This fuel type is also known as liquid alternative fuel. Figure 4.11 demonstrates the difference between the processes.

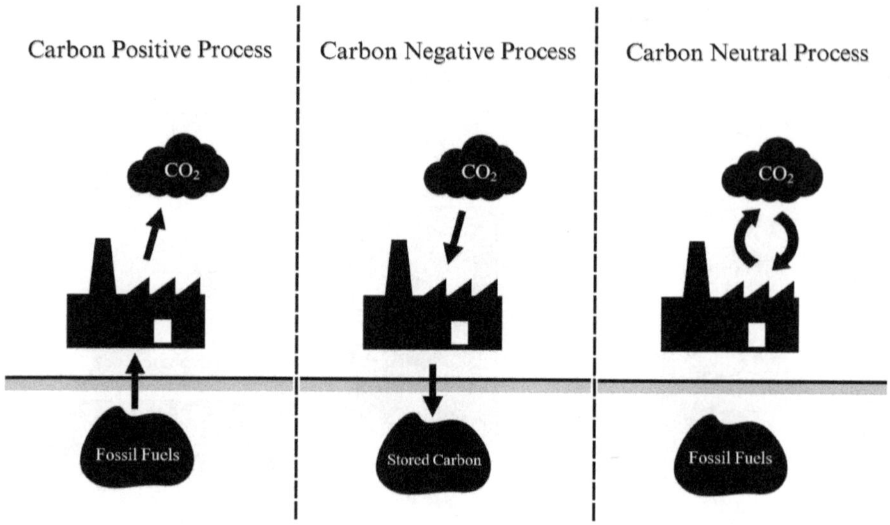

FIGURE 4.11 Carbon positive, negative, and neutral processes.

4.3.1 Electro-Methane

Methane is the most harmful greenhouse gas. For instance, methane is 25 times more potent in the atmosphere than carbon dioxide. Also, it is responsible for around 20% of global warming. Around one-third of methane emissions worldwide are related to fossil fuel extraction, utilization, and transport. Hence, avoiding the creation of methane from fossil fuels and relying completely on electro-methane created from green hydrogen and captured carbon emissions are necessary for the net-zero transition [32,33]. Methane is important due to its suitability for a huge variety of applications. Methane can be used as fuel for internal combustion engines, heating fuel, chemical production feedstock, refrigerant, and many other applications [34–36]. Further, methane makes up around 85% of compressed natural gas and more than 95% of liquid natural gas, which, in turn, are very valuable resources that cannot be replaced entirely in the near future [37]. Considering all-natural gases, including methane, ethane, propane, and butane, methane has the highest energy density (= 55.5 MJ/kg), followed by ethane (= 53.2 MJ/kg), then propane (= 50.3 MJ/kg), and finally, butane (= 49.5 MJ/kg) [38]. As mentioned, there are no chemical differences between conventional methane and electro-methane since the difference is all in the production stage.

Electro-methane is produced via the Sabatier process, in which methane and water are produced from a reaction between hydrogen and carbon dioxide at elevated temperatures and pressure in the presence of a catalyst. This process was discovered by the French chemist Paul Sabatier in the early 1900s, and since then, it has been utilized for the chemical creation of methane. The following chemical reaction governs the Sabatier process [39]:

$$CO_2 + 4H_2 \rightleftharpoons CH_4 + 2H_2O \tag{4.12}$$

The Sabatier process occurs at very high temperatures that go up to (300°C–400°C) and a pressure of around 3 MPa in the presence of a nickel catalyst. Like the Haber–Bosch process, not all reactants' moles are converted into products. Hence, the methanation efficiency of the Sabatier process must be considered, which is typically less than 80% in practice [40].

Example 4.5: Methane for Cooking

A single household consumes 30 L of propane for cooking per week. Assuming that each liter weighs 0.51 kg. If the propane is replaced by pure methane, calculate how much methane is required per week and how much hydrogen is needed to produce that amount. Considering that the energy density of propane is 50.3 MJ/kg, and 55.5 MJ/kg for methane, the methanation efficiency is 78%, $MW_{CH_4} = 16.04$ g/mol, $MW_{H_2} = 2.016$ g/mol, $MW_{CO_2} = 44.01$ g/mol.

Solution:

In order to determine how much energy is required on a weekly basis, we must multiply the energy density of propane by the total weekly consumed mass of propane:

$$\text{Total consumed mass} = 0.51\ \text{kg/L} \times 30\ \text{L} = 15.3\ \text{kg}$$

$$\text{Required energy} = 15.3\ \text{kg} \times 50.3\ \text{MJ/kg} = 769.59\ \text{MJ}$$

The required mass of methane can be found by dividing the required energy by the energy density of methane:

$$\text{Required mass of CH}_4 = \frac{\text{Required energy}}{\text{Energy density of CH}_4} = \frac{769.59}{55.5} = 13.86\ \text{kg}$$

Considering the molar weight of methane (= 16.08 g/mol), the number of methane moles can be calculated as follows:

$$n_{CH_4} = \frac{m_{CH_4}}{MW_{CH_4}} = \frac{13{,}860}{16.08} = 861.94\ \text{mol}$$

Considering the stoichiometry of the reaction (equation 4.12), the theoretical number of moles of hydrogen and carbon dioxide should be as follows:

$$1\ \text{mol CH}_4\text{:}4\ \text{mol H}_2\text{:}1\ \text{mol CO}_2$$

Hydrogen:

$$861.94\ \text{mol CH}_4\text{:x mol H}_2$$

$$x = 861.94 \times 4 = 3447.76\ \text{mol H}_2$$

However, only 78% of hydrogen moles were converted to methane.

Hence,

$$\text{Actual moles } H_2 = \frac{\text{Theoretical moles } H_2}{\text{Methanation efficiency}} = \frac{3447.76}{0.78} = 4420.2 \text{ mol } H_2$$

Carbon dioxide:

$$861.94 \text{ mol } CH_4 : x \text{ mol } CO_2$$

$$x = 861.94 \times 1 = 861.94 \text{ mol } CO_2$$

Similar to hydrogen:

$$\text{Actual moles } CO_2 = \frac{\text{Theoretical moles } CO_2}{\text{Methanation efficiency}} = \frac{861.94}{0.78} = 1105 \text{ mol } CO_2$$

Finally, to find the required mass of hydrogen and carbon dioxide, the following must be applied:

$$m_{H_2} = \frac{n_{H_2}}{MW_{H_2}} = \frac{4420.2}{2.016} = 2192.5 \text{ g} = 2.19 \text{ kg } H_2$$

$$m_{CO_2} = \frac{n_{CO_2}}{MW_{CO_2}} = \frac{1105}{44.01} = 25.1 \text{ g} = 0.025 \text{ kg } CO_2$$

Methane can be converted back to hydrogen and carbon dioxide through a steam–methane reforming process, which reacts with steam at elevated pressure between 3 and 25 bar in the presence of a catalyst. In this case, the hydrogen produced cannot be considered green since the process produces carbon emissions. Carbon capture systems can be used to limit the impact of carbon emissions. Thus, the produced hydrogen is blue [41].

4.3.2 Synthetic Gas

Synthetic gas, also known as "syngas," is a mixture of hydrogen and carbon monoxide. Syngas can be defined as the building block for the production of liquid fuels through the Power-to-Liquid chain. Syngas can be produced through many methods, including natural gas reforming, coal gasification, partial oxidation of coal or oil, gasification of municipal solid waste, plasma gasification, pyrolysis, biomass gasification, and direct solid oxide electrolysis of carbon dioxide (SOECD) [42–48]. SOECD is the most environmental method among all alternatives since it relies on clean and recycled feedstock only. The chemical reactions that take place at the anode and the cathode during the production of syngas through SOECD are as follows [49]:

Cathode:

$$CO_2 + 2e^- \rightarrow CO + O^{2-} \tag{4.13a}$$

$$H_2O + 2e^- \rightarrow H_2 + O^{2-} \tag{4.13b}$$

Anode:

$$2O^{2-} \rightarrow O_2 + 2e^- \tag{4.14}$$

The application of SOEC units in syngas production holds great promise for economical and efficient production of syngas from fossil fuels, natural gas, coal, and biomass. In SOECs, the carbon-based feedstock is transformed into syngas through an electrochemical reaction that operates at high temperatures and includes the oxidation of carbon dioxide and/or water vapor. The efficiency of SOECs in producing syngas exceeds that of conventional methods due to lower cost, reduced carbon emissions, and greater flexibility in the desired feedstock. There are, however, some engineering problems that must be solved before SOEC-based syngas production can be done commercially.

The utilized carbon emission constituents must be captured in c systems, and the utilization of artificially manufactured carbon emissions must not be allowed. Figure 4.12 demonstrates the process of integration of syngas production through SOEC.

In the SOEC's operation, carbon dioxide and steam are introduced on the cathode side of the cell, where they undergo reduction and electrolysis to generate carbon monoxide, hydrogen gas, and oxygen ions (equations 4.13a and 4.13b). The oxygen ions subsequently move through the solid oxide electrolyte to the anode side of the cell, where they, with other oxygen ions, form oxygen gas molecules (equation 4.14). By controlling the feedstock's composition and the operational parameters, the resulting syngas product can be modified. Syngas is reported to have approximately

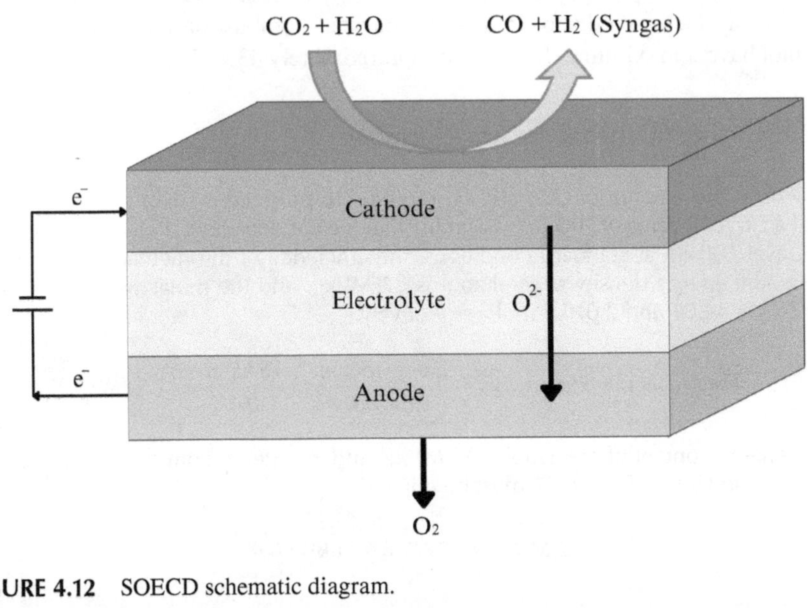

FIGURE 4.12 SOECD schematic diagram.

half the energy density of natural gas [50]. Nevertheless, it stands to reason that it can be employed as fuel for myriad uses. Some of the applicative components of syngas fuel include its burning for thermal energy generation, burning for electric power generation, and power generation via fuel cells. Alas, these are too carbon-heavy and cannot be done without capturing carbon technologies, which, in most cases, can be quite sophisticated and expensive. Thus, syngas assumes a crucial position in the hydrogen economy because it is able to produce multiple liquid fuels that serve as substitutes for fossil fuels, including methanol, diesel, jet fuel, and gasoline [51].

4.3.3 Electro-Methanol

Methanol is the simplest alcohol, consisting of one oxygen, one hydrogen, and one carbon atom. Moreover, it possesses the greatest hydrogen-to-carbon ratio out of all the liquid fuels [52]. Electro-methanol and its derivatives are regarded as one of the cleanest fuels for multipurpose applications. With little modifications to the current infrastructure, electro-methanol can be utilized as fuel in aviation, marine, and road transport [53].

Electro-methanol holds an extremely beneficial potential since it can help decrease global transportation emissions by 95%. Having room temperature and ambient pressure makes the storage and lifetime of electro-methanol very long, which is another advantage [54]. It is produced from carbon-dioxide-bearing syngas in the presence of a copper/zinc oxide/aluminium oxide catalyst at high pressure (50–100 bar) and high temperature (200°C–300°C). The following chemical reaction controls the production of methanol:

$$CO_2 + 3H_2 \rightleftharpoons CH_3OH + H_2O \tag{4.15}$$

Notably, the CO_2 hydrogenation process requires around 1.4 kg of CO_2 to produce 1 kg of electro-methanol [55]. Also, internal combustion engines running on methanol have a maximum efficiency of approximately 43% [56].

Example 4.6: Methanol for Transportation

Calculate the amount of CO_2 and H_2 required to produce methanol sufficient to fuel a car for a drive of 100 km. Assuming that the car consumes 15 kWh of energy to travel 100 km at standard conditions, the efficiency of the methanol engine is 40%, the energy density of methanol is 22 MJ/kg, and the molar weight of CO_2 and H_2 is 44.01 and 2.016 g/mol, respectively.

$$\text{Energy input} = \text{Fuel energy} = \frac{\text{Energy output}}{\text{Efficiency}} = \frac{15\ \text{kWh}}{0.4} = 37.5\ \text{kWh}$$

The energy content of methanol is 22 MJ/kg, and in order to convert MJ to kWh, a conversion factor of 1:0.2777 must be applied:

$$22\ \text{MJ/kg} \times 0.2777 = 6.11\ \text{kWh/kg}$$

Hence, the amount of methanol required for the operation of the car for 100 km is as follows:

$$\frac{37.5\text{ kWh}}{6.11\text{ kWh/kg}} = 6.137\text{ kg}$$

Each 1 kg of methanol requires 1.4 kg of CO_2 to be produced. Therefore, the amount of CO_2 emissions in kg is as follows:

$$6.137 \times 1.4 = 8.59\text{ kg of }CO_2$$

The number of equivalent CO_2 moles can be calculated as the following:

$$CO_2\text{ moles} = \frac{\text{Mass of }CO_2}{MW_{CO_2}} = \frac{8.59 \times 1000}{44.01} = 195.18\text{ mol}$$

Considering the stoichiometry of the chemical reaction in (equation 4.15), the number of hydrogen moles can be determined by utilizing the ratio of carbon dioxide moles to hydrogen moles of 1:3.

$$H_2\text{ moles} = CO_2\text{ moles} \times 3 = 195.18 \times 3 = 585.54\text{ mol}$$

Finally, the mass of hydrogen can be calculated using the following equation:

$$H_2\text{ mass} = H_2\text{ moles} \times MW_{H_2} = \frac{585.54 \times 2.016}{1000} = 1.18\text{ kg of }H_2$$

As a petrochemical feedstock and fuel, methanol has a very wide variety of applications. For instance, methanol is used to produce paints, car parts, medical equipment, and many other products. Also, it can be used as a hydrogen carrier to generate electricity via fuel cells. Further, it can be used as a combustion fuel in numerous applications, including space heating, transportation, and cooking. Methanol is currently available at more than 100 ports worldwide. Hence, its infrastructure is similar to that of ammonia in maturity and readiness. In 2019, China launched the "100 M" project to fuel vehicles with pure methanol. The 100 M fleet is expected to expand up to 50,000 vehicles in the near future. On the other hand, a fuel called "A20" was developed in Italy in 2019, composed of 15% methanol, 5% ethanol, and 80% gasoline. The A20 fleet comprises 500 vehicles traveling around 50,000 km annually [57–59].

4.3.4 Synthetic Petroleum

Petroleum products, including gasoline, diesel, kerosene, petroleum waxes, paraffins, and gases, can also be synthesized from syngas feedstock [60]. Each and every petroleum fuel has its unique importance and application. Hence, it is extremely challenging to replace all petroleum fuels in the near future. Thus, reducing emissions

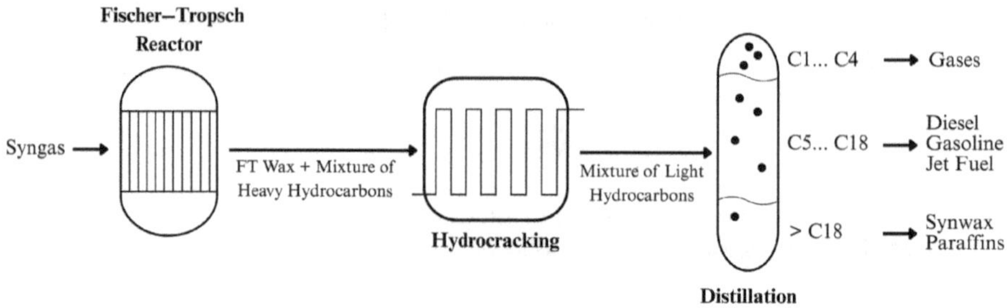

FIGURE 4.13 Petroleum fuels; the production process.

resulting from extracting crude oil from underground and emitting its carbon content into the atmosphere to net zero is the current goal for decision-makers worldwide [61]. The synthesis of petroleum fuels from syngas is carried through the Fischer–Tropsh process, followed by additional upgrading processes. The Fischer–Tropsch process was developed by two German chemists, Franz Fischer and Hans Tropsch, to produce long-chain hydrocarbons (heavy hydrocarbons) and petroleum waxes from syngas. The process occurs at around 200°C–300°C and a pressure between 10 and 40 bar [62].

The by-products of the Fischer–Tropsch are considered low-grade petroleum products. Thus, hydrocracking is necessary to obtain high-grade fuels. Hydrocracking is a process in which heavy hydrocarbons are subjected to high temperatures and pressures of 290°C–425°C and 80–205 bar, respectively [63]. The process typically involves four main components: a catalytic reactor, gas separator, scrubber for sulfur removal, and product fractionator. After the hydrocracking process, the mixture of light hydrocarbons is then distilled to produce high-grade petroleum fuels [64,65]. The complete process is shown in the following simplified schematic diagram (Figure 4.13).

4.4 CHAPTER SUMMARY

This chapter explored the diverse landscape of hydrogen derivatives and their potential to bridge industrial sectors under the banner of sustainability. Focusing on key examples, it examined green ammonia's pivotal role in decarbonizing food production and energy storage, contrasting the established Haber–Bosch process with emerging decomposition techniques. Further, it delved into electrofuels, highlighting their potential as sustainable alternatives to traditional fossil fuels, with specific emphasis on methane and methanol production via the Sabatier process and syngas conversion. The chapter details the production of syngas from captured CO_2 and green hydrogen through the SOECD process, followed by its utilization in methanol synthesis and the Fischer–Tropsch process. This analysis showcases how syngas can be transformed into a wide array of refined fuels, including diesel, gasoline, and kerosene, through catalytic synthesis, hydrocracking, and distillation, ultimately illustrating the transformative potential of hydrogen derivatives in achieving a sustainable energy future.

SELF-ASSESSMENT QUESTIONS

1. What is the most important derivative of green hydrogen, and what are its potential uses in the food and energy industries?
2. How is ammonia produced industrially, and what are the methods for decomposing it?
3. What are electro-fuels, and how are they different from conventional fuels?
4. Describe the process for producing electro-methane.
5. Why is syngas considered an important hydrogen derivative, and how is it produced?
6. Explain the process of producing methanol from syngas.
7. What is the Fischer–Tropsch process, and what are its products?
8. How are long-chain hydrocarbons and waxes from the Fischer–Tropsch process upgraded for refining?
9. What is the final step in producing high-quality fuels such as diesel, gasoline, and kerosene from hydrocarbons?
10. List some of the ways hydrogen can be utilized in its pure form.
11. Why is understanding hydrogen derivatives crucial for developing the global hydrogen economy?
12. What are some of the hydrogen derivatives mentioned in the chapter, and what are their applications?
13. Why is green ammonia considered important in the context of green hydrogen?
14. Compare the volumetric energy density of green ammonia and pure hydrogen.
15. What is the primary use of ammonia today, and what are the future prospects for its use in other sectors?
16. According to the International Energy Agency (IEA), what is the significance of green ammonia in achieving net zero emissions (NZE) by 2050?
17. Describe the Haber–Bosch process for producing green ammonia.
18. Who invented the Haber–Bosch process, and what is its significance in ammonia production?
19. Explain the schematic of the typical Haber–Bosch process and its components.
20. What is the ideal ratio of nitrogen to hydrogen in the Haber–Bosch process, and why is high pressure required?
21. What is the role of the catalyst bed in the Haber–Bosch process, and what is it mainly composed of?
22. Explain the reversible reaction governing ammonia production and the influence of temperature and pressure on it.
23. How does the Haber–Bosch process utilize Chatelier's principle to favor ammonia production?
24. Why are energy-capturing devices such as heat exchangers used in the Haber–Bosch process?
25. Describe the process of ammonia separation in the condenser and the fate of unreacted nitrogen and hydrogen.

26. How is the produced liquid ammonia further cooled, and what type of cooling system is typically used for large-scale applications?
27. What are the common measurement units for ammonia production?
28. What is the fractional conversion factor in ammonia production, and what is its typical range at nominal operating conditions?
29. Why is recycling of unreacted hydrogen and nitrogen gases necessary in the Haber–Bosch process?
30. Explain the concept of the extent of reaction and how it is calculated.
31. How can the extent of reaction be used to calculate the number of reacted and produced moles of a substance?
32. What equation can be used to assess the concentration of a substance before and after a chemical reaction?
33. How is the production rate of ammonia calculated in MTPD?
34. What are the two main categories of chemical analyses used to obtain accurate values for the ammonia production rate?
35. Give examples of theoretical methods used to analyze the production rate of ammonia.
36. What are the different methods for storing ammonia after its production?
37. Describe the conditions for storing gaseous and liquid ammonia.
38. What are the advantages of using ammonia as an energy carrier compared with hydrogen in its pure form?
39. Why is storing and transferring hydrogen in its gaseous phase inefficient?
40. How can ammonia be broken down into hydrogen and nitrogen after transportation?
41. Describe the process of ammonia electrolysis and its energy requirements compared with water electrolysis.
42. Explain the thermal decomposition of ammonia for producing hydrogen and nitrogen.
43. What is the decomposition efficiency in ammonia decomposition, and how is it calculated?
44. How can the amount of undecomposed ammonia be calculated in the decomposition process?
45. What are the applications of ammonia in energy generation and other industries?
46. What is the efficiency range of ammonia when used in an internal combustion engine?
47. What are the benefits of using ammonia as a green fuel for internal combustion engines?
48. How can ammonia be utilized in fuel cell applications for electricity generation?
49. What are the major fertilizers that rely on ammonia directly or indirectly?
50. Why is understanding ammonia's role in the hydrogen supply chain crucial for green hydrogen and power-to-X initiatives?
51. An ammonia production plant uses 150 moles of nitrogen gas per second to produce ammonia. The single-pass conversion factor is 12%. Calculate the number of moles of hydrogen gas required per second.

52. For the ammonia plant in Problem 1, calculate the extent of the reaction.
53. For the ammonia plant in Problem 1, calculate the output molar flow rate of ammonia from a single pass, assuming the reaction takes 1 second to complete.
54. An ammonia production plant consumes 500 moles of nitrogen gas per second. The molar weights are 28.014, 2.016, and 17.031 g/mol for nitrogen gas, hydrogen, and ammonia, respectively. Assuming a total conversion factor of 98%, calculate the production rate of ammonia in MTPD.
55. An ammonia decomposition plant consumes 700 TPD of ammonia with a decomposition efficiency of 95%. Calculate the production rate of hydrogen in moles per second, considering the following molar weights: $MW_{H_2} = 2.016$ g/mol, $MW_{NH_3} = 17.031$ g/mol.
56. A single household consumes 25 L of propane for heating per week. Assuming that each liter weighs 0.51 kg and the energy density of propane is 50.3 MJ/kg, calculate the weekly energy consumption in MJ.
57. If the household in Problem 6 switches to methane, which has an energy density of 55.5 MJ/kg, how much methane would be required per week to meet their energy needs?
58. Calculate the mass of hydrogen required to produce the amount of methane needed in Problem 7, assuming a methanation efficiency of 80%. Use the following molar weights: $MW_{CH_4} = 16.04$ g/mol, $MW_{H_2} = 2.016$ g/mol.
59. Calculate the amount of CO_2 and H_2 required to produce enough methanol to fuel a car for a 100 km drive. Assume the car consumes 12 kWh of energy to travel 100 km, the methanol engine efficiency is 42%, the energy density of methanol is 22 MJ/kg, and the molar weights of CO_2 and H_2 are 44.01 and 2.016 g/mol, respectively.
60. A car consumes 8 L of gasoline/100 km. If this car were to switch to electro-methanol, how much methanol would be required for the same distance, assuming that the energy density of gasoline is 44 MJ/kg and the energy density of methanol is 22 MJ/kg?

REFERENCES

1. Aminudin, Muhammad Akmal, Siti Kartom Kamarudin, Boon Han Lim, et al. 2023. "An overview: Current progress on hydrogen fuel cell vehicles." *International Journal of Hydrogen Energy* 48 ((11): 4371–4388. https://doi.org/10.1016/j.ijhydene.2022.10.156.
2. Cavaliere, Pasquale. 2023. "Hydrogen applications." In *Water Electrolysis for Hydrogen Production*, pp. 653–727. Cham: Springer International Publishing. https://doi.org/10.1007/s11814-024-00264-5.
3. Chavez-Angel, Emigdio, Alejandro Castro-Alvarez, Nicolas Sapunar, et al. 2023. "Exploring the potential of green hydrogen production and application in the antofagasta region of Chile." *Energies* 16 (11): 4509. https://doi.org/10.3390/en16114509.
4. Qazi, Umair Yaqub. 2022. "Future of hydrogen as an alternative fuel for next-generation industrial applications; challenges and expected opportunities." *Energies* 15 (13): 4741. https://doi.org/10.3390/en15134741.

5. Chatterjee, Sudipta, Rajesh Kumar Parsapur, and Kuo-Wei Huang. 2021. "Limitations of ammonia as a hydrogen energy carrier for the transportation sector." *ACS Energy Letters* 6 (12): 4390–4394. https://doi.org/10.1021/acsenergylett.1c02189.
6. Mazloomi, Kaveh and Chandima Gomes. 2012. "Hydrogen as an energy carrier: Prospects and challenges." *Renewable and Sustainable Energy Reviews* 16 (5): 3024–3033. https://doi.org/10.1007/s10669-023-09932-z.
7. Pearson, Andy. 2005. "Carbon dioxide—New uses for an old refrigerant." *International Journal of Refrigeration* 28 (8): 1140–1148. https://doi.org/10.1016/j.ijrefrig.2005.09.005.
8. Faria, Jimmy A. 2021. "Renaissance of ammonia synthesis for sustainable production of energy and fertilizers." *Current Opinion in Green and Sustainable Chemistry* 29: 100466. https://doi.org/10.1016/j.cogsc.2021.100466.
9. International Energy Agency. 2021. "Ammonia Technology Roadmap: Towards More Sustainable Nitrogen Fertiliser Production." OECD Publishing. Available at: https://www.iea.org/reports/ammonia-technology-roadmap.
10. Humphreys, John, Rong Lan and Shanwen Tao. 2021. "Development and recent progress on ammonia synthesis catalysts for Haber–Bosch process." *Advanced Energy and Sustainability Research* 2 (1): 2000043. https://doi.org/10.1002/aesr.202000043.
11. Lin, Liangrui, Zihan Lin, Qi Xia, et al. 2024. "Full-spectrum solar energy utilization for green ammonia production via solid oxide electrolysis cell coupled with Haber-Bosch process." *Energy Conversion and Management* 310: 118488. https://doi.org/10.1016/j.enconman.2024.118488.
12. Lee, Jae Seong, Gao-Feng Han, and Jong-Beom Baek. 2023. "Mechanochemical ammonia synthesis: Old is new again." *ChemSusChem* 16 (22): e202300459. https://doi.org/10.1002/cssc.202300459.
13. Zheng, Jianyun, Li Jiang, Yanhong Lyu, et al. 2022. "Green synthesis of nitrogen-to-ammonia fixation: Past, present, and future." *Energy & Environmental Materials* 5 (2): 452–457. https://doi.org/10.1002/eem2.12192.
14. Binoy, Mariya and P. A. Soloman. 2021. "Heat integration of synthesis gas production part of Haber-Bosch process of ammonia manufacture." In *Proceedings of the Yukthi 2021—The International Conference on Emerging Trends in Engineering*, GEC Kozhikode, Kerala, India, pp. 1–6. SSRN. https://dx.doi.org/10.2139/ssrn.3991462.
15. Verleysen, Kevin, Alessandro Parente, and Francesco Contino. 2021. "How sensitive is a dynamic ammonia synthesis process? Global sensitivity analysis of a dynamic Haber-Bosch process (for flexible seasonal energy storage)." *Energy* 232: 121016. https://doi.org/10.1016/j.energy.2021.121016.
16. Denis Constales, Gregory S. Yablonsky, Dagmar R. D'hooge, et al. 2017. "Chapter 3—Complex reactions: Kinetics and mechanisms—Ordinary differential equations—Graph theory." In *Advanced Data Analysis & Modelling in Chemical Engineering*, pp. 35–82. Elsevier. https://doi.org/10.1016/C2011-0-08276-5.
17. Dmitry Murzin and Tapio Salmi. 2005. "Chapter 10—Kinetic modeling." In *Catalytic Kinetics*, pp. 419–468. Elsevier Science. https://doi.org/10.1016/B978-0-444-51605-3.X5000-X.
18. Parker, Vernon D. 2005. "Is the single transition-state model appropriate for the fundamental reactions of organic chemistry?" *Pure and Applied Chemistry* 77 (11): 1823–1833. https://doi.org/10.1039/D3CP00380A.
19. Binous, Housam, Ahmed Aheed, and Mohammed M. Hossain. 2016. "Haber process and steam-coal gasification: Two standard thermodynamic problems elucidated using two distinct approaches." *Computer Applications in Engineering Education* 24 (1): 58–70. https://doi.org/10.1007/s13399-023-04167-z.
20. Mohamed Rashidi and Nur Amirah. 2021. "Mixing Profile of Nitrogen and Hydrogen Gases in Monolithic Channel through Computational Fluid Dynamics." PhD diss., Universiti Teknologi PETRONAS. https://utpedia.utp.edu.my/id/eprint/22777.

21. Paramanantham, Salai Sargunan S., Boris Brigljević, Aleksey Ni et al. 2023. "Numerical simulation of ball milling reactor for novel ammonia synthesis under ambient conditions." *Energy* 263: 125754. https://doi.org/10.1016/j.energy.2022.125754.
22. Sam Mannan. 2007. "Chapter 22—Storage." In *Lees' Loss Prevention in the Process Industries*, pp. 22/1–22/78. Butterworth-Heinemann. ISBN: 9780080962313.
23. Aziz, Muhammed, Agung Tri Wijayanta, and Asep Bayu Dani Nandiyanto. 2020. "Ammonia as effective hydrogen storage: A review on production, storage and utilization." *Energies* 13 (12): 3062. https://doi.org/10.3390/en13123062.
24. Sun, Hui Ying, Guang-Rui Xu, Fu-Min Li, et al. 2020. "Hydrogen generation from ammonia electrolysis on bifunctional platinum nanocubes electrocatalysts." *Journal of Energy Chemistry* 47: 234–240. https://doi.org/10.1016/j.jechem.2020.01.035.
25. Lucentini, Ilaria, Xènia Garcia, Xavier Vendrell, et al. 2021. "Review of the decomposition of ammonia to generate hydrogen." *Industrial & Engineering Chemistry Research* 60 (51): 18560–18611. https://doi.org/10.1021/acs.iecr.1c00843.
26. Cardoso, João Sousa, Valter Silva, Rodolfo C. Rocha, et al. 2021. "Ammonia as an energy vector: Current and future prospects for low-carbon fuel applications in internal combustion engines." *Journal of Cleaner Production* 296: 126562. https://dx.doi.org/10.1016/j.jclepro.2021.126562.
27. Zarebska, Agata, Daniel Romero Nieto, Knud Villy Christensen, et al. 2015. "Ammonium fertilizers production from manure: A critical review." *Critical Reviews in Environmental Science and Technology* 45 (14): 1469–1521. https://dx.doi.org/10.1080/10643389.2014.955630.
28. Klimczyk, Marta, Anna Siczek, and Lech Schimmelpfennig. 2021. "Improving the efficiency of urea-based fertilization leading to reduction in ammonia emission." *Science of the Total Environment* 771: 145483. https://doi.org/10.1016/j.scitotenv.2021.145483.
29. Davidson, F. Todd, Kazunori Nagasawa, and Michael E. Webber. 2017. "Electrofuels." *Mechanical Engineering* 139 (9): 30–35. https://doi.org/10.1115/1.2017-Sep-1.
30. Chen, Lin, Goodluck Msigwa, Mingyu Yang, et al. 2022. "Strategies to achieve a carbon neutral society: A review." *Environmental Chemistry Letters* 20 (4): 2277–2310. https://doi.org/10.1007/s10311-022-01435-8.
31. US Congress. 2025. "Is Biopower Carbon Neutral?" Available at: https://www.congress.gov/crs-product/R41603.
32. Grahn, Maria, Elin Malmgren, Andrei D. Korberg, et al. 2022. "Review of electrofuel feasibility—Cost and environmental impact." *Progress in Energy* 4 (3): 032010. https://doi.org/10.1088/2516-1083/ac7937.
33. Reay, Dave, Pete Smith, and Andre van Amstel, eds. 2010. *Methane and Climate Change*. Routledge. ISBN 9781138866935.
34. US Environmental Protection Agency. 2022. "Importance of Methane." Available at: https://www.epa.gov/gmi/importance-methane#:~:text=Methane%20is%20the%20second%20most,trapping%20heat%20in%20the%20atmosphere.
35. Klell, Manfred, Helmut Eichlseder, and Markus Sartory. 2012. "Mixtures of hydrogen and methane in the internal combustion engine—Synergies, potential and regulations." *International Journal of Hydrogen Energy* 37 (15): 11531–11540. https://doi.org/10.1016/j.ijhydene.2012.03.067.
36. Franz, Robert, Evgeny A. Uslamin, and Evgeny A. Pidko. 2021. "Challenges for the utilization of methane as a chemical feedstock." *Mendeleev Communications* 31 (5): 584–592. https://doi.org/10.1016/j.mencom.2021.09.002.
37. Moein, Peyman, Mehran Sarmad, Mahmoud Khakpour, et al. 2016. "Methane addition effect on a dual nitrogen expander refrigeration cycle for LNG production." *Journal of Natural Gas Science and Engineering* 33: 1–7. https://doi.org/10.1016/j.jngse.2016.04.061.

38. Longinos, Sotirios Nik and Mahmut Parlaktuna. 2020. "The effect of experimental conditions on methane (95%)—Propane (5%) hydrate formation." *Energies* 13 (24): 6710. https://doi.org/10.3390/en13246710.
39. Bin, Gong, Liu Xuan, and Qin Guan. 2014. "A Lattice Boltzmann model for multi-component vapor-liquid two phase flow." *Petroleum Exploration and Development* 41 (5): 695–702. https://doi.org/10.1016/S1876-3804(14)60083-5.
40. Guerra, Luís Filipe, Silvia Rossi, João Rodrigues, et al. 2018. "Methane production by a combined Sabatier reaction/water electrolysis process." *Journal of Environmental Chemical Engineering* 6 (1): 671–676. https://doi.org/10.1016/j.jece.2017.12.066.
41. Lisbona, Pilar, Manuel Bailera, Begoña Peña, et al. 2020. "Integration of CO_2 capture and conversion." In *Advances in Carbon Capture*, pp. 503–522. Woodhead Publishing. https://doi.org/10.1016/B978-0-12-819657-1.00022-0.
42. Pashchenko, Dmitry, 2019. "Experimental investigation of synthesis gas production by methane reforming with flue gas over a $NiO–Al_2O_3$ catalyst: Reforming characteristics and pressure drop." *International Journal of Hydrogen Energy* 44 (14): 7073–7082. https://doi.org/10.1016/j.ijhydene.2019.01.250.
43. Gür, Mesut and Engin Deniz Canbaz. 2020. "Analysis of syngas production and reaction zones in hydrogen oriented underground coal gasification." *Fuel* 269: 117331. https://doi.org/10.1016/j.fuel.2020.117331.
44. Naeem, Nida, Asif Hussain Khoja, Faaz Ahmed Butt, et al. 2022. "Partial oxidation of methane over biomass fly ash (BFA)-supported Ni/CaO catalyst for hydrogen-rich syngas production." *Research on Chemical Intermediates* 48 (5): 2007–2034. https://doi.org/10.1007/s11164-022-04685-x.
45. Mehdi, Muzaffar, Syed Ali Ammar Taqvi, Asif Ahmed Shaikh, et al. 2023. "Aspen plus simulation model of municipal solid waste gasification of metropolitan city for syngas production." *Fuel* 344: 128128. https://doi.org/10.1016/j.fuel.2023.128128.
46. Zhou, Nan, Junwen Zhou, Leilei Dai, et al. 2020. "Syngas production from biomass pyrolysis in a continuous microwave assisted pyrolysis system." *Bioresource Technology* 314: 123756. https://doi.org/10.1016/j.biortech.2020.123756.
47. Dang, Qi, Xiaoqi Zhang, Yuling Zhou, et al. 2021. "Prediction and optimization of syngas production from a kinetic-based biomass gasification process model." *Fuel Processing Technology* 212: 106604. https://doi.org/10.1016/j.fuproc.2020.106604.
48. Vecten, Simon, Michael Wilkinson, Nuno Bimbo, et al. 2021. "Hydrogen-rich syngas production from biomass in a steam microwave-induced plasma gasification reactor." *Bioresource Technology* 337: 125324. https://doi.org/10.1016/j.biortech.2021.125324.
49. Choe, Changgwon, Seunghyun Cheon, Jiwon Gu, et al. 2022. "Critical aspect of renewable syngas production for power-to-fuel via solid oxide electrolysis: Integrative assessment for potential renewable energy source." *Renewable and Sustainable Energy Reviews* 161: 112398. https://doi.org/10.1016/j.rser.2022.112398.
50. Schäfer, Dominik, Qingping Fang, Ludger Blum, et al. 2019. "Syngas production performance and degradation analysis of a solid oxide electrolyzer stack." *Journal of Power Sources* 433: 126666. https://doi.org/10.1016/j.jpowsour.2019.05.072.
51. Moghadam, Reza Alipour, Suzana Yusup, Yoshimitsu Uemura, et al. 2014. "Syngas production from palm kernel shell and polyethylene waste blend in fluidized bed catalytic steam co-gasification process." *Energy* 75: 40–44. https://doi.org/10.1016/j.energy.2014.04.062.
52. Rauch, Reinhard, Jitka Hrbek, and Hermann Hofbauer. 2014. "Biomass gasification for synthesis gas production and applications of the syngas." *Wiley Interdisciplinary Reviews: Energy and Environment* 3 (4): 343–362. https://doi.org/10.1002/9781118957844.ch7.
53. Kundu, Arunabha, Yong Gun Shul, and Dong Hyun Kim. 2007. "Methanol reforming processes." In *Advances in Fuel Cells*, vol. 1, pp. 419–472. Elsevier Science. https://doi.org/10.1016/S1752-301X(07)80012-3.

54. Dalena, Francesco, Alessandro Senatore, Alessia Marino, et al. 2018. "Methanol production and applications: An overview." In *Methanol*, pp. 3–28. Elsevier. https://doi.org/10.1016/B978-0-444-63903-5.00001-7.
55. Hobson, Charlie. 2018. "Renewable Methanol Report." methanol.org. edited by Carlos Márquez. Methanol Institute. Available at: https://www.methanol.org/wp-content/uploads/2019/01/MethanolReport.pdf
56. Do, Thai Ngan and Jiyong Kim. 2019. "Process development and techno-economic evaluation of methanol production by direct CO_2 hydrogenation using solar-thermal energy." *Journal of* CO_2 *Utilization* 33: 461–472. https://doi.org/10.1016/j.jcou.2019.07.003.
57. Brusstar, Matthew, Mark Stuhldreher, David Swain, et al. 2002. "High efficiency and low emissions from a port-injected engine with neat alcohol fuels." *SAE Transactions*, pp. 1445–1451. https://doi.org/10.4271/2002-01-2743.
58. Basile, Angelo and Francesco Dalena, eds. 2017. *Methanol: Science and Engineering*. Elsevier. ISBN: 978-0-444-63903-5.
59. Zhao, Kai. 2019. "A Brief Review of China's Methanol Vehicle Pilot and Policy." Methanol Institute. Available at: https://www.methanol.org/wp-content/uploads/2019/03/A-Brief-Review-of-Chinas-Methanol-Vehicle-Pilot-and-Policy-20-March-2019.pdf.
60. Yellapragada, Datta Bharadwaz and Govinda Rao Budda. 2021. "Methanol—A sustainable fuel for SI engine." In *Methanol: A Sustainable Transport Fuel for SI Engines*, pp. 103–137. Singapore: Springer Singapore. https://doi.org/10.1007/978-981-16-1224-4_5.
61. dos Santos, Ronaldo Gonçalves and Andre Cardoso Alencar. 2020. "Biomass-derived syngas production via gasification process and its catalytic conversion into fuels by Fischer Tropsch synthesis: A review." *International Journal of Hydrogen Energy* 45 (36): 18114–18132. https://doi.org/10.1016/j.ijhydene.2019.07.133.
62. International Energy Agency. 2021. "Net Zero by 2050: A Roadmap for the Global Energy Sector—Event." Available at: https://www.iea.org/events/net-zero-by-2050-a-roadmap-for-the-global-energy-system.
63. Dry, Mark E. 2002. "The fischer–tropsch process: 1950–2000." *Catalysis Today* 71 (3–4): 227–241. https://doi.org/10.1016/S0920-5861(01)00453-9.
64. Weitkamp, Jens. 2012. "Catalytic hydrocracking—Mechanisms and versatility of the process." *ChemCatChem* 4 (3): 292–306. https://doi.org/10.1002/cctc.201100315.
65. Kinsara, Rami A. and Ayhan Demirbas. 2016. "Upgrading of crude oil via distillation processes." *Petroleum Science and Technology* 34 (14): 1300–1306. https://doi.org/10.1080/10916466.2016.1200080.

5 Hydrogen Storage

5.1 INTRODUCTION

For various reasons, green hydrogen storage is an essential stage in the supply chain. For instance, as previously discussed in the green hydrogen production chapter, green hydrogen is produced via water electrolysis powered by renewable energy, which in turn has an intermittent behavior on a temporal basis [1]. This necessitates integrating green hydrogen storage systems within the supply chain to sustain a reliable hydrogen supply. Moreover, hydrogen is an extremely light gas with a volumetric density of 0.09 kg/m^3 at standard conditions. Therefore, large volumes of hydrogen must be available to obtain sufficient quantities of hydrogen on a mass basis. This can be avoided by compressing and storing the produced green hydrogen before transporting it to the desired locations [2].

Green hydrogen is mainly stored as an energy carrier or chemical feedstock for various industries. For instance, producing green hydrogen via water electrolysis utilizing excess renewable energy is among the most promising techniques for renewable energy storage, especially on a seasonal basis. The stored green hydrogen can then be utilized in fuel cell systems to generate electricity when there is a lack or an absence of renewable energy generation, thus ensuring a stable supply of electricity while minimizing the reliance on fossil fuels for electricity generation. On the other hand, numerous industries depend on hydrogen as a chemical feedstock, including oil refinery, steel reforming, fertilizers production, and many more [3].

5.2 PHYSICAL-BASED HYDROGEN STORAGE

Hydrogen has a low volumetric density of 0.09 kg/m^3 and a low energetic density per volume of 3 kWh/m^3 at standard atmospheric conditions [2]. This fact prevents hydrogen from being stored at atmospheric pressure and temperature without the need for enormous storage volumes. This can be resolved by altering the physical conditions of hydrogen through compression or liquefaction to increase its density. The alteration of hydrogen's physical conditions is referred to as physical-based hydrogen storage, which can be carried out via three main methods: compressed hydrogen storage, liquified hydrogen storage, and cryo-compressed hydrogen storage, as illustrated in Figure 5.1 [4].

5.2.1 Compressed Hydrogen Storage

Compressed hydrogen storage technologies are the most common and mature technologies for storing hydrogen [5]. In fact, around 80% of hydrogen worldwide is stored and transported via compressed storage technologies [6]. As the name implies, this technology relies on increasing the pressure of hydrogen gas in high-pressure reservoirs. This significantly increases hydrogen density, allowing large quantities to be stored in a given volume [7].

DOI: 10.1201/9781003639190-5

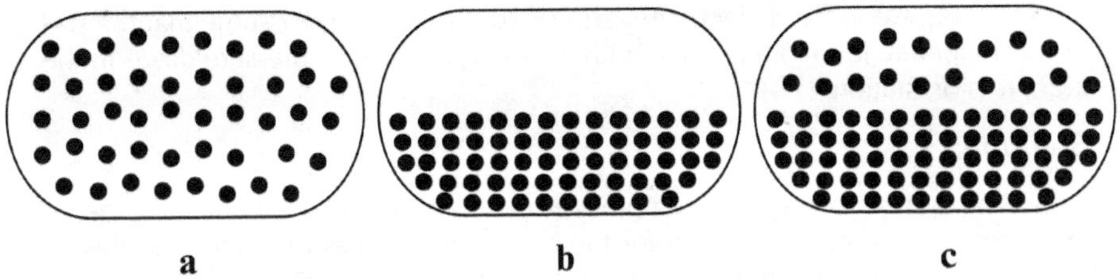

FIGURE 5.1 Physical-based hydrogen storage: (a) compressed hydrogen storage. (b) Liquefied hydrogen storage. (c) Cryo-compressed hydrogen storage.

After being produced, green hydrogen passes through a compressor before storage in vessels or underground cavities [8]. Hydrogen compressors must possess high-pressure ratios, minimal leakage, and high energy efficiency [9]. The compression ratio can be mathematically expressed as follows:

$$r_p = \frac{P_{out}}{P_{in}} \tag{5.1}$$

where r_p is the pressure ratio, P_{out} is the output pressure from the compressor, and P_{in} is the input pressure of hydrogen entering the compressor. The compression ratio in hydrogen compressors can be as high as 20:1, which means that the pressure of hydrogen can be multiplied by a factor of 20 when passing through a single compression stage, minimizing the necessary number of compressors that are required to produce highly pressurized hydrogen [10]. Moreover, hydrogen compression is an energy-intensive process with typical energy requirements between 1.7 and 6.4 kWh/kg, depending on the specific characteristics of compression systems and the final pressure and quantity of compressed hydrogen [11]. On the other hand, the specific cost for hydrogen compression is roughly \$1.34/kg [12]. These requirements must always be considered when designing a hydrogen compression station, as they can hugely influence the feasibility of such projects.

Example 5.1: Designing a Compression Station

An ammonia production plant requires 250 kg of 600 bar pressurized hydrogen daily. This demand can be fulfilled by a nearby water electrolysis plant that produces green hydrogen at a pressure of 3 bar. Hence, the ammonia plant aims to incorporate a hydrogen compression station within its facilities. Determine (a) the number and type of compressors and (b) the investment and annual operational costs of the compression station. The available compressors in the market have the following characteristics:

Compressor	Compression Ratio	Energy Consumption (kWh/kg)	Market Price	Annual O&M Cost
A	5:1	2.7	\$60,000	10.2% of market price
B	10:1	4.2	\$85,000	11.3% of market price
C	15:1	4.8	\$100,000	12.8% of market price
D	20:1	5.9	\$132,000	14.1% of market price

Also, consider a fixed electricity tariff of \$0.085/kWh and assume that the station will operate for 345 days/year with no leakage losses or pressure drops in the compression station.

Solution:

a. The number of required compressors can be determined by obtaining the required compression ratio for the hydrogen compression station. In this case, hydrogen is fed at a pressure of 3 bar and is required to be compressed to 600 bar. Calculating the compression ratio using equation 5.1:

$$r_p = \frac{P_{out}}{P_{in}} = \frac{600}{3} = 200:1$$

This means that hydrogen's inlet pressure must be multiplied by 200 to achieve the desired output pressure, which cannot be achieved via a single compressor.

Therefore, several compressors must be chosen to ensure a total compression ratio of 200:1. In other words, an output pressure of 600 bar is required, considering an inlet pressure of 3 bar. This can be solved by trial and error for each type of compressor accordingly.

Solving for the outlet pressure for the first compression stage using equation 5.1 for each compressor,

Compressor(s)	Compression Ratio	First Stage Output Pressure (bar)
A	5:1	15
B	10:1	30
C	15:1	45
D	20:1	60

As the results imply, compressors A, B, and D can be used for compressing hydrogen in the first stage, while compressor C cannot be used in this stage as it produces hydrogen at a pressure of 45 bar, which is not a divisor for 600. Solving for the outlet pressure for the first compression stage using equation 5.1 for each compressor,

First Stage Compressor	Second Stage Inlet Pressure (bar)	Second Stage Compressor	Second-Stage Compression Ratio	Second Stage Output Pressure (bar)
		B	10:1	150
		C	15:1	225
A	15	D	20:1	300
B	30	A	5:1	150
		C	15:1	450
		D	20:1	600
D	60	A	5:1	300
		B	10:1	600
		C	15:1	900

As shown in the results, considering the available compressors in the market, there is only one possible arrangement of compressors to compress hydrogen from 3 to 600 bar. This is to use compressors B and D in a two-stage compression station, regardless of which compressor comes first.

b. The required initial investment can be simply obtained by summing the market prices for the utilized compressors. In this case, the total market price for compressors B and D equals:

$$\text{Total market price} = \$85{,}000 + \$132{,}000 = \$217{,}000$$

In practice, there are more costs to consider when designing a compression station. These include the cost of potential land expansion, logistics and installation costs, and many more.

In order to calculate the annual costs for the compression station, the required Operation and Maintenance (O&M) costs and the cost of electricity must be obtained and summed for both compressors.

The annual O&M cost can be obtained for each compressor as follows:

$$\text{Annual O\&M cost} = \text{O\&M\% of market price} \times \text{market price (\$)}$$

$$\text{Annual O\&M cost, B} = 11.3\% \times \$85{,}000 = \$9605$$

$$\text{Annual O\&M cost, D} = 14.1\% \times \$132{,}000 = \$18{,}612$$

$$\text{Total O\&M cost} = \$9605 + \$18{,}612 = \$28{,}217/\text{year}$$

Moreover, the annual cost of electricity consumed within the station under an electricity tariff of \$0.085/kWh can be calculated as follows:

$$\begin{aligned}\text{Annual cost of electricity} &= \text{Total daily electricity consumption (kWh/day)} \\ &\times \text{operational period (day/year)} \\ &\times \text{electricity tariff (\$/kWh)}\end{aligned}$$

Where the daily electricity consumption for each compressor can be obtained from the following equation:

$$\begin{aligned}\text{Daily electricity consumption (kWh/day)} &= \text{Specific electricity consumption (kWh/kg)} \\ &\times \text{daily quantity of compressed hydrogen (kg/day)}\end{aligned}$$

$$\text{Daily electricity consumption, B} = 4.2\ \text{kWh/kg} \times 250\ \text{kg/day} = 1050\ \text{kWh/day}$$

$$\text{Daily electricity consumption, D} = 5.9\ \text{kWh/kg} \times 250\ \text{kg/day} = 1475\ \text{kWh/day}$$

$$\begin{aligned}\text{Total daily electricity consumption} &= 1050\ \text{kWh/day} + 1475\ \text{kWh/day} \\ &= 2525\ \text{kWh/day}\end{aligned}$$

$$\text{Annual cost of electricity} = 2525 \text{ kWh/day} \times 345 \text{ day/year} \times 0.085 \text{ \$/kWh}$$

$$= \$74{,}045/\text{year}$$

Finally, the total annual operational costs for the compression can be obtained as follows:

$$\text{Annual operational cost} = \text{Total O\&M (\$/year)} + \text{Cost of electricity (\$/year)}$$

$$= \$102{,}262/\text{year}$$

5.2.1.1 Compressed Hydrogen Vessels

As mentioned earlier, compressed gaseous hydrogen can be stored in storage vessels or underground cavities. There are five types of hydrogen storage vessels, namely Type I, II, III, IV, and V. Each of the five types has its characteristics as follows:

- **Type I:** These storage vessels are typically made of aluminum or steel (Figure 5.2a) and can withstand pressures of between 175 and 200 bar. Also, they are the heaviest hydrogen storage vessels, with a volumetric

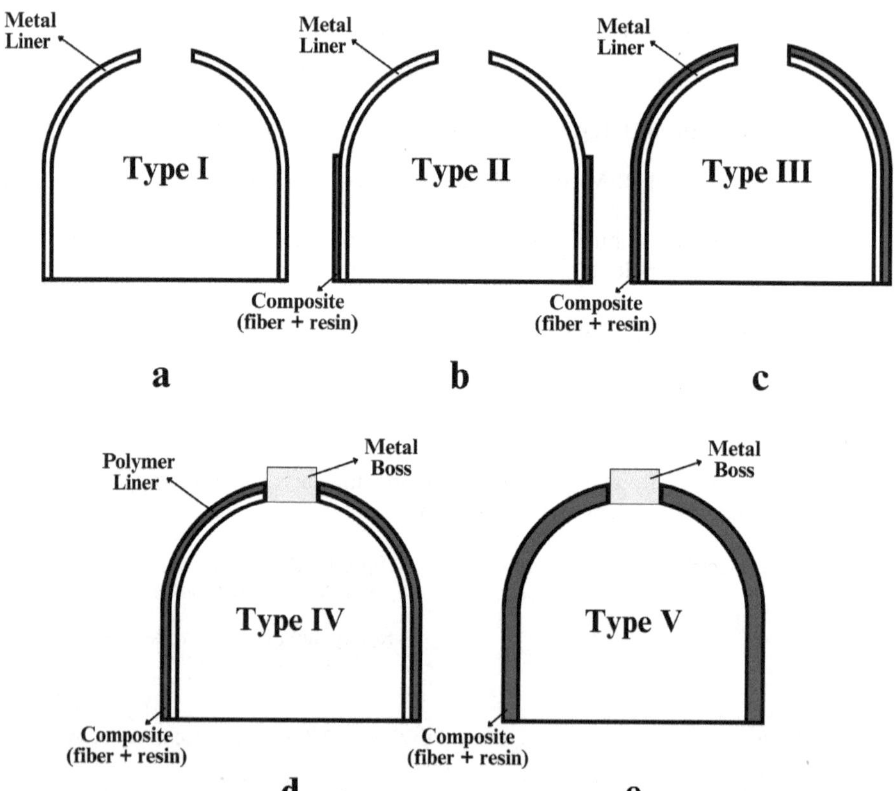

FIGURE 5.2 Compressed hydrogen storage vessels: (a) Type I. (b) Type II. (c) Type III. (d) Type IV. (e) Type V.

density of around 1.36 kg/L. Additionally, they are the least expensive option among all alternatives for storage vessels, costing approximately $83/kg. Their abundant availability, ease of manufacturing, and low cost have made this type of storage vessel become the predominant container for hydrogen storage in many industrial applications. However, their relatively heavy weight and low storage density of 1 wt.% have limited them to stationary applications. It is worth mentioning that the storage density refers to the percentage of hydrogen's weight out of the total tank's weight [12–14].

- **Type II:** Storage vessels of this type are made of aluminum or steel wrapped with a composite material of glass or carbon fiber along with resin around the mid-section of the cylinder (Figure 5.2b). Typically, these vessels can withstand a pressure between 263 and 300 bar. Furthermore, Type II vessels typically cost 50% more than Type I vessels (= $124.5/kg) but weigh 30%–40% less than them (0.82–0.95 kg/L). However, they are still considered heavy tanks and have a storage density of between 1 and 2 wt.%, making them only suitable for stationary industrial applications [12–14].
- **Type III:** These vessels are made of aluminum or steel and fully wrapped with a composite material made of glass or carbon fiber and resin (Figure 5.2c). They can endure a wide range of high pressures ranging from 350 to 700 bar. Furthermore, they weigh around (0.30–0.45 kg/L), making them approximately half as heavy as Type II vessels but twice the price (= $249/kg). Moreover, they can hold 25%–75% more hydrogen mass than Type I and II vessels (1.25–3.5 wt.%), making them suitable for vehicular hydrogen applications [12,13].
- **Type IV:** This type of storage vessel is typically made of a polymer liner consisting of high-density polyethylene (HDPE) covered by a carbon fiber composite layer and supported by a metal boss. Hence, they are the lightest commercially available storage vessels for compressed hydrogen storage, with a volumetric density of around 0.30 kg/L in best designs, making them slightly lighter than Type III tanks. Further, these vessels can withstand a pressure of up to 1000 bar, enabling them to store hydrogen on a volumetric basis effectively. Moreover, they can store hydrogen at a storage density of around 5.5 wt.% [12–14]. Also, vessels under this type have a typical cost of approximately equivalent to Type III vessels (= $249/kg), making them the most suitable option for vehicular hydrogen storage and transportation.

As illustrated in Figure 5.3, a typical Type IV hydrogen storage vessel consists of an HPDE layer covered by a carbon fiber composite layer. These layers increase the vessel's strength in resisting the mechanical stress resulting from high hydrogen pressure. Moreover, the vessel is equipped with a protection dome that plays a significant safety role by isolating the vessel from the surrounding environment and providing additional mechanical resistance. Furthermore, a pressure relief device is utilized to decrease the pressure within the vessel in the case of pressure accumulation. Last but not least, the mechanical valve controls the direction and magnitude of hydrogen's flow out of and into the storage vessel [15].

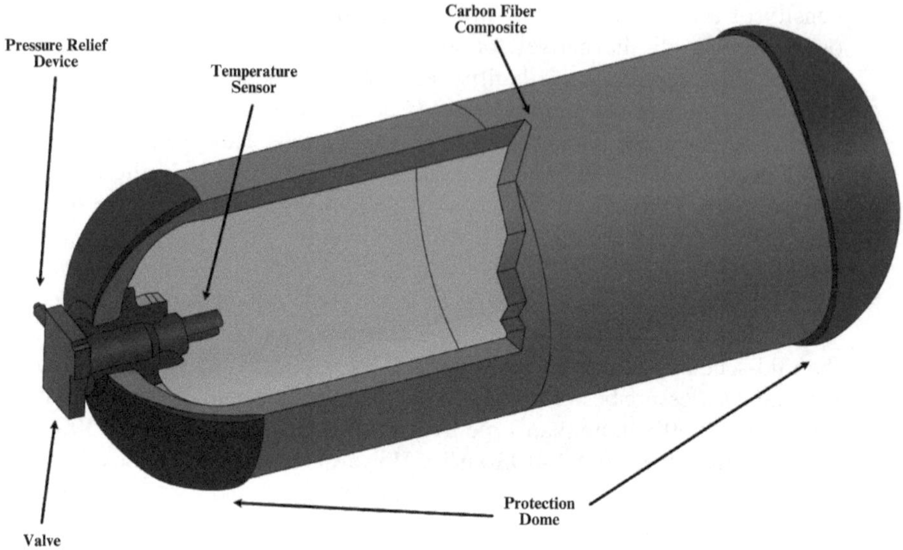

FIGURE 5.3 Type IV compressed hydrogen storage vessel.

- **Type V:** These vessels are still under development and are not yet commercially available. They ought to be fully made of a carbon fiber composite without a liner and supported by a metal boss. Moreover, they are 20%–30% lighter than Type IV tanks, with an approximate density of between 2.1 and 2.4 kg/L, and can theoretically withstand up to 1000 bar pressures. Although they show a significant potential for increasing the gravimetric and volumetric density of hydrogen storage, significant research is still required to fully uncover their applicability in commercial settings [16].

When handling theoretical calculations relevant to the sizing of hydrogen storage tanks, it is essential to be familiar with basic physical principles, including the ideal gas law, volumetric density, and storage density. Each of these can be mathematically expressed as follows: Ideal gas law [17]:

$$PV = nRT \tag{5.2}$$

where P is the pressure of gas in units of Pascals (Pa), V is the volume occupied by the gas in cubic meters (m^3), n is the gas number of moles (mol), R is the universal gas constant in Joules per Kelvin multiplied by moles (= 8.3144 J/K mol). T is the gas temperature in Kelvin (K).

Moreover, it is important to recall that the number of moles can be calculated through the following formula:

$$n = \frac{m}{MW} \tag{5.3}$$

where n is the number of moles (mol), m is the mass of a substance in grams (g), and MW is the molecular weight of a substance in grams per mole (g/mol). Furthermore, the storage density can be expressed as follows [18]:

$$\text{wt.\%} = \frac{m}{M_t} \tag{5.4}$$

where wt.% is the storage density percentage, m is the mass of stored substance in kilograms, and M_t is the mass of the storage tank in kilograms. Finally, we must recall that the volumetric density for storage vessels is given by the following simple equation [17]:

$$\rho_s = \frac{M_t}{V_t} \tag{5.5}$$

where ρ_s is the storage density, which is often given in units of kilograms per liter (kg/L), M_t is the mass of the storage tank in kilograms, and V_t is the volume of the storage tank(s) in liters (L).

Example 5.2: Selection of Stationary Storage Tanks

A hydrogen production plant produces 120,000 L of gaseous green hydrogen per day and compresses it to a pressure of 200 bar to be stored in storage vessels on-site at a temperature of 25°C. If the available storage tanks in the market possess the following characteristics:

Tank	Maximum Pressure (bar)	Volumetric Density (kg/L)	Storage Density (wt.%)	Market Price ($/kg)
I	200	1.36	1.0	83
II	275	0.92	2.0	124.5
III	700	0.43	3.0	230
IV	950	0.30	5.5	250

Determine the most suitable type of tank(s) for on-site hydrogen storage and calculate the tanks' mass, volume, and total cost.

Solution:

Considering that hydrogen is compressed to 200 bar after being produced, all the available tank types (I, II, III, and IV) are capable of withstanding the pressure limit. However, for a pressure of 200 bar in a stationary application, Types I and II are the strongest candidates due to their lower cost and lower pressure rating, resulting in investment savings and higher storage efficiency. Hence, these two types must be compared for on-site hydrogen storage.

In order to obtain the total volume, weight, and cost of each hydrogen storage alternative, we must calculate the mass of hydrogen to be stored using equations 5.2 and 5.3.

To find the number of hydrogen moles, we can apply the ideal gas law from equation 5.2:

$$PV = nRT$$

Further, to find the mass of hydrogen in kilograms (kg), the following modified form of equation 5.3 applies:

$$m = \frac{n \times MW}{1000}$$

Where MW is the molecular weight of hydrogen (= 2.0157 g/mol).

The pressure, volume, and temperature in the problem are all given in inconsistent units. Therefore, we must convert the values to apply the ideal gas law.

Recall:

$1\ m^3 = 1000\ L$

$1\ bar = 100{,}000\ Pa$

$$T_K = T_C + 273.15$$

Therefore,

- **Pressure:**

$$200\ \text{bar} \times \frac{100{,}000\ \text{Pa}}{1\ \text{bar}} = 20 \times 10^6\ \text{Pa}$$

- **Volume:**

$$120{,}000\ \text{L} \times \frac{1\ \text{m}^3}{1000\ \text{L}} = 120\ \text{m}^3$$

- **Temperature:**

$$T_K = 25 + 273.15 = 298.15\ \text{K}$$

To find the number of produced hydrogen gas moles, we substitute the converted values into the ideal gas law as follows:

$$PV = nRT$$

$$n = \frac{PV}{RT}$$

$$n = \frac{20 \times 10^6 \times 120}{8.3144 \times 298.15}$$

$$n = 968.15 \times 10^3\ \text{mol}$$

This translates into a mass of hydrogen of:

$$m = \frac{n \times MW}{1000}$$

$$m = \frac{968.15 \times 10^3 \times 2.0157}{1000}$$

$$m = 1951.5 \text{ kg}$$

Therefore, the hydrogen production plant produces around 1.95 tons of green hydrogen per day. To calculate the total volume of hydrogen storage tanks as well as their total cost, the required mass of tanks must be obtained. This can be calculated through equation 5.4:

$$wt.\% = \frac{m}{M_t}$$

- **Type I:**

$$0.01 = \frac{1951.5}{M_t}$$

$$M_t = 195{,}150 \text{ kg}$$

- **Type II:**

$$0.02 = \frac{1951.5}{M_t}$$

$$M_t = 97{,}575 \text{ kg}$$

The results show that Type II tanks weigh 50% less than Type I tanks due to their higher storage density. Moreover, the total market price for each type of tank can be formulated by multiplying the specific cost per kg by the total mass of each type.

- **Type I:**

$$\text{Total price} = M_t \times \text{Specific cost}$$

$$\text{Total price} = 195{,}150 \times \$83 = \$16.2 \text{ million}$$

- **Type II:**

$$\text{Total price} = M_t \times \text{Specific cost}$$

$$\text{Total price} = 97{,}575 \times \$124.5 = \$12.1 \text{ million}$$

Finally, the total required volume for each type of storage tanks can be calculated through equation 5.5 as follows:

$$\rho_s = \frac{M_t}{V_t}$$

Recalling that each 1 m^3 corresponds to 1000 L, the volumetric density becomes 1360 kg/m^3 for Type I storage tanks and 920 kg/m^3 for Type II storage tanks. Substituting these values in equation 5.5, we get:

- **Type I:**

$$1360 = \frac{195{,}150}{V_t}$$

$$V_t = 143.5\ m^3$$

- **Type II:**

$$920 = \frac{97{,}575}{V_t}$$

$$V_t = 106.1\ m^3$$

As the results imply, Type II storage tanks will be lighter weight, less mass, and cheaper to purchase. This is mainly due to their higher storage density and lower volumetric density, making them more suitable for storing hydrogen on-site.

Compressed hydrogen storage is among the most practical and cost-effective methods to store hydrogen for small-scale applications with a levelized cost of storage (LCOS) of \$0.19/kg, which is expected to drop down to \$0.17/kg in the near future [19]. However, it fails to deliver sufficient quantities of hydrogen for large-scale applications due to the volumetric constraints posed by the physical characteristics of gaseous hydrogen [20]. Hence, in order to effectively store compressed hydrogen for large-scale applications, extensive engineering efforts have been dedicated to exploring innovative solutions that account for the volume of gaseous hydrogen. Among these solutions is storing compressed hydrogen naturally or man-made underground. Man-made cavities include salt caverns, hard rock caverns, and abandoned mines, while natural cavities are saline aquifers and depleted oil/gas reservoirs [21]. Generally, underground hydrogen storage technologies offer various advantages over other storage methods, including higher safety and lower risk of accidents, smaller on-land footprint, and lower investment costs [22].

5.2.1.2 Salt Caverns

Salt caverns are artificial man-made cavities formed in underground salt deposits (Figure 5.4). They are created by the dissolution of rock salt through leaching, which refers to injecting highly pressurized freshwater to create a cavity within rock salt suitable for storing hydrogen and many other gases [23,24].

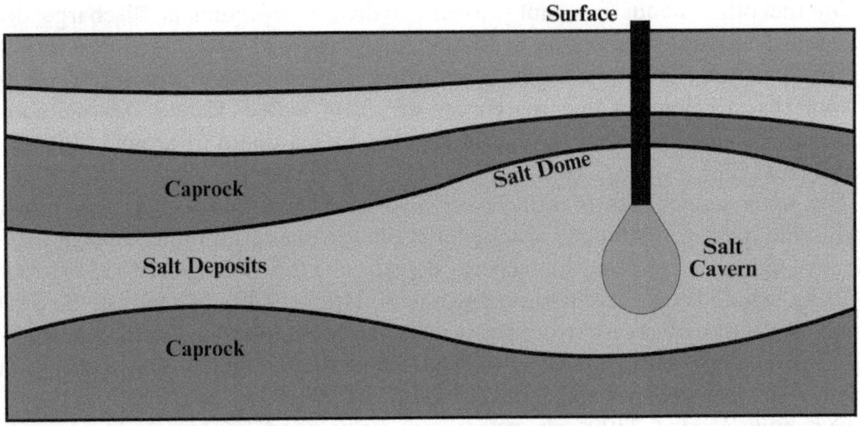

FIGURE 5.4 Salt cavern for underground hydrogen storage.

As shown in Figure 5.4, salt caverns are embedded within a salt dome connected to a salt deposit and surrounded by caprock. Further, salt caverns possess a low permeability due to the tightness of salt deposits and the low porosity of the surrounding caprock, resulting in a hydrogen leakage rate of less than 1% [25]. Typically, salt caverns have a height of between 300 and 500 m and a diameter of 50–150 m, making them capable of storing 50,000–1,000,000 m^3 of hydrogen [26].

The pressure of hydrogen stored in a salt cavern mainly depends on the cavern's depth and the lithostatic pressure caused by the caprock above the cavern. Lithostatic pressure refers to the pressure caused by the weight of the caprock layers above a specific level under the ground. Alas, no specific mathematical equations can calculate hydrogen pressure within salt caverns. Hence, geologists and engineers rely on geophysical logs and numerical simulation tools [27]. However, hydrogen's maximum pressure within a salt cavern is typically 30%–80% of the lithostatic pressure value at the depth of the cavern roof [28].

Salt caverns are cyclic storage reservoirs that can be filled and emptied 10–12 times a year. Hydrogen can be filled and discharged into and from salt caverns using variable-pressure and constant-pressure techniques [26].

In variable-pressure injection and discharge, hydrogen is removed at different pressures, resulting in a significant pressure drop in the cavern. This leads to salt creep, damaging the structure of the salt cavern. Salt creep refers to the plastic (permanent) deformation of salt walls enclosing the cavern due to the accumulative stress resulting from rapid variations in pressure. This phenomenon becomes more crucial when storing hydrogen in salt caverns near the maximum allowable depth of 2000 m. This is because the pressure drop in a piping system rises as the length of the system increases. To avoid salt creep, a portion of hydrogen must always be kept within the cavern to act as a cushion gas, alleviating the pressure difference when filling and discharging the cavern. This portion typically contributes to one-third of the total mass of hydrogen stored in the salt cavern, especially in high-depth caverns [26].

On the other hand, constant-pressure hydrogen injection at discharge does not require any cushion gas, as hydrogen is injected with saturated brine into the cavern. The addition of saturated brine to the cavern reduces pressure drop and ensures that all pores within the cavern are fully sealed. Hence, this method is more suitable for hydrogen storage in caverns with a depth of around 2000 m or slightly less [26].

The main advantages of salt cavern hydrogen storage are that they provide large-scale hydrogen storage capacity, effective tightness, minimal leakage due to effective tightness, relatively high energy density, and low hydrogen LCOS of around $0.23/kg, which is expected to drop down to $0.11/kg [19,29]. In contrast, the disadvantages of salt caverns for hydrogen storage can be summarized in the operational and environmental burden relevant to brine management and the geographical limitation of suitable salt rock and deposits worldwide [30].

Currently, around 2300 salt caverns are being used to store energy carriers, including natural gas, petroleum gas, and other liquid hydrocarbons. Around 2000 of these caverns are in North America, while the remaining 300 are in Germany [31]. However, plenty of salt deposits worldwide can be utilized for underground hydrogen storage. Many countries worldwide do not stand above salt deposits suitable for hydrogen storage. On the other hand, North America, Western and Eastern, possess an abundance of salt deposits, giving them the highest potential for utilizing salt caverns for underground hydrogen storage. Considering the Middle East and North Africa Region, some countries have varying potential for underground hydrogen storage in salt caverns. The most promising potential is in Algeria and Iran, followed by Morocco, Oman, Iraq, Yemen, and Jordan [31].

Despite the availability of suitable salt deposits for hydrogen storage in various places worldwide, only a few are being utilized for large-scale hydrogen storage. As of 2022, only four salt caverns were being utilized for this purpose worldwide. Specifically, there are three salt caverns in the United States of America (USA) and one salt cavern in the United Kingdom (UK) [32]. Each has the characteristics demonstrated in Table 5.1 [32,33].

As shown in Table 5.1, the salt cavern "Spindletop" in Texas, USA, has the maximum pressure rating, energy content, and cavern overall volume and depth among all other existing salt caverns. This is mainly due to the significant development in underground hydrogen storage technologies in recent decades, allowing for more effective design of salt caverns to withstand higher pressures and larger volumes of hydrogen [33].

5.2.1.3 Hard Rock Caverns

Hydrogen can be stored in rock caverns, excavated underground chambers in hard rock formations. These caverns can be lined rock caverns (LRC) or unlined rock caverns (URC). LRC is the most advanced and superior type of rock cavern since it is sealed with a special lining system made of steel composites or plastics to achieve a tight storage space with minimal hydrogen leakage and contamination. At the same time, URC relies on the tightness of natural hard rock formations, resulting in higher leakage rates [34]. Further, the steel or plastic lining is supported by two layers of concrete to ensure its structural integrity. This, along with the host hard rock, can

TABLE 5.1
Existing Large-Scale Salt Caverns for Hydrogen Storage Worldwide [33,34]

Salt Cavern	Teesside	Clemens Dome	Moss Bluff	Spindletop
Location	Middleborough, UK	Texas, USA	Texas, USA	Texas, USA
Establishment Year	1972	1983	2007	2017
Depth of cavern (m)	350–450	850	823	1128
Cavern height (m)	15–40	300	579	518
Cavern diameter (m)	70	49	60	76
Cavern volume (m^3)	70,000	580,000	566,000	906,000
Storage pressure (bar)	45–50	70–135	55–152	68–202
Energy stored (GWh)	25	92	120	>120

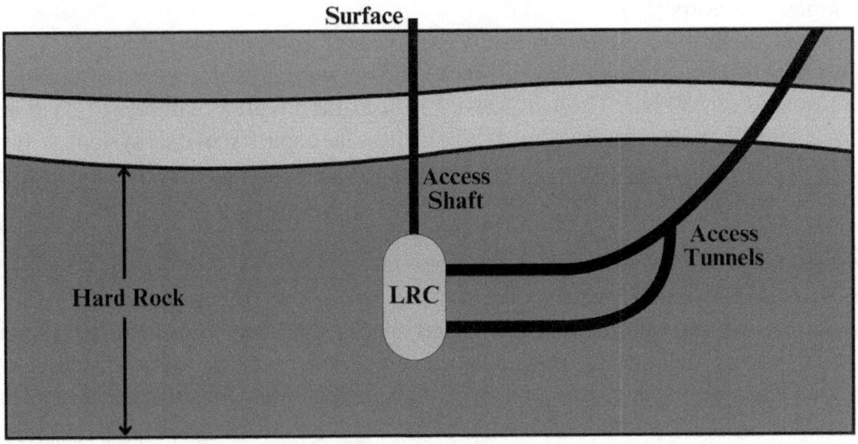

FIGURE 5.5 Lined rock cavern system for hydrogen storage.

withstand significant stresses affecting the LRC, avoiding the need for thick lining walls in LRC systems [35]. Commonly, hard rock caverns have a depth of 100–200 m below ground level, a 35–40 m diameter, and a cavern height of 60–100 m [36]. Moreover, the pressure rating of LRC systems can go as high as 250 bar in the best cases [37]. Figure 5.5 illustrates a typical LRC system [36].

As shown in Figure 5.5, an LRC system mainly consists of the LRC storage cavity, access tunnels, access shaft, and surface facility. The surface facility houses several systems for operating and controlling the LRC system. The access shafts and tunnels are underground passages supported by a thick layer of concrete, and they allow workers, heavy machinery, and equipment to travel from the surface facility to the LRC cavity to conduct maintenance activities. Last, the LRC storage cavity is installed within the hard rock formation and supported by several layers of concrete to ensure its structural integrity [36].

Hydrogen storage in LRC systems provides significant advantages over other underground hydrogen storage methods. These include the superior tightness of LRC cavities due to their layered design incorporating a steel or plastic liner supported by two layers of concrete. Also, the stored hydrogen can be extracted from LRC systems and used directly without any purification steps, unlike in many other systems where hydrogen can be mixed with brine, water, or other contaminants. Moreover, LRC systems can be applied worldwide as this method does not rely on specific underground formations such as salt caverns, aquifers, or oil and gas reservoirs. On the other hand, the effective design and operation of LRC systems require substantive financial investments, especially in installing and regularly inspecting the concrete supporting layers. The average LCOS of hydrogen in rock caverns is around \$0.71/kg, making it almost three times more expensive than hydrogen storage in salt caverns. However, this value is expected to drop to \$0.23/kg [19]. Altogether, LRC systems can potentially become the most common method to store hydrogen underground, especially in countries with little or no salt deposits or depleted aquifers and oil or gas reservoirs [37].

To date, there are no large-scale LRC systems in operation. However, a sole pilot project for an LRC system in Southern Sweden can store 100 m^3 of hydrogen at a pressure of 250 bar. The LRC cavity is 50 m high with a diameter of 35 m and is located 30 m below ground level. The full-scale capacity of the system will be 10,000–12,000 m^3 of hydrogen, which will be utilized to power a nearby iron production plant with an energy content of 100 GWh per discharging cycle [38].

5.2.1.4 Abandoned Mines

The mineable resources within minerals and coal mines get depleted after a long excavation, leaving behind vast unutilized spaces that can be used to store compressed hydrogen and other gases. Further, employing existing tunnels, shafts, and large underground spaces can greatly minimize the initial investments and effort required to excavate an underground hydrogen storage infrastructure [39]. However, several modifications must be made to ensure that hydrogen is stored effectively with minimal leakage and geochemical contamination. These modifications include surrounding the internal walls of abandoned mines with a lining medium like that used in LRC systems [40]. Also, several aspects must be considered when renovating an abandoned mine for hydrogen storage. These include the availability of storage spaces, tightness of spaces, rock mass stability, maximum hydrogen storage pressure, degree of fracture development, and storage duration [40].

Hydrogen storage in abandoned mines is very similar to rock caverns, especially LRC systems. This is because most mines worldwide are surrounded by hard rock mass; see Figure 5.4 for more information. Also, isolating hydrogen storage spaces in abandoned mines is almost identical to creating an LRC storage space. However, one key difference between the two methods is that abandoned mines typically contain minerals and/or carbon leftovers, increasing the risk of geochemical contamination and resulting in purity inferiority if not carefully dealt with through proper isolation. Further, specialized LRC and URC systems provide more effective storage spaces since they are excavated only to store hydrogen gas in the first place [36]. To date, the storage of different gases, including hydrogen, in abandoned mines is still under investigation, with very few pilot projects being implemented worldwide. The most successful pilot project is located in Hokkaido, Japan, where an abandoned coal mine has been modified to incorporate a storage system for storing compressed air at a pressure of 8 megapascal (MPa). The storage space is made of a 57 m tunnel with a diameter of 6 m and is lined with synthetic rubber inside a layer of concrete [41].

It is worth mentioning that hydrogen storage in URC systems and in abandoned unlined mines must be carried out deeply underground (i.e., more than 1000 m deep) to ensure sufficient tightness of the storage space and prevent contamination from groundwater leaking into the storage space through cracks in the rock walls [41].

5.2.1.5 Saline Aquifers

Saline aquifers are deep geological formations of porous rocks containing salt water saturated with brine. These formations present significant potential for storing hydrogen gas in large quantities with minimal investment. The use of saline aquifers for storing gases has been known, thoroughly investigated, and implemented for decades. More than 85 saline aquifers currently store natural gas worldwide [42,43].

Compressed hydrogen storage in saline aquifers is implemented by injecting pressurized hydrogen into porous rocks enclosed by upper and lower layers of impermeable caprock Figure 5.6. The porous rock formations are typically located at depths around 450–2000 m below ground. When hydrogen is injected into the porous layer, it displaces the saturated brine, filling the pores downward. This creates a layer of hydrogen gas at the top of the porous rock formation, which does not mix with brine due to the difference in density. When hydrogen is discharged from the saline aquifer, some cushion hydrogen gas is left in the aquifer to prevent the sudden pressure drop and to maintain a separation layer between brine and hydrogen storage cavities [42,43].

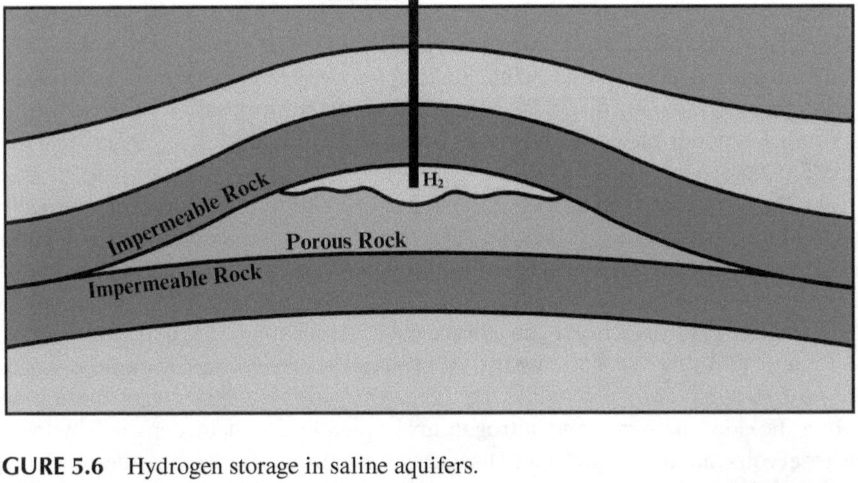

FIGURE 5.6 Hydrogen storage in saline aquifers.

The volume of the stored hydrogen mainly depends on the volume of the saline aquifer and the amount of saturated brine enclosed within the aquifer. While the pressure of hydrogen is typically maintained at 200 bar [37]. Currently, saline aquifers are limited to storing natural gas, methane, and carbon dioxide [44]. However, there is significant potential for saline aquifers to become one of the most prevailing technologies for large-scale hydrogen storage in the future. Saline aquifers present an opportunity for large-scale hydrogen storage in the future. Saline aquifers present an opportunity for large-scale, cost-effective, and long-term hydrogen storage with numerous advantages over other storage technologies. These include the geographical availability of saline aquifers worldwide, the minimal adverse environmental impacts, and the minimal risk of lethal accidents [45]. Current estimates for LCOS in saline aquifers stand at around \$1.90/kg and are expected to decrease to \$1.07/kg [19]. Further, an ideal saline aquifer must possess numerous characteristics to be suitable for hydrogen storage, including tectonic stability, adequate storage capacity, sufficient porosity and permeability, and structural integrity of upper and lower impermeable caprock [45]. On the other hand, there are still some challenges facing the mass implementation of hydrogen storage in saline aquifers, especially due to the uncertain nature of saline aquifers. Being located hundreds of meters below ground and spread along vast areas, the precise analysis of saline aquifers regarding their tectonic stability is still quite challenging [45]. Moreover, the abundance of brine in saline aquifers creates a rich microbial environment, increasing the risk of hydrogen contamination and necessitating hydrogen purification before use [46]. Also, the uncertain nature of pores/fluid interactions within the saline aquifer underlines a technical challenge when estimating the possible volume of stored hydrogen [45]. Finally, the porous nature of saline aquifers corresponds to significant losses in hydrogen quantities that can be 22% in the best cases [47].

5.2.1.6 Depleted Gas or Oil Reservoirs

Hydrocarbon reservoirs are geological traps naturally made to store oil or natural gas. These traps are porous in nature and usually surrounded by an impermeable rock layer similar to saline aquifers. In fact, a depleted or about-to-be-depleted hydrocarbon reservoir possesses low pressure and a high water saturation, especially in the zone where oil or gas existed before depletion. The pores of the reservoir get filled with saturated water and brine, making the depleted reservoir part of a saline aquifer. However, depleted reservoirs are differentiated from saline aquifers through the smaller amount of water and brine and a larger amount of oil or gas residuals trapped within their pores [48,49].

Like the principles of hydrogen storage in saline aquifers, pressurized hydrogen is injected into a porous rock formation surrounded by impermeable rock. The porous rock formation in depleted hydrocarbon reservoirs typically contains salt water and oil and/or gas residuals. The pressurized hydrogen displaces brine and gas/oil residuals downward, creating a hydrogen storage cavity at the top of the depleted reservoir. See Figure 5.7. In the case of a hydrocarbon-depleted reservoir, the cushion gas can be pure hydrogen, carbon dioxide, methane, nitrogen, or a mixture of these gases. Carbon dioxide, methane, and nitrogen are typically found in depleted hydrocarbon reservoirs naturally, and their usage as cushion gases reduces the amount of

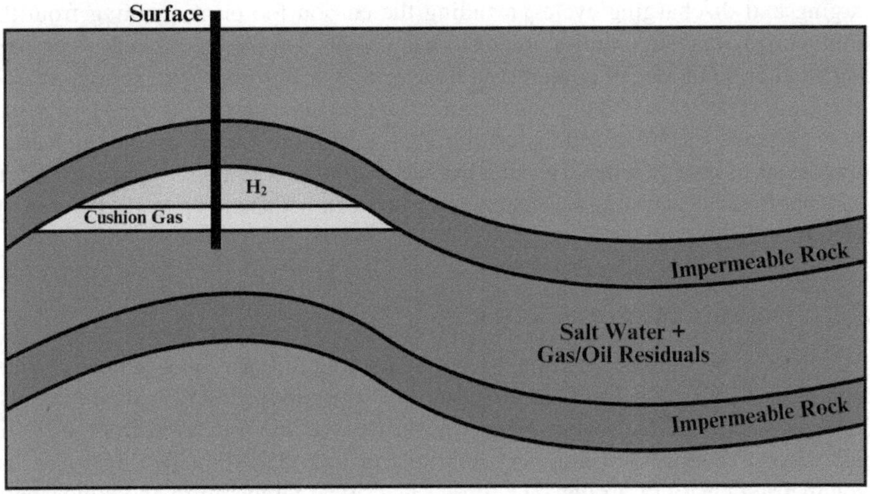

FIGURE 5.7 Hydrogen storage in depleted gas/oil reservoirs.

hydrogen loss by allowing for a larger portion of hydrogen to act as a cyclic working gas [50]. Previous studies report that cushion gas in depleted oil/gas reservoirs makes up around 50%–60% of the total stored hydrogen, while this share goes as high as 80% in saline aquifers [51]. However, using carbon dioxide, nitrogen, and/or methane as cushion gases can compromise the purity of the stored hydrogen [51].

Depleted hydrocarbon reservoirs are considered to be among the most stable underground storage formations due to their proven resistance over very long periods of time. Also, they are well-researched, explored, and geologically defined by engineers and experts worldwide, facilitating more precise and certain analyses before their hydrogen storage utilization [52]. Similar to hydrogen storage in saline aquifers, the volume of the stored hydrogen depends on the spacing between pores and the amount of existing brine and residual oil/gas between pores. Further, storing hydrogen at a pressure of around 200 bar is advised in the case of depleted hydrocarbon aquifers [37]. Moreover, the current LCOS estimates for hydrogen storage in depleted hydrocarbon reservoirs are identical to that of saline aquifers with a value of $1.90/kg, expected to drop to $1.07/kg with technological advancements [19].

Altogether, storing hydrogen in underground formations is a very promising method for large-scale hydrogen quantities. Further, it is considered to be the safest method to store compressed hydrogen since storage spaces are located deep underground, minimizing the risk of hydrogen ignition and/or explosion. In terms of cost, storing large quantities of hydrogen in compressed vessels is not economically feasible. Hence, exploring the potential of underground hydrogen storage can boost the hydrogen economy worldwide by providing cost-effective methods with varying costs depending on the underground storage technology. Environmentally, underground cavities can be used for very long periods and a significant number of charging and discharging cycles, avoiding the carbon footprint resulting from the manufacturing and decommissioning storage vessels. On the other hand, when handling smaller quantities of compressed hydrogen, it is advised to use storage vessels as these are superior to underground storage cavities in terms of hydrogen purity, and they experience less hydrogen losses over short periods of time. Last but not least, it is important to note that underground hydrogen storage is limited to pressures of up to 250 bar in the best cases, while storage vessels can withstand up to 1000 bar in the near future [53].

5.2.2 Liquefied Hydrogen Storage

Hydrogen can be stored in its liquid form by cooling it down to its critical condensation temperature of −253°C at a pressure of 1 bar [54]. Liquid hydrogen is also known as "cryogenic hydrogen," where the term cryogenic is derived from the Latin word "kryos," simply meaning very cold or freezing [55]. Liquefied hydrogen has a volumetric density of around 71 kg/m^3 at its critical temperature and atmospheric pressure (= 1 bar), which is 1.8 times higher than the volumetric density of compressed hydrogen at a pressure of 700 bar (= 39.2 kg/m^3). Correspondingly, liquid hydrogen has a higher volumetric energy density of 2.36 kWh/L in comparison with 0.78 kWh/L in the case of compressed hydrogen [56]. Hence, hydrogen storage in its liquid phase presents a substantial potential to the hydrogen supply chain, especially by enabling hydrogen transport via long-haul trucks and ships [57].

Hydrogen liquefaction can be carried out through numerous thermodynamics theoretical cycles, including Linde–Hampson, refrigerant liquefaction, Claude, expander liquefaction, Brayton, and magnetic refrigeration cycles [58]. These theoretical cycles act as the scientific and engineering bases for all practical liquefaction plants worldwide. Initially, hydrogen gas is fed to the cryogenic liquefaction system at a pressure between 1 and 30 bar and a temperature between 25°C and 40°C. In cases where the hydrogen pressure is below 20 bar, a feed compressor is used to increase the pressure to around 20–80 bar. Thereafter, the gaseous hydrogen stream enters the cold box, which sometimes contains a set of subsequent heat exchangers and a precooling system. See Figure 5.8.

If the cold box is equipped with a precooling system, it initially decreases the temperature of hydrogen to around −193°C to ease the remaining cooling stages.

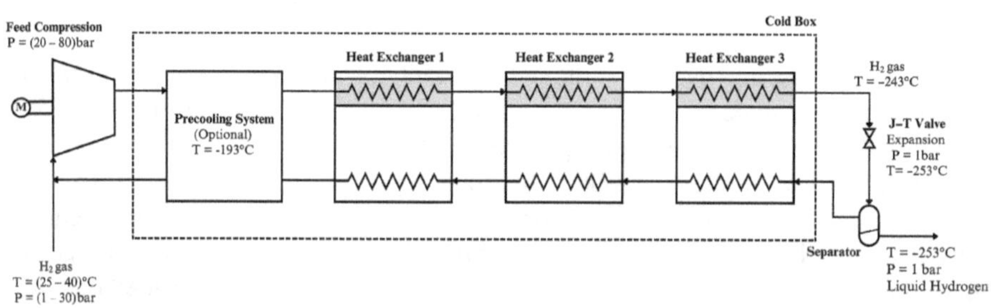

FIGURE 5.8 Hydrogen liquefication process.

Afterward, hydrogen gas enters a series of heat exchangers in which it gets cooled to around −243°C. After leaving the cold box, hydrogen is expanded in a Joule–Thomson (J–T) valve adiabatically to drop its pressure to around 1 bar and its temperature to the critical temperature of −253°C. At this stage, a large portion of the hydrogen's stream is condensed and liquefied, while some cooled hydrogen remains in its gaseous phase. Therefore, a separator separates liquid from gaseous hydrogen after cooling and expanding. Liquid hydrogen is then obtained and stored in specialized tanks. At the same time, the uncondensed hydrogen is recycled back into the system to benefit from its low temperature for cyclic cooling and to cool it once again until it fully condenses [58].

After being liquefied, hydrogen can be stored in specialized tanks that can withstand and preserve its cryogenic characteristics. The tank has many components to ensure hydrogen is maintained safely in its liquid phase. Typically, such tanks are double-walled and insulated by a vacuum. Also, they include several other components that aim to reduce heat transfer into the tank and to keep liquid hydrogen within the tank at its critical temperature. These include the inner and outer shields and a continuously operating heat exchanger. Moreover, the tank has several valves, such as the shut-off and safety valves. The shut-off valve isolates the liquid hydrogen inside the tank, especially during maintenance and emergencies. On the other hand, the safety valve mainly prevents pressure build-up by continuously venting gaseous hydrogen out of the tank [59].

Even with extensive thermal insulation, liquid hydrogen can heat up slightly over time, gasifying small amounts of liquid hydrogen in what is referred to as "boil-off" losses. These losses typically range between 0.3% and 3% per day [60]. Another drawback of cryogenic hydrogen storage is the relatively low efficiency of hydrogen liquefaction. Specifically, the specific energy consumption for hydrogen liquefaction stands at around 15.2 kWh/kg in best cases, which is equivalent to 45.6% of hydrogen's lower heating value (= 33.33 kWh/kg). In comparison, only 10% of hydrogen's lower heating value is consumed when compressing hydrogen and storing it in its gaseous phase [61]. These factors and the complex design requirements have made cryogenic hydrogen storage an extremely costly method to store hydrogen with a current LCOS of $4.57/kg, making it the most expensive method for hydrogen storage and 24 times more expensive than hydrogen storage in compressed vessels. This value is anticipated to drop to $0.95/kg in the best possible cases [19]. The total boil-off mass losses can be calculated as follows [62]:

$$m_{\text{boil-off}} = m_{H_2} - m_{H_2} \times (1 - \text{boil-off\%})^n \tag{5.7}$$

where $m_{\text{boil-off}}$ is the mass lost due to boil-off losses in kg, m_{H_2} is the initial mass of stored hydrogen in kg, boil-off% is the daily boil-off losses percentage in kg/day, and n is the number of days for hydrogen storage [62].

Example 5.3: Liquid Hydrogen Storage for Transport

A hydrogen production plant produces 120 m³ of gaseous green hydrogen per day at a pressure of 20 MPa and a temperature of 298.15 K. The plant aims to liquefy the produced hydrogen daily to be transported via ships to a foreign consumer. Considering that the specific energy consumption for liquefaction is 15.2 kWh/kg, the LCOS for liquid hydrogen is \$4.57/kg, and the volumetric density of hydrogen is 71 kg/m³. Recalling that the molecular weight (MW) of hydrogen gas is 2.0157 g/mol and the universal ideal gas constant (R) is 8.3144 J/K mol. Also, assuming that the hydrogen boil-off losses are 2%/day and that a shipment leaves the plant daily and reaches its destination within 8 days.

Calculate:

A. The energy consumption for hydrogen liquefaction.
B. The volume of hydrogen post liquefaction.
C. The daily cost for hydrogen storage.
D. The percentage of hydrogen lost due to boil-off losses.

Solution:

Initially, the mass of produced hydrogen on a daily basis must be obtained by applying equations 5.2 and 5.3, respectively. We get

$$PV = nRT$$

$$n = \frac{PV}{RT}$$

$$n = \frac{20 \times 10^6 \times 120}{8.3144 \times 298.15}$$

$$n = 968.15 \times 10^3 \text{ mol}$$

Also,

$$m = \frac{n \times MW}{1000}$$

$$m = \frac{968.15 \times 10^3 \times 2.0157}{1000}$$

$$m = 1951.5 \text{ kg}$$

A. Considering that the specific energy consumption for hydrogen liquefaction is 15.2 kWh/kg, we can simply calculate the total energy consumption per day by multiplying the specific energy consumption by the total mass to be liquefied on a daily basis. We get

$$\text{Daily energy consumption (kWh/day)} = 1951.5 \text{ kg/day} \times 15.2 \text{ kWh/kg}$$

$$= 29.67 \text{ MWh/kg}$$

The amount of energy required for hydrogen liquefaction is tremendous, resulting in significant losses in the overall useful energy.

B. Considering that liquid hydrogen has a volumetric density of 71 kg/m³, the required volume to store hydrogen on a daily basis can be calculated by dividing the total daily mass of liquid hydrogen over the volumetric density of liquid hydrogen as follows:

Note: in Example 5.2, the required daily volumes for compressed hydrogen storage vessels to store 1951.5 kg of hydrogen were 143.5 m³ for Type I tanks and 106.1 m³ for Type II tanks. This means that storing hydrogen in its liquid phase corresponds to a volume that is 19% and 26% of the required volume to store hydrogen in Type I and Type II compressed vessels, respectively.

C. The daily cost for liquid hydrogen storage can be calculated by multiplying the total daily mass of hydrogen by the LCOS per kg as follows:

$$\text{Hydrogen storage cost (\$/day)} = 1951.5\ \text{kg/day} \times 4.57\ \text{\$/kg} = \$8918/\text{day}$$

D. Considering a boil-off loss percentage of 2%/day and a shipping period of 8 days, the mass lost due to boil-off can be calculated through (equation 5.7) as follows:

$$m_{\text{boil-off}} = m_{H_2} - m_{H_2} \times (1 - \text{boil-off\%})^n$$

$$m_{\text{boil-off}} = 1951.5 - 1951.5 \times (1 - 0.02)^8$$

$$m_{\text{boil-off}} = 291.2\ \text{kg}$$

Calculating the loss percentage by dividing the mass lost due boil-off over the original stored mass of liquid hydrogen, we get:

$$\text{Percentage of lost mass } (\%) = \frac{291.2}{1951.5}\% = 14.9\%$$

We can clearly see that boil-off losses account for a significant loss of hydrogen due to continuous evaporation and venting. This makes hydrogen storage in its liquid phase inefficient when transporting hydrogen over long periods.

Despite the financial challenges surrounding cryogenic hydrogen storage, the storage density in cryogenic hydrogen systems can go as high as 19.6 wt.%, allowing for effective hydrogen transportation with minimal weight requirements. For instance, a cryogenic hydrogen tanker can transport around 5000 kg of hydrogen, which is five times the capacity of common compressed hydrogen tanks [63]. Hence, justifying the need for such technology, especially for hydrogen transportation within short periods of time via ships or trucks [64].

5.2.3 Cryo-Compressed Hydrogen Storage

Cryogenic-compressed hydrogen storage, abbreviated cryo-compressed storage, refers to hydrogen storage at a low temperature (cryogenic) of around −253°C and high pressure between 250 and 300 bar [65]. This developing method of hydrogen storage promises to alleviate the constraints of both compressed and liquid hydrogen storage by storing hydrogen at a higher volumetric density and with minimal or no boil-off losses [66]. The volumetric density in cryo-compressed storage systems can go as high as 87 kg/m^3 at a pressure of 240 bar, which is 1.2 greater than the volumetric density of liquid hydrogen systems and 2.1 times greater than that of compressed hydrogen storage [66]. Indeed, the energy consumption in cryo-compressed systems is higher than that of compressed and liquefied hydrogen systems since, in this method, hydrogen is first liquefied to its critical temperature and then compressed to 250–300 bar.

The specially designed tanks where cryo-compressed hydrogen is stored simultaneously allow hydrogen storage in different phases. Specifically, the tank can hold liquid, gaseous, and supercritical hydrogen [67]. Supercritical hydrogen is an intermediate state between gas and liquid, which occurs when hydrogen's pressure is below 8 bar and its temperature is between −237.15°C and −239.15°C [68]. The hydrogen storage volume of the tank is typically made of a Type III composite overwrapped pressure vessel similar to those used for compressed hydrogen storage. This vessel is enclosed by a vacuum enclosure consisting of multilayer insulation to minimize heat transfer into the tank. Further, the tank is equipped with two major heat exchangers dedicated to maintaining the temperature of the coolant and the stored hydrogen at their cryogenic temperature values. This tank type also has various control and monitoring components embedded within an auxiliary systems unit. These sensors and controllers ensure that hydrogen is maintained within the tank at its desired temperature and pressure with minimal leakage. In case of any hazards or leakage, the secondary vacuum module acts spontaneously to stall the storage system. The module consists of a shut-off valve, which shuts off all ports and electric systems in case of any hazards, and a safety valve, which vents out hydrogen in cases of pressure buildup [69].

To date, the most successful prototype for a cryo-compressed hydrogen storage system was developed by the automobile company BMW in 2011 for use in a BMW fuel cell electric vehicle. The system could withstand a pressure of up to 350 bar and a temperature of −240°C.

Further, it could store 8 kg of cryo-compressed hydrogen with a storage gravimetric density of 5.4 wt.%, which is much less than that of liquid hydrogen systems [70]. Cryo-compressed systems are still under research and development. Hence, there are no clear estimates on LCOS and their applicability for large-scale hydrogen storage. However, their ability to store different phases of hydrogen at once, as well as the proven storage efficiency with minimal leakage losses, still makes the ongoing research on cryo-compressed systems justified, especially for aircraft and vehicular applications [71].

5.3 MATERIAL-BASED HYDROGEN STORAGE

Material-based hydrogen storage depends on utilizing specific materials that can store hydrogen reversibly. Reversible chemical storage involves utilizing a chemical reaction that can be reversed, allowing for the creation and breaking of chemical bonds between elements or substances. Furthermore, hydrogen storage materials must possess certain characteristics to store hydrogen effectively. These include good reversibility, high storage capacities at desired operational pressure and temperature, and rapid kinetics (bonding and releasing hydrogen molecules quickly). The latter is important to ensure that hydrogen is charged and discharged from storage materials quickly, facilitating a responsive storage mechanism, which is crucial in many applications such as vehicular and aircraft [72].

As previously discussed, the drawbacks of physical-based storage technologies are mainly related to the limited storage capacity, high energy requirements, safety concerns as in compressed vessels storage, geographical constraints as in underground storage, high leakage rates as in liquid storage, and complexity of design to withstand high pressure and temperature as in cryo-compressed storage [73]. Therefore, researchers worldwide sought new technologies to achieve higher hydrogen storage capacities while operating under low-to-moderate pressure and temperature. The mainly developing material-based technologies are expected to have higher storage capacities at room temperature and moderate pressures while having significantly lower hydrogen leakage rates and being safer than physical-based storage technologies [74]. The classification of material-based storage technologies is shown in Figure 5.9 [74].

As shown in Figure 5.9, material-based hydrogen storage technologies are mainly categorized into adsorption and absorption technologies [74]. In chemistry,

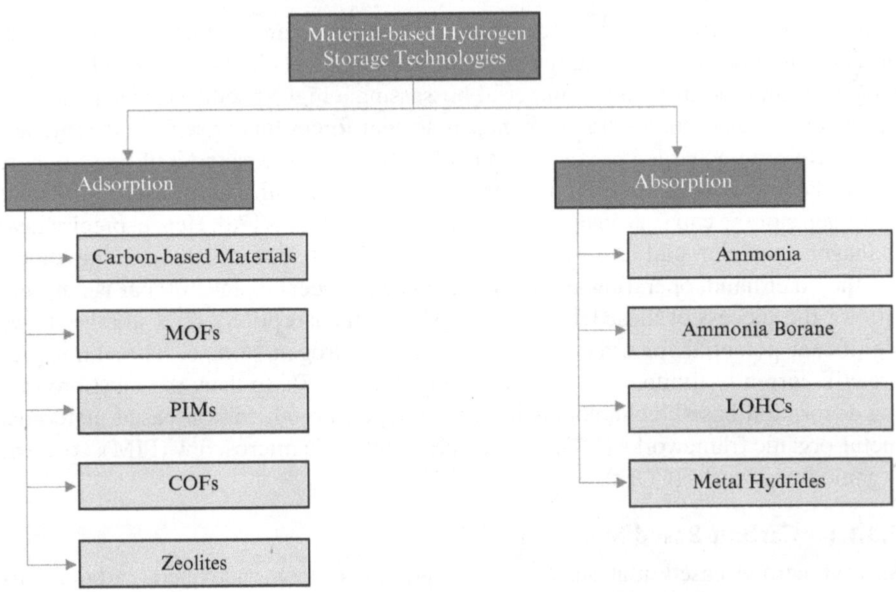

FIGURE 5.9 Material-based hydrogen storage technologies.

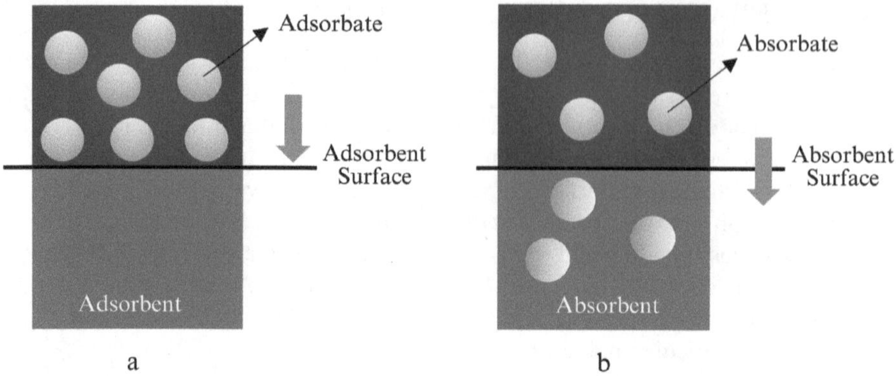

FIGURE 5.10 (a) Adsorption process. (b) Absorption process.

adsorption refers to the process in which the ions or molecules of a substance adhere to the surface of another substance. At the same time, absorption involves the transfer of ions or molecules of a substance into another substance. In both processes, a sorbate is a substance that adheres to the surface (adsorbate) or transfers through another substance (absorbate). On the other hand, the substance that the adsorbate adheres to its surface is called an adsorbent, and the substance through which the absorbate transfers is called an absorbent. The illustrations in Figure 5.10 define the nature of each process along with the above-mentioned terminology [75].

5.3.1 Adsorption Hydrogen Storage

In adsorption hydrogen storage technologies, hydrogen can be stored on the surface of a porous solid by exploiting the physical van der Waals bonds between hydrogen molecules and the surface of a material possessing a high specific surface area. Van der Waals bonds or forces are weak intermolecular forces that arise from the interactions between uncharged atoms or molecules. Due to the weakness of these bonds, relatively low temperatures and high pressures are required to ensure sufficient hydrogen storage capacity in adsorption-based technologies [76]. Hence, precise heat management is a crucial factor in determining the effectiveness of these technologies. On the other hand, operating at a pressure range between 10 and 100 bar is a necessity for the success of these technologies. Although adsorption-based storage shows significant potential for effective high-capacity hydrogen storage, its technologies are still currently limited to bench-scale prototypes [77]. To date, the most promising adsorption-based technologies include storage through carbon-based materials, metal-organic frameworks (MOF), polymers of intrinsic microscopy (PIM), covalent organic frameworks (COFs), and zeolites [74].

5.3.1.1 Carbon-Based Materials

Several carbon-based materials such as porous carbon, activated carbon, carbon nanostructures, and graphene have numerous hydrogen storage properties [78]. For instance, porous carbon (Figure 5.11) obtained via the

FIGURE 5.11 Porous carbon sponge-like structure.

carbonization of rayon fibers can be developed to obtain a specific surface area of up to 3144 m^2g. Utilizing such a material can allow for a hydrogen gravimetric storage density of between 0.9 wt.% and 7.01 wt.% at an operating temperature of 77 K (= −196.15°C) and an operating pressure of 40 bar [78,79]. On the other hand, activated carbon obtained via biomass carbonization is a well-known and thoroughly researched material for hydrogen adsorption storage. The specific surface area in activated carbon technology can be between 2000 and 3100 m^2/g [80]. Also, the gravimetric hydrogen storage density at a temperature of 77 Kelvin and a pressure of between 1 and 20 bar typically ranges between 1 and 7 wt.%, respectively [74,78].

Also, carbon nanostructures with a highly specific surface area are promising materials for high-capacity hydrogen storage. Carbon nanotubes are among the most important carbon nanostructures for hydrogen storage. These nanotubes are made of graphite in a cylindrical fullerenes form (a type of carbon wherein its molecules consist of a large number of carbon atoms connected by single and double bonds, for example: C60) and possess superior adsorption capacities, especially when combined with metals. Further, nanotubes come in different forms, including single-layered, double-layered, and multiple-layered nanotubes. Depending on the size and curvature of the tube as well as the type of combined metal, the gravimetric storage density in carbon nanotubes can range between 5 and 9.8 wt.% [78]. For instance, carbon nanotubes combined with aluminum can provide a gravimetric storage density of

6.15 wt.%, while when combined with scandium, their gravimetric storage density can reach 9.8 wt.% [81–83]. Last but not least, graphene holds the highest potential among carbon-based materials for high-capacity hydrogen storage. Pure or material-enhanced graphene sheets are stacked together to form a cage-like structure that can adsorb hydrogen with high gravimetric storage density values [78]. For instance, when graphene is combined with aluminum, they can provide a gravimetric storage density of up to 13.8 wt.% [84,85], while a boron-doped graphene combined with Potassium can provide an extraordinary gravimetric storage density of up to 20 wt.% [86].

5.3.1.2 Metal–Organic Frameworks

MOFs are porous crystalline materials with a large surface area of up to 4800 m^2/g [74]. MOFs consist of metal ions (clusters) that act as joints, bound by organic linkers (ligands) that link the MOF structure (Figure 5.12) [87]. More than 90,000 known MOFs are expected to grow to over 500,000 [88]. Among all MOF species, some strong candidates have shown auspicious characteristics for hydrogen storage.

These include MOF-5, which is made of $Zn_4O(BDC)_3$ [89], where BDC is 1,4-benzenedicarboxylate. MOF-5 is the most mature type of MOF, as it has been developing since 2003 [90]. MOF-5 has a specific storage surface area of 3362 m^2/g and is capable of storing hydrogen at both ambient and cryogenic temperatures and operating at ambient pressure. Under ambient temperature and a pressure of 20 bar, MOF-5 can provide a gravimetric hydrogen storage density of around 1 wt.%, while under cryogenic temperature and a pressure of 1 bar, the gravimetric density of MOF-5 rises to 4.5 wt.% [74]. On the other hand, among the currently developing MOFs, MIL-101(Cr) holds the best-anticipated characteristics for hydrogen storage. MIL-101(Cr) is a

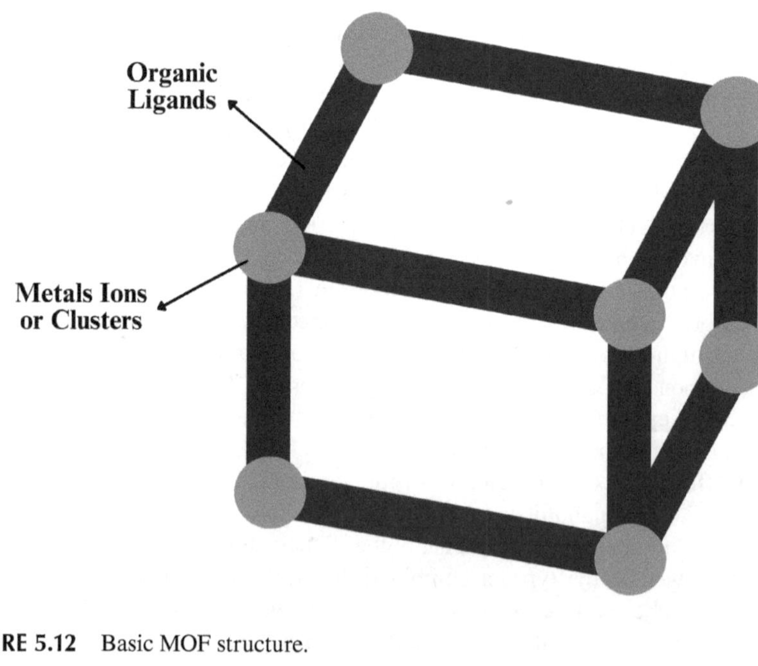

FIGURE 5.12 Basic MOF structure.

chromium-based MOF made of metal chromium ions and terephthalic acid ligands. It has an extremely high specific surface area of 6240 m^2/g and a gravimetric hydrogen storage density of up to 17.6 wt.% under cryogenic temperature and a pressure of 80 bar [90,91], making it far more superior than any other known type of MOFs. However, the current LCOS for MOF-based storage systems is extremely high. For instance, the LCOS for hydrogen via MOF-5 currently stands at around $490/kg, impairing the adoption of such a technology for practical hydrogen storage [74].

5.3.1.3 Polymers of Intrinsic Microscopy

PIMs are amorphous (noncrystalline) organic microporous materials that can store hydrogen via adsorption. Typically, polymers have a weak attraction for hydrogen molecules, but specific PIMs have shown a promising potential for hydrogen storage via adsorption. These PIMs are PIM-1 and PIM-7, which could store hydrogen in small capacities at a cryogenic temperature and a pressure of 1 bar. Specifically, PIM-1 can achieve a gravimetric hydrogen storage density of between 0.95 and 1.04 wt.%, while PIM-7 can store hydrogen at a gravimetric storage density of 1 wt.% [92]. This can be justified by the weak bonds between hydrogen and polymers in general and the low specific storage surface values for PIM-1 and PIM-7. At the same time, it yielded a maximum of 850 and 680 m^2/g, respectively [92]. However, the low cost of polymers and the abundance of preparation materials are urging the scientific community to develop novel PIM types that can provide larger hydrogen storage capacities at reasonably low costs [92].

5.3.1.4 Covalent Organic Frameworks

COFs are crystalline porous organic polymers (Figure 5.13) with a permanent porosity and a highly ordered structure. They are different from other polymers in that

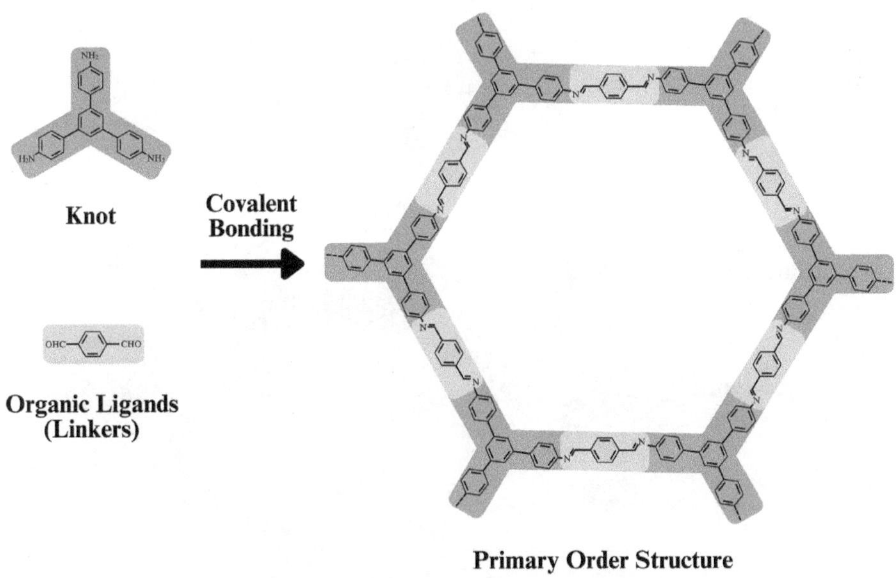

FIGURE 5.13 Covalent organic framework structures.

they are structurally designable, controllable, and highly manageable [93]. Similar to MOFs, COFs have a wide range of species that can be used for hydrogen storage via adsorption.

Among these, COF-5 has been proven to be the best alternative with a specific storage surface area of 1670 m^2/g, a hydrogen gravimetric storage density of up to 5.5 wt.% at cryogenic temperatures, and a pressure of around 100 bar [94,95]. Even if COFs do not compete with superior MOFs such as MIL-101(Cr), they still show a very promising potential for hydrogen storage since they are relatively cheaper, easier to fabricate, and can be operated more precisely, all due to their highly ordered polymer structure [93].

5.3.1.5 Zeolites

Zeolites are a class of highly ordered microporous, three-dimensional crystalline solids of aluminum silicate (Figure 5.14). Zeolites have been known for being good adsorbents for various gases, such as hydrogen and carbon dioxide, and effective moisture adsorbents [74]. Zeolites can store hydrogen via adsorption in two main ways. The first of which occurs at cryogenic temperatures and moderate pressure. For instance, NaX zeolites can store hydrogen at a gravimetric storage density of up to 2.55 wt.% at cryogenic temperature and a pressure of around 40 bar. The other method for hydrogen storage in zeolites is called "encapsulation," in which hydrogen is forced into the pores of zeolites at very high pressure (around 900 bar) and high temperature (up to 350°C). The major benefit of this method is that the stored hydrogen can remain confined at room temperature, allowing for stable hydrogen storage without minimal concerns regarding hydrogen release at standard atmospheric

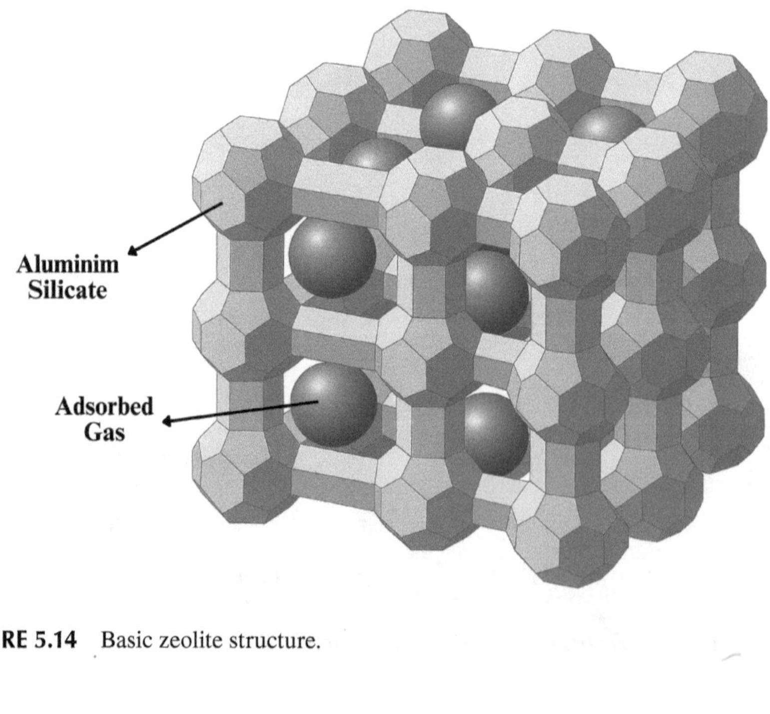

FIGURE 5.14 Basic zeolite structure.

conditions [74]. Although zeolites do not provide high hydrogen storage capacities, they still possess significant potential for hydrogen storage due to their relatively low cost, ease of fabrication, and technical maturity [96].

5.3.2 Absorption Hydrogen Storage

As previously mentioned, hydrogen storage through adsorption only requires creating and destroying weak Van der Waals bonds. In contrast, hydrogen storage via absorption technologies involves chemically reacting hydrogen with its host absorbent. Thus, exploiting strong chemical bonds results in a more stable hydrogen storage, but at the cost of more energy requirements [97]. The stability of hydrogen storage in absorption technologies enables hydrogen storage for longer periods under a wider range of temperatures and pressures [97]. Hence, extensive research is being dedicated worldwide to explore the potential of such technologies in assisting a widespread adoption of the green hydrogen economy. To date, the most promising absorption-based storage technologies include ammonia–borane, liquid organic hydrogen carriers (LOHC), metal boron–nitrogen–hydrogen compounds, and metal hydrides [74].

5.3.2.1 Ammonia

Hydrogen storage in ammonia is one of the most promising and widely researched technologies in the green hydrogen theme. It is important to understand that hydrogen storage in ammonia refers to the process of creating ammonia in the first place by reacting hydrogen and nitrogen via the Haber–Bosch reaction (refer to Chapter 4 for more details) [98]. Once ammonia is created, it can act as a hydrogen carrier rather than being used as fuel or a feedstock for chemical applications [99]. In this regard, ammonia shows very promising characteristics including that ammonia has a well-established global infrastructure relevant to storage, liquefication, and transport [100]. Further, ammonia has a high hydrogen content of around 17.65 wt.%, corresponding to a significantly large storage capacity of around 10.7 kg of hydrogen for every 100 L of ammonia [74,99]. When compared with liquid hydrogen, ammonia must be cooled down only to −33°C to be liquefied, resulting in less energy requirements for ammonia liquefication. Also, liquid ammonia boil-off rates are limited to below 0.025%/day, preserving the stored hydrogen for longer periods of time and facilitating long-distance transport via ships [101]. Ammonia decomposition into hydrogen and nitrogen can be carried out through various methods, such as ammonia electrolysis and ammonia thermal decomposition (refer to Chapter 4 for more details). Both processes are technically mature and can provide gaseous hydrogen with very high purity [99,102].

5.3.2.2 Ammonia Borane

Ammonia borane (NH_3BH_3), abbreviated AB, is a chemical hydride consisting of ammonia and boron hydride joined by a boron–nitrogen chemical bond (Figure 5.15) [103]. Similar to ammonia, the storage of hydrogen within AB is implemented by creating AB in the first place and then decomposing it to obtain hydrogen once again [97]. AB possesses exceptional hydrogen storage characteristics: its significantly high hydrogen content of up to 19.6 wt.%, its high stability under varying ambient

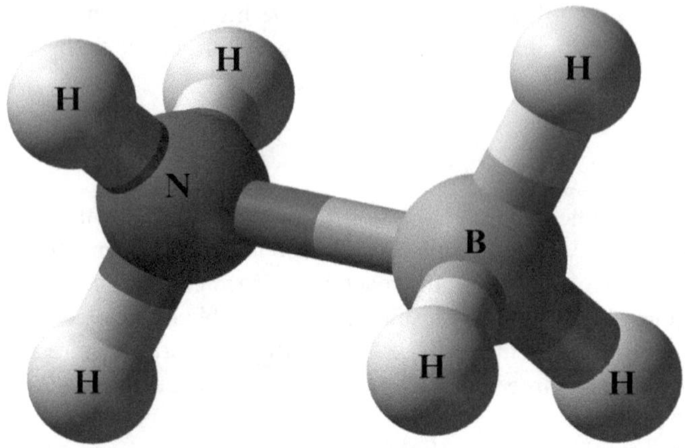

FIGURE 5.15 Ammonia borane chemical structure.

conditions, its nontoxicity, and its high solubility in common solvents [104]. The dehydrogenation (extraction of hydrogen) of AB can be achieved in the presence of a suitable catalyst at room temperature, thus minimizing the energy requirements for hydrogen extraction [104]. However, there is ongoing research to develop suitable catalysts with a long lifetime and reasonable costs to fulfill this task [104]. Furthermore, the thermal decomposition of AB can be achieved at a temperature between 120°C and 180°C, which is almost half of that required for the thermal decomposition of ammonia, resulting in a better overall efficiency [97]. However, there are still some challenges regarding hydrogen storage in AB, including the lack of technically mature large-scale AB synthesis technologies as well as the fact that several materials involved in AB dehydrogenation are classified as highly corrosive and explosive, compromising the safety of the technology, especially for on-board storage applications [97,105].

5.3.2.3 Liquid Organic Hydrogen Carriers

Liquid organic hydrogen carriers (LOHCs) are a group of chemical compounds that can bond with hydrogen and store it at high capacities covalently. These compounds include many alternatives, including methanol, methyl cyclohexane–toluene, cycloalkenes, formaldehyde, formic acid, and many more [106]. LOHCs can be hydrogenated and dehydrogenated cyclically (Figure 5.16) due to the highly reversible nature of their bonding with hydrogen [106].

For several reasons, methanol is among the LOHCs with the highest potential for hydrogen storage. The first is that methanol has a high hydrogen content of around 12.5 wt.%, which corresponds to a volumetric hydrogen storage density of around 99 kg/m^3. Moreover, methanol exists in a liquid phase at ambient room temperature, corresponding to a high volumetric density without liquefication or compression. This fact makes methanol among the most promising hydrogen carriers for on-board hydrogen storage for medium to long-range transport [107–109]. The dehydrogenation of methanol can be carried out in various ways, including steam reforming, partial oxidation, and thermal decomposition. The most efficient method among the

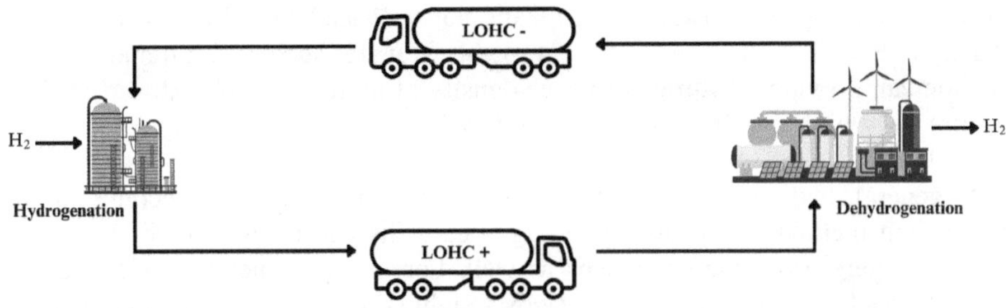

FIGURE 5.16 LOHCs hydrogenation and dehydrogenation.

three is steam reforming, as it produces three moles of hydrogen per one mole of decomposed methanol, while partial oxidation and thermal decomposition produce two moles of hydrogen per one mole of methanol. Methanol steam reforming typically occurs at a temperature between 230°C and 330°C with the assistance of a Cu/ZnO/Al_2O_3 catalyst [109]. With ongoing research and development, hydrogen storage in LOHCs holds a significant potential to contribute to the advancement of the hydrogen economy by offering a reliable method for storing and transporting hydrogen in a continuous manner for extended periods.

5.3.2.4 Metal Hydrides

Metal hydrides comprise one or more metal cations (positively charged metal ions) and one or more hydride ions (H^-). Several metallic-based materials can absorb hydrogen under low temperatures and moderate pressures. Specifically, metal hydrides made of elements such as lithium, boron, nitrogen, aluminum, and magnesium have a significant potential to be utilized as hydrogen storage materials [74]. Upon pressurization, metals form a strong bond with hydrogen, leading to a stable metal hydride that can be utilized for hydrogen storage. Metal hydrides can be in the form of liquids or powders stored in tanks at a pressure of 10 bar. When the temperature of metal hydrides increases from 120°C to 200°C and their pressure is reduced, hydrogen is released. Once emptied, a metal hydride tank can be recharged without high pressure or low temperatures. Global research has been focused on the utilization of magnesium-based alloys as they possess superior hydrogenation capabilities, have a hydrogen gravimetric storage density of up to 7.6 wt.%, have a relatively lightweight in comparison with other hydrides, resist heat very effectively, can be recycled and reused, and are available at low cost [74].

Moreover, several other metallic hydrides can compete with magnesium alloys, especially in terms of hydrogen storage densities. For instance, lithium borohydride ($LiBH_4$) can have a hydrogen storage gravimetric density of 18.5 wt.% at room temperature. However, $LiBH_4$ has an extremely difficult hydrogenation process, a very high energy of activation, and a very high decomposition temperature [74]. Hence, researchers have discussed various modifications to $LiBH_4$ to alleviate some of its drawbacks. For example, the utilization of silicon dioxide with $LiBH_4$ lowers its decomposition to around 200°C but leads to reducing the storage density to 13.5 wt.% [74]. The same applies to all proposed modifications, which can enhance the characteristics of $LiBH_4$ by all means but will lead to a drop in its storage density to 8 wt.% in the same cases [74]. Last, Alanates such as lithium aluminum hydride can provide a hydrogen storage density of up to 10.6 wt.%. However, they require very high pressures and temperatures, making them thermodynamically unstable [74].

In general, hydrogen storage in metal hydrides can potentially become one of the leading methods for hydrogen storage due to its enormous storage capacities. However, there are some challenges as well. For instance, metal hydride storage technologies have a low energy density by weight (kWh/kg), making metal hydride storage tanks four times heavier than gasoline tanks. This bulkiness of metal hydride tanks prevents them from contributing to clean transportation solutions. Moreover, the cost of some expensive metals involved in metal hydride systems raises the overall system cost substantially, compromising their economic feasibility. Also, many metal hydride systems possess a very slow hydrogen charging and discharging response, preventing them from being used in systems where a highly responsive hydrogen flow is necessary. Moreover, the drop in efficiency of metal hydride systems over long periods and a large number of cycles is still a challenge to date [110].

5.4 CHAPTER SUMMARY

This chapter explored hydrogen storage, which is pivotal for the green hydrogen supply chain. It addressed supply–demand variability and hydrogen's low volumetric density. The chapter explored physical and material-based storage technologies crucial for effective transport and delivery. Physical storage involved compression (up to 700 bar), liquefaction (–253°C), and cryo-compression (250–300 bar, –253°C). Storage was in high-pressure vessels, underground cavities, or specialized insulated tanks. Material-based storage utilized adsorption (weak van der Waals bonds) or absorption (stronger chemical bonds) for hydrogen interaction with host materials. Underground storage offered the highest capacity (up to 1,000,000 m^3), while material-based methods provided superior gravimetric density (around 20 wt.%). However, physical storage was currently more cost-effective, and advancements aimed to reduce material-based storage costs and enhance capacities. The chapter delved into the characteristics of each storage technology, highlighting their significance in the evolving green hydrogen landscape.

SELF-ASSESSMENT QUESTIONS

1. What is the main challenge in transporting gaseous green hydrogen produced via water electrolysis?
2. What are the two main classifications of hydrogen storage technologies?
3. Describe the three methods of physical-based hydrogen storage.
4. Where is compressed hydrogen typically stored?
5. What are the characteristics of vessels used to store liquid and cryo-compressed hydrogen?
6. Explain the difference between adsorption and absorption in material-based hydrogen storage.

7. Which hydrogen storage technology has the highest capacity?
8. Which hydrogen storage technologies have the highest gravimetric storage density?
9. Compare the cost of physical hydrogen storage to material hydrogen storage.
10. What is the main reason for integrating green hydrogen storage systems into the supply chain?
11. What are the two main purposes for storing green hydrogen?
12. How does storing green hydrogen help in renewable energy storage?
13. List some industries that depend on hydrogen as a chemical feedstock.
14. What is the volumetric density of hydrogen at standard atmospheric conditions?
15. What are the three main methods of physical-based hydrogen storage?
16. Which hydrogen storage technology is the most common and mature?
17. What percentage of hydrogen worldwide is stored using compressed storage technologies?
18. How does compressed hydrogen storage increase hydrogen density?
19. What are the desired characteristics of hydrogen compressors?
20. Define the compression ratio in hydrogen compressors.
21. What are the typical energy requirements for hydrogen compression?
22. What is the approximate specific cost for hydrogen compression?
23. What are the five types of compressed hydrogen storage vessels?
24. Describe the composition and characteristics of Type I hydrogen storage vessels.
25. How do Type II vessels differ from Type I vessels in terms of composition, pressure, and cost?
26. What are the advantages of Type III vessels over Type I and Type II vessels?
27. Explain the composition and characteristics of Type IV hydrogen storage vessels.
28. What are the key components of a Type IV hydrogen storage vessel?
29. Describe the current status and potential advantages of Type V vessels.
30. What physical principles are essential for sizing hydrogen storage tanks?
31. What is the levelized cost of storage (LCOS) for compressed hydrogen storage?
32. Why does compressed hydrogen storage fail to deliver sufficient quantities of hydrogen for large-scale applications?
33. What are the advantages of underground hydrogen storage?
34. List the different types of man-made and natural underground cavities used for hydrogen storage.
35. How are salt caverns created for hydrogen storage?
36. What factors determine the pressure of hydrogen stored in a salt cavern?
37. Explain the difference between variable pressure and constant pressure techniques for filling and discharging hydrogen in salt caverns.
38. What are the advantages and disadvantages of salt cavern hydrogen storage?
39. Where are the majority of salt caverns currently used for energy storage located?

40. Which countries have the highest potential for utilizing salt caverns for underground hydrogen storage?
41. A hydrogen production plant produces 500 kg of gaseous green hydrogen per day at a pressure of 5 bar. The plant aims to compress the hydrogen to 700 bar for storage in Type III vessels. Calculate the required number of compressors and the total compression ratio, assuming the available compressors have a compression ratio of 20:1.
42. A fuel cell electric vehicle (FCEV) requires a Type IV hydrogen storage vessel with a storage density of 5 wt.% to store 4 kg of hydrogen. Calculate the total weight of the tank.
43. A city plans to store 500,000 m^3 of hydrogen in an underground salt cavern. The lithostatic pressure at the cavern roof is 200 bar. Determine the maximum pressure at which hydrogen can be stored in the cavern.
44. A liquid hydrogen storage tank has a boil-off rate of 1%/day. Calculate the mass of hydrogen lost over 30 days if the initial mass of stored hydrogen is 1000 kg.
45. A cryo-compressed hydrogen storage system operates at a temperature of −253°C and a pressure of 300 bar. Compare the volumetric density of hydrogen in this system to that of liquid hydrogen storage at −253°C and 1 bar, and compressed hydrogen storage at 25°C and 700 bar.
46. A metal–organic framework (MOF) material has a specific surface area of 4000 m^2/g. Calculate the maximum amount of hydrogen that can be adsorbed on 1 kg of this material, assuming a gravimetric storage density of 10 wt.%.
47. A zeolite material can store hydrogen at a gravimetric storage density of 2 wt.% at cryogenic temperature and a pressure of 40 bar. Calculate the mass of hydrogen that can be stored in a 100-L tank filled with this zeolite material.
48. An ammonia borane (AB) storage system has a hydrogen content of 19.6 wt.%. Calculate the mass of AB required to store 10 kg of hydrogen.
49. A liquid organic hydrogen carrier (LOHC) system uses methanol as the carrier. Calculate the volume of methanol required to store 5 kg of hydrogen, given that methanol has a hydrogen content of 12.5 wt.% and a density of 792 kg/m^3.
50. A metal hydride storage system uses a magnesium-based alloy with a hydrogen gravimetric storage density of 7.6 wt.%. Calculate the mass of the alloy required to store 2 kg of hydrogen.

REFERENCES

1. Capurso, Tiziana, Michele Stefanizzi, Marianna Torresi, et al. 2022. "Perspective of the role of hydrogen in the 21st century energy transition." *Energy Conversion and Management* 251: 114898. https://doi.org/10.1016/j.enconman.2021.114898.
2. Zsiborács, Henrik, Norbert Baranyai, Andrea Vincze, et al. 2019. "Intermittent renewable energy sources: The role of energy storage in the European power system of 2040." *Electronics* 8 (7): 729. https://doi.org/10.3390/electronics8070729.

3. Rasul, M. G., Mohammad Hazrat, Mohammad Sattar, et al. 2022. "The future of hydrogen: Challenges on production, storage and applications." *Energy Conversion and Management* 272: 116326. https://doi.org/10.1016/j.enconman.2022.116326.
4. Tarhan, Cansu, and Metin Akif Çil. 2021. "A study on hydrogen, the clean energy of the future: Hydrogen storage methods." *Journal of Energy Storage* 40: 102676. https://doi.org/10.1016/j.est.2021.102676.
5. Zheng, Jie, Xiaohua Liu, Ping Xu, et al. 2012. "Development of high pressure gaseous hydrogen storage technologies." *International Journal of Hydrogen Energy* 37 (1): 1048–1057. https://doi.org/10.1016/j.ijhydene.2011.02.125.
6. Xu, Ping, Jie Zheng, and P. F. Liu. 2009. "Finite element analysis of burst pressure of composite hydrogen storage vessels." *Materials & Design* 30 (7): 2295–2301. https://doi.org/10.1016/j.matdes.2009.03.006.
7. Zhang, Qiang, Hua Xu, Xu Jia, et al. 2020. "Design of a 70MPa type IV hydrogen storage vessel using accurate modeling techniques for dome thickness prediction." *Composite Structures* 236: 111915. https://doi.org/10.1016/j.compstruct.2020.111915
8. Sdanghi, Gérald, Géraldine Maranzana, Ammar Celzard, et al. 2019. "Review of the current technologies and performances of hydrogen compression for stationary and automotive applications." *Renewable and Sustainable Energy Reviews* 102: 150–170. https://doi.org/10.1016/j.rser.2018.11.028.
9. Franco, Alessandro, and Caterina Giovannini. 2024. "Hydrogen gas compression for efficient storage: balancing energy and increasing density." *Hydrogen* 5(2): 293–311. https://doi.org/10.3390/hydrogen5020017
10. Kurz, Rainer., Berend Winkelmann, Sascha Freund, et al. 2022. "Transport and storage." In *Machinery and Energy Systems for the Hydrogen Economy*, pp. 215–249. Elsevier. https://doi.org/10.1016/B978-0-323-90394-3.00003-5.
11. Khan, Mohd Adnan, Cameron Young, Catherine Mackinnon, et al. 2021. "The techno-economics of hydrogen compression." *Transition Accelerator Technical Briefs* 1 (1): 1–36. Available at: https://transitionaccelerator.ca/wp-content/uploads/2023/04/TA-Technical-Brief-1.1_TEEA-Hydrogen-Compression_PUBLISHED.pdf.
12. Langmi, Henrietta W., Nicolaas Engelbrecht, Phillimon M. Modisha, et al. 2022. "Hydrogen storage." In *Electrochemical Power Sources: Fundamentals, Systems, and Applications*, pp. 455–486. Elsevier. https://doi.org/10.1016/B978-0-12-819424-9.00006-9.
13. Qureshi, Fazil, Mohammad Yusuf, Mohd Arham Khan, et al. 2023. "A state-of-the-art review on the latest trends in hydrogen production, storage, and transportation techniques." *Fuel* 340: 127574. https://doi.org/10.1016/j.fuel.2023.127574.
14. Shin, Hyun Kyu, and Sung Kyu Ha. 2023. "A review on the cost analysis of hydrogen gas storage tanks for fuel cell vehicles." *Energies* 16 (13): 5233. https://doi.org/10.3390/en16135233.
15. U.S Department of Energy—Fuel Cell Technologies Office. 2017. "Hydrogen Storage." Available at: https://www.energy.gov/sites/prod/files/2017/03/f34/fcto-h2-storage-fact-sheet.pdf.
16. Air, Alexander, Md. Shamsuddoha, and B. G. Prusty. 2023. "A review of Type V composite pressure vessels and automated fibre placement based manufacturing." *Composites Part B: Engineering* 253: 110573. https://doi.org/10.1016/j.compositesb.2023.110573.
17. Cengel, Yunus, and Michael Boles. 2002. "Thermodynamics: An engineering approach." *Sea* 1000 (8862): 287–293. ISBN10: 0071150714.
18. Sapre, Sonal, Mihir Vyas, and Kanchan Pareek. 2021. "Impact of refueling parameters on storage density of compressed hydrogen storage tank." *International Journal of Hydrogen Energy* 46 (31): 16685–16692. https://doi.org/10.1016/j.ijhydene.2020.08.136.
19. BloombergNEF. 2020. "Hydrogen Economy Outlook—Key Messages." Available at: https://data.bloomberglp.com/professional/sites/24/BNEF-Hydrogen-Economy-Outlook-Key-Messages-30-Mar-2020.pdf.

20. Barthélémy, Henri, Miriam Weber, and Florence Barbier. 2017. "Hydrogen storage: Recent improvements and industrial perspectives." *International Journal of Hydrogen Energy* 42 (11), 7254–7262. https://doi.org/10.1016/j.ijhydene.2016.03.178.
21. Tarkowski, Radosław. 2019. "Underground hydrogen storage: Characteristics and prospects." *Renewable and Sustainable Energy Reviews* 105: 86–94. https://doi.org/10.1016/j.rser.2019.01.051.
22. Tarkowski, Radosław, Beata Uliasz-Misiak, and Piotr Tarkowski. 2021. "Storage of hydrogen, natural gas, and carbon dioxide–geological and legal conditions." *International Journal of Hydrogen Energy* 46 (38): 20010–20022. https://doi.org/10.1016/j.ijhydene.2021.03.131.
23. Donadei, Sabine, and Gregor-Sönke Schneider. 2022. "Compressed air energy storage." In *Storing Energy*, pp. 141–156. Elsevier. https://doi.org/10.1016/B978-0-12-824510-1.00034-9.
24. Geoscience Australia. 2023. "Australia's Hydrogen Production Potential." Available at: https://www.ga.gov.au/scientific-topics/energy/resources/hydrogen/australias-hydrogen-production-potential.
25. Lord, Andrew, Peter Kobos, and David Borns. 2014. "Geologic storage of hydrogen: Scaling up to meet city transportation demands." *International Journal of Hydrogen Energy* 39 (28): 15570–15582. https://doi.org/10.1016/j.ijhydene.2014.07.121.
26. Thiyagarajan, Sasikumar, Hossein Emadi, Arshad Hussain, et al. 2022. "A comprehensive review of the mechanisms and efficiency of underground hydrogen storage." *Journal of Energy Storage* 51: 104490. https://doi.org/10.1016/j.est.2022.104490.
27. Zhang, Hao, Peng Wang, Qiqi Wanyan, et al. 2023. "Sensitivity analysis of operation parameters of the salt cavern under long-term gas injection-production." *Scientific Reports* 13 (1): 20012. https://doi.org/10.1038/s41598-023-47352-w.
28. Hui, Shangli, Shanqing Yin, Xingchen Pang, et al. 2023. "Potential of salt caverns for hydrogen storage in Southern Ontario, Canada." *Mining* 3 (3): 399–408. https://doi.org/10.3390/mining3030024.
29. Liu, Wei, Qian Li, Chenglong Yang, et al. 2023. "The role of underground salt caverns for large-scale energy storage: A review and prospects." *Energy Storage Materials* 63: 103045. https://doi.org/10.1016/j.ensm.2023.103045.
30. Tackie-Otoo, Benard, and Mohammad Haq. 2024. "A comprehensive review on geo-storage of H_2 in salt caverns: Prospect and research advances." *Fuel* 356: 129609. https://doi.org/10.1016/j.fuel.2023.129609.
31. Donadei, Sabine, and Gregor-Sönke Schneider. 2016. "Compressed air energy storage in underground formations." In *Storing Energy*, pp. 113–133. Elsevier. https://doi.org/10.1016/B978-0-12-803440-8.00006-3.
32. Williams, Jonathan, Jonathan Williamson, David Parkes, et al. 2022. "Does the United Kingdom have sufficient geological storage capacity to support a hydrogen economy? Estimating the salt cavern storage potential of bedded halite formations." *Journal of Energy Storage* 53: 105109. https://doi.org/10.1016/j.est.2022.105109.
33. Małachowska, Anna, Natalia Łukasik, Joanna Mioduska, et al. 2022. "Hydrogen storage in geological formations—The potential of salt caverns." *Energies* 15 (14): 5038. https://doi.org/10.3390/en15145038.
34. Patanwar, Yogesh, H. M. Kim, Deb Deb, et al. 2023. "Underground storage of hydrogen in lined rock caverns: An overview of key components and hydrogen embrittlement challenges." *International Journal of Hydrogen Energy* 50: 116–133. https://doi.org/10.1016/j.ijhydene.2023.08.342.
35. Damasceno, Daniella. 2022. "Modeling Aspects of Reliability-Based Design of Lined Rock Caverns." Doctoral dissertation, KTH Royal Institute of Technology. ISBN 978-91-8040-404-4.
36. Masoudi, Masoud, Hassanpouryouzband Ali, Helge Hellevang, et al. 2024. "Lined rock caverns: A hydrogen storage solution." *Journal of Energy Storage* 84: 110927. https://doi.org/10.1016/j.est.2024.110927.

37. Londe, Leonardo. 2021. "Four ways to store large quantities of hydrogen." In *Abu Dhabi International Petroleum Exhibition and Conference*, Abu Dhabi, UAE, p. D022S189R001. SPE. https://doi.org/10.2118/208178-MS.
38. Damasceno, Daniella, Joakim Spross, and Fredrik Johansson. 2023. "Rock mass response for lined rock caverns subjected to high internal gas pressure." *Journal of Rock Mechanics and Geotechnical Engineering* 15 (1): 119–129. https://doi.org/10.1016/j.jrmge.2022.03.006.
39. Peppas, Alexandros, Spyros Kottaridis, and Charalambos Politi. 2023. "Potential application of RES and underground H_2 storage in abandoned mines." *Materials Proceedings* 15 (1): 30. https://doi.org/10.3390/materproc2023015030.
40. Liu, Wenlong, and Peijun Pei. 2021. "Evaluation of the influencing factors of using underground space of abandoned coal mines to store hydrogen based on the improved ANP method." *Advances in Materials Science and Engineering* 2021: 1–9. https://doi.org/10.1155/2021/7506055.
41. Lalanne, Pascal, and Patrick Byrne. 2019. "Large-scale pumped thermal electricity storages—Converting energy using shallow lined rock caverns, carbon dioxide and underground pumped-hydro." *Applied Sciences* 9 (19): 4150. https://doi.org/10.3390/app9194150.
42. Heinemann, Nial, J. Scafidi, Gordon Pickup, et al. 2021. "Hydrogen storage in saline aquifers: The role of cushion gas for injection and production." *International Journal of Hydrogen Energy* 46 (79): 39284–39296. https://doi.org/10.1016/j.ijhydene.2021.09.174.
43. Trent, John 2023. "The uncertain bright future of underground hydrogen storage." *Journal of Petroleum Technology*. Available at: https://jpt.spe.org/the-uncertain-bright-future-of-underground-hydrogen-storage.
44. Kumar, Sandeep, J. Foroozesh, Karl Edlmann, et al. 2020. "A comprehensive review of value-added CO_2 sequestration in subsurface saline aquifers." *Journal of Natural Gas Science and Engineering* 81: 103437. https://doi.org/10.1016/j.jngse.2020.103437.
45. Raad, Seyed, Yuriy Leonenko, and H. Hassanzadeh. 2022. "Hydrogen storage in saline aquifers: Opportunities and challenges." *Renewable and Sustainable Energy Reviews* 168: 112846. https://doi.org/10.1016/j.rser.2022.112846.
46. Strobel, Gion, Birger Hagemann, Thiago Huppertz, et al. 2020. "Underground bio-methanation: Concept and potential." *Renewable and Sustainable Energy Reviews*, 123, 109747. https://doi.org/10.1016/j.rser.2020.109747.
47. Sáinz-García, Álvaro, Elena Abarca, Victòria Rubí, et al. 2017. "Assessment of feasible strategies for seasonal underground hydrogen storage in a saline aquifer." *International Journal of Hydrogen Energy* 42 (26): 16657–16666. https://doi.org/10.1016/j.ijhydene.2017.05.076.
48. Muhammed, Nasiru, Md Bashirul Haq, Dhafer Al Shehri, et al. 2023. "Hydrogen storage in depleted gas reservoirs: A comprehensive review." *Fuel* 337: 127032. https://doi.org/10.1016/j.fuel.2022.127032.
49. Madirisha, Mamokete, and Bidemi Ikotun. 2024. "Hydrogen storage in depleted geological hydrocarbon reservoirs: Enhancing wellbore integrity and sustainability with geopolymer cement—Review." *Journal of Energy Storage* 84: 110834. https://doi.org/10.1016/j.est.2024.110834.
50. Hemme, Claudia, and Wolfgang Van Berk. 2018. "Hydrogeochemical modeling to identify potential risks of underground hydrogen storage in depleted gas fields." *Applied Sciences* 8 (11): 2282. https://doi.org/10.3390/app8112282.
51. Wallace, Robert, Zhiming Cai, Huilin Zhang, et al. 2021. "Utility-scale subsurface hydrogen storage: UK perspectives and technology." *International Journal of Hydrogen Energy* 46 (49): 25137–25159. https://doi.org/10.1016/j.ijhydene.2021.05.034.
52. Lysyy, Mykhailo, Magnus Fernø, and Geir Ersland. 2021. "Seasonal hydrogen storage in a depleted oil and gas field." *International Journal of Hydrogen Energy* 46 (49): 25160–25174. https://doi.org/10.1016/j.ijhydene.2021.05.030.

53. Elberry, Alaa, Jayant Thakur, Anna Santasalo-Aarnio, et al. 2021. "Large-scale compressed hydrogen storage as part of renewable electricity storage systems." *International Journal of Hydrogen Energy* 46 (29): 15671–15690. https://doi.org/10.1016/j.ijhydene.2021.02.080.
54. Qiu, Yong, Haibo Yang, Liming Tong, et al. 2021. "Research progress of cryogenic materials for storage and transportation of liquid hydrogen." *Metals* 11 (7): 1101. https://doi.org/10.3390/met11071101.
55. Zohuri, Bahman, and Bijan Zohuri. 2018. "Cryogenic technologies." In *Hybrid Energy Systems: Driving Reliable Renewable Sources of Energy Storage*, pp. 39–88. Springer Nature. https://doi.org/10.1007/978-3-319-70721-1_2.
56. Al Ghafri, Saeed, Sean Munro, Umberto Cardella, et al. 2022. "Hydrogen liquefaction: A review of the fundamental physics, engineering practice and future opportunities." *Energy & Environmental Science* 15 (7): 2690–2731. https://doi.org/10.1039/D2EE00099G.
57. Aziz, Muhammad 2021. "Liquid hydrogen: A review on liquefaction, storage, transportation, and safety." *Energies* 14 (18): 5917. https://doi.org/10.3390/en14185917.
58. Zhang, Tao, Jun Uratani, YujieHuang, et al. 2023. "Hydrogen liquefaction and storage: Recent progress and perspectives." *Renewable and Sustainable Energy Reviews* 176: 113204. https://doi.org/10.1016/j.rser.2023.113204.
59. Rao, Amol, Feng Yin, and Henk Werij. 2020. "Energy transition in aviation: The role of cryogenic fuels." *Aerospace* 7 (12): 181. https://doi.org/10.3390/aerospace7120181.
60. Morales-Ospino, Rafael, Ammar Celzard, and Vanessa Fierro. 2023. "Strategies to recover and minimize boil-off losses during liquid hydrogen storage." *Renewable and Sustainable Energy Reviews* 182: 113360. https://doi.org/10.1016/j.rser.2023.113360.
61. Kumar, Anshul, Nilay Mukhopadhyay, and Thakur Yadav. 2024. "Hydrogen storage in high entropy alloys." In *Towards Hydrogen Infrastructure*, pp. 133–164. Elsevier. https://doi.org/10.1016/B978-0-323-95553-9.00007-8.
62. Ghaffari-Tabrizi, Fardin, Jan Haemisch, and Daniela Lindner. 2022. "Reducing hydrogen boil-off losses during fuelling by pre-cooling cryogenic tank." *Hydrogen* 3 (2): 255–269. https://doi.org/10.3390/hydrogen3020015.
63. Hosseini, Seyed Ehsan. 2023. "Hydrogen storage and delivery challenges." In *Fundamentals of Hydrogen Production and Utilization in Fuel Cell Systems*, pp. 237–254. Elsevier. https://doi.org/10.1016/C2020-0-03183-X.
64. Ahluwalia, Rajesh Hee Seok Roh, Jae-Kwon Peng, et al. 2023. "Liquid hydrogen storage system for heavy duty trucks: Configuration, performance, cost, and safety." *International Journal of Hydrogen Energy* 48 (35): 13308–13323. https://doi.org/10.1016/j.ijhydene.2022.12.152.
65. Wang, Hongsheng, Yang Zhao, Xiaorong Dong, et al. 2022. "Thermodynamic analysis of low-temperature and high-pressure (cryo-compressed) hydrogen storage processes cooled by mixed-refrigerants." *International Journal of Hydrogen Energy* 47 (67): 28932–28944. https://doi.org/10.1016/j.ijhydene.2022.06.193.
66. Langmi, Henrietta, Nicolaas Engelbrecht, Phillimon Modisha, et al. 2022. "Hydrogen storage." In *Electrochemical Power Sources: Fundamentals, Systems, and Applications*, pp. 455–486. Elsevier. https://doi.org/10.1016/B978-0-12-819424-9.00006-9.
67. Ahluwalia, Rajesh, Jae-Kwon Peng, and Thanh Hua. 2016. "Cryo-compressed hydrogen storage." In *Compendium of Hydrogen Energy*, pp. 119–145. Woodhead Publishing. https://doi.org/10.1016/B978-1-78242-362-1.00005-5.
68. Ibrahimoglu, Burak, Yavuz Sarikaya, Aykut Veziroglu, et al. 2021. "Supercritical phases of hydrogen." *International Journal of Hydrogen Energy* 46 (65): 32762–32767. https://doi.org/10.1016/j.ijhydene.2021.07.032.
69. Sharma, Pooja, Aditya Burolia, Ansuman Mandal, et al. 2024. "Commercially available resources for physical hydrogen storage and distribution." In *Towards Hydrogen Infrastructure*, pp. 225–256. Elsevier. https://doi.org/10.1016/B978-0-323-95553-9.00010-8.

70. Godula-Jopek, Anna. 2015. "Hydrogen storage options including constraints and challenges." In *Hydrogen Production: Electrolysis*, pp. 273–310. Wiley. https://doi.org/10.1002/9783527676507.ch7.
71. Moreno-Blanco, Jorge, Guillaume Petitpas, Francisco Espinosa-Loza, et al. 2019. "The storage performance of automotive cryo-compressed hydrogen vessels." *International Journal of Hydrogen Energy* 44 (31): 16841–16851. https://doi.org/10.1016/j.ijhydene.2019.04.189.
72. Rusman, Nor Aishah Ahmad, and Maznah Dahari. 2016. "A review on the current progress of metal hydrides material for solid-state hydrogen storage applications." *International Journal of Hydrogen Energy* 41 (28): 12108–12126. https://doi.org/10.1016/j.ijhydene.2016.05.244.
73. Usman, Muhammad Rizwan. 2022. "Hydrogen storage methods: Review and current status." *Renewable and Sustainable Energy Reviews* 167: 112743. https://doi.org/10.1016/j.rser.2022.112743.
74. Hassan, Ibrahim, Hamdy Ramadan, Mohammed Saleh, et al. 2021. "Hydrogen storage technologies for stationary and mobile applications: Review, analysis and perspectives." *Renewable and Sustainable Energy Reviews* 149: 111311. https://doi.org/10.1016/j.rser.2021.111311.
75. Zbair, Mohamed, and Simona Bennici. 2021. "Survey summary on salts hydrates and composites used in thermochemical sorption heat storage: A review." *Energies* 14 (11): 3105. https://doi.org/10.3390/en14113105.
76. Prayogo, Genki Imam, Homin Shin, Abdelghani Benali, et al. 2021. "Importance of van der Waals interactions in hydrogen adsorption on a silicon-carbide nanotube revisited with vdW-DFT and quantum Monte Carlo." *ACS Omega* 6 (38): 24630–24636. https://doi.org/10.1021/acsomega.1c03318.
77. Halter, Dominik , Ryan Klein, Michael Boreen, et al. 2020. "Self-adjusting binding pockets enhance H_2 and CH_4 adsorption in a uranium-based metal–organic framework." *Chemical Science* 11 (26): 6709–6716. https://doi.org/10.1039/D0SC02394A.
78. Jastrzębski, Krzysztof, and Piotr Kula. 2021. "Emerging technology for a green, sustainable energy-promising materials for hydrogen storage, from nanotubes to graphene—A review." *Materials* 14 (10): 2499. https://doi.org/10.3390/ma14102499.
79. Wang, Guanzhong, De-Bin Ji, Hongjun Wu, et al. 2021. "Superior capacitive storage behavior of porous carbon electrode with high mass loading." *Advanced Powder Technology* 32 (12): 4677–4686. https://doi.org/10.1016/j.apt.2021.10.013.
80. Sdanghi, Gérald, Roberto Canevesi, Ammar Celzard, et al. 2020. "Characterization of carbon materials for hydrogen storage and compression." *C* 6 (3): 46. https://doi.org/10.3390/c6030046.
81. Zheng, Yulin, Xian Chen, Nedgine Joseph, et al. 2022. "Occurrences and impacts of engineered nanoparticles in soils and groundwater." In *Emerging Contaminants in Soil and Groundwater Systems*, pp. 165–204. Elsevier. https://doi.org/10.1016/B978-0-12-824088-5.00005-7.
82. Seenithurai, Sonai, R. Kodi Pandyan, S. Vinodh Kumar, et al. 2014. "Al-decorated carbon nanotube as the molecular hydrogen storage medium." *International Journal of Hydrogen Energy* 39 (23): 11990–11998. https://doi.org/10.1016/j.ijhydene.2014.05.184.
83. Durgun, Engin, Salim Ciraci, and Taner Yildirim. 2008. "Functionalization of carbon-based nanostructures with light transition-metal atoms for hydrogen storage." *Physical Review B* 77 (8): 085405. https://doi.org/10.1103/PhysRevB.77.085405.
84. Öztürk, Zeynel. 2021. "Lithium decoration characteristics for hydrogen storage enhancement in novel periodic porous graphene frameworks." *International Journal of Hydrogen Energy* 46 (21): 11804–11814. https://doi.org/10.1016/j.ijhydene.2021.01.073.

85. Ghosh Anupama, Kovvali Siva Subrahmanyam, Katla Sai Krishna, et al. 2008. "Uptake of H_2 and CO_2 by graphene." *The Journal of Physical Chemistry* C 112 (40): 15704–15707. https://doi.org/10.1021/jp805802w.
86. Tokarev, Alexey, Aleksey Avdeenkov, Henrietta Langmi, et al. 2015. "Modeling hydrogen storage in boron-substituted graphene decorated with potassium metal atoms." *International Journal of Energy Research* 39 (4): 524–528. https://doi.org/10.1002/er.3268.
87. Baumann Avery, David Burns, Bingqian Liu, et al. 2019. "Metal-organic framework functionalization and design strategies for advanced electrochemical energy storage devices." *Communications Chemistry* 2 (1): 86. https://doi.org/10.1038/s42004-019-0184-6.
88. Jones Christopher. 2022. "Metal–organic frameworks and covalent organic frameworks: Emerging advances and applications." *Jacs Au* 2 (7): 1504–1505. https://doi.org/10.1021/jacsau.2c00376.
89. Aghajanloo, Mostafa, Abbas Rashidi, and Mohammad Moosavian. 2014. "Synthesis of zinc-organic frameworks nano adsorbent and their application for methane adsorption." *Journal of Chemical Engineering & Process Technology* 5 (5): 1000203. https://doi.org/10.4172/2157-7048.1000203.
90. Zhang, Xiaolin, Peng Liu, and Yongli Zhang. 2023 "The application of MOFs for hydrogen storage." *Inorganica Chimica Acta* 557: 121683. https://doi.org/10.1016/j.ica.2023.121683.
91. Niknam, Esmaeil, Farhad Panahi, Fatemeh Daneshgar, et al. 2018. "Metal–organic framework MIL-101 (Cr) as an efficient heterogeneous catalyst for clean synthesis of benzoazoles." *ACS Omega* 3 (12): 17135–17144. https://doi.org/10.1021/acsomega.8b02309.
92. Ramimoghadam, Delaram, Edward Gray, and Chris Webb. 2016. "Review of polymers of intrinsic microporosity for hydrogen storage applications." *International Journal of Hydrogen Energy* 41 (38): 16944–16965. https://doi.org/10.1016/j.ijhydene.2016.07.134.
93. Geng, Keyu, Ting He, Ruoyang Liu, et al. 2020. "Covalent organic frameworks: Design, synthesis, and functions." *Chemical Reviews* 120 (16): 8814–8933. https://doi.org/10.1021/acs.chemrev.9b00550.
94. Tan, Kian, Suvendu Ghosh, Ziqi Wang, et al. 2023. "Covalent organic frameworks." *Nature Reviews Methods Primers* 3 (1): 1. https://doi.org/10.1038/s43586-022-00181-z.
95. Zhu, Lin, and Y. B. Zhang. 2017. "Crystallization of covalent organic frameworks for gas storage applications." *Molecules* 22 (7): 1149. https://doi.org/10.3390/molecules22071149.
96. Feied, Charles. 2004. "Novel Antimicrobial Surface Coatings and the Potential for Reduced Fomite Transmission of SARS and Other Pathogens." Available at: https://graphics8.nytimes.com/images/blogs/freakonomics/pdf/FeiedAntimicrobialSurfaces.pdf.
97. Kumar, Anshul, Ponnan Muthukumar, Pankaj Sharma, et al. 2022. "Absorption based solid state hydrogen storage system: A review." *Sustainable Energy Technologies and Assessments* 52: 102204. https://doi.org/10.1016/j.seta.2022.102204.
98. Humphreys, J., Rong Lan, and Shanwen Tao. 2021. "Development and recent progress on ammonia synthesis catalysts for Haber–Bosch process." *Advanced Energy and Sustainability Research* 2 (1): 2000043. https://doi.org/10.1002/aesr.202000043.
99. Aziz, Muhammad, Agung Wijayanta, and A. B. D. Nandiyanto. 2020. "Ammonia as effective hydrogen storage: A review on production, storage and utilization." *Energies* 13 (12): 3062. https://doi.org/10.3390/en13123062.
100. MacFarlane, Douglas, Pavel Cherepanov, J. Choi, et al. 2020. "A roadmap to the ammonia economy." *Joule* 4 (6): 1186–1205. https://doi.org/10.1016/j.joule.2020.04.004.

101. Serpell, Oliver, Zoubida Hsain, Andrea Chu, et al. 2023. "Ammonia's Role in a Net-Zero Hydrogen Economy." *Kleinman Center for Energy Policy*, March, 17. Available at: https://kleinmanenergy.upenn.edu/research/publications/ammonias-role-in-a-net-zero-hydrogen-economy/.
102. Sun, Hanyu, Guorui Xu, Fengmei Li, et al. 2020. "Hydrogen generation from ammonia electrolysis on bifunctional platinum nanocubes electrocatalysts." *Journal of Energy Chemistry* 47: 234–240. https://doi.org/10.1016/j.jechem.2020.01.035.
103. Demirci, Umit. 2017. "Ammonia borane, a material with exceptional properties for chemical hydrogen storage." *International Journal of Hydrogen Energy* 42 (15): 9978–10013. https://doi.org/10.1016/j.ijhydene.2017.01.154.
104. Akbayrak, Saim, and S. Özkar. 2018. "Ammonia borane as hydrogen storage materials." *International Journal of Hydrogen Energy* 43 (40): 18592–18606. https://doi.org/10.1016/j.ijhydene.2018.02.190.
105. Ramachandran, Veeraraghavan, Hiren Mistry, Arshin Kulkarni, et al. 2014. "Ammonia-mediated, large-scale synthesis of ammonia borane." *Dalton Transactions* 43 (44): 16580–16583. https://doi.org/10.1039/C4DT02467B.
106. Choi, Jeung, Héctor Valdés, Dirk Pingen, et al. 2018. "Metal pincer catalysts in aqueous media: Approaches with water as solvent, reagent, and molecular hydrogen storage." In *Pincer Compounds*, pp. 273–294. Elsevier. https://doi.org/10.1016/B978-0-12-812931-9.00013-X.
107. Mission Innovation. 2024. "Liquid Organic Hydrogen Carriers." Available at: https://mission-innovation.net/our-work/mission-innovation-breakthroughs/liquid-organic-hydrogen-carriers/.
108. Wojnarowski, Paweł, and Dominika Janiga. 2024. "Enhanced system for hydrogen storage and conversion into green methanol in a geothermal environment." *International Journal of Hydrogen Energy* 52: 834–844. https://doi.org/10.1016/j.ijhydene.2023.10.133.
109. Andersson, Johan, and Stefan Grönkvist. 2019. "Large-scale storage of hydrogen." *International Journal of Hydrogen Energy* 44 (23): 11901–11919. https://doi.org/10.1016/j.ijhydene.2019.03.063.
110. Sheffield, John, Kenneth Martin, and Richard Folkson. 2014. "Electricity and hydrogen as energy vectors for transportation vehicles." In *Alternative Fuels and Advanced Vehicle Technologies for Improved Environmental Performance*, pp. 117–137. Woodhead Publishing. https://doi.org/10.1533/9780857097422.1.117.

6 Hydrogen Transportation

6.1 INTRODUCTION

Over time, attention has been drawn to green hydrogen for its purpose of aiding in global decarbonization growth. Remarkable efforts have been made to study hydrogen's complete value chain, starting from production, passing through storage and delivery, and finishing with the endpoint use. Green hydrogen depends on wind, water, and photovoltaic (PV) sources. On the other hand, not all the rest of the districts have sustainable energy sources. Therefore, the storage and energy carriers that transport power from sources to end users have to be upgraded. Additionally, as outlined in [1], the supply of green hydrogen on the European market is expected to increase sharply in a matter that, at some point, would exceed the capacity of the internal supply of green hydrogen. That results in the necessity of establishing monitoring frameworks for intercountry transportation and trade of hydrogen.

On a mass basis, hydrogen has a comparably high energy density compared with other fuels. Its specific energy is also extremely high, with each kilogram of hydrogen containing an approximate total of 120 megajoules (MJ) of energy.

Nonetheless, if one looks at volumetric energy density, hydrogen's energy density is lower than those other fuels. Hydrogen is a low-density gas with a poor volumetric energy density, thus requiring large amounts of space to store meaningful energy. To overcome this limitation, hydrogen's energy density can be increased through compression or liquefaction, and as a result, reasonable amounts of hydrogen can be stored or transported.

6.2 HYDROGEN TRANSPORTATION METHODS

The provision of hydrogen mobility encompasses the transportation of hydrogen from the point of production to the point of usage. The infrastructure required at this stage may comprise pipelines, vessels, liquefaction plants, compressors, vehicles, and warehouses. As shown in Figure 6.1, three methods are commonly used in the delivery of hydrogen.

The main conventional methods for transporting hydrogen rely on using compressed gaseous hydrogen transportation by trucks or cryogenic liquid H_2, which is considered costly and slows the economic feasibility of hydrogen as a fuel. Therefore, the utilization of pipelines has gained more interest since they are considered a promising solution for large-scale and cost-effective hydrogen deployment that can link production hubs to various hydrogen applications.

The cost factor of hydrogen delivery depends on several aspects, such as transportation distance, the amount of hydrogen needed to be transmitted, and its nature, whether liquid hydrogen, compressed hydrogen, or liquid organic hydrogen carrier (LOHC).

DOI: 10.1201/9781003639190-6

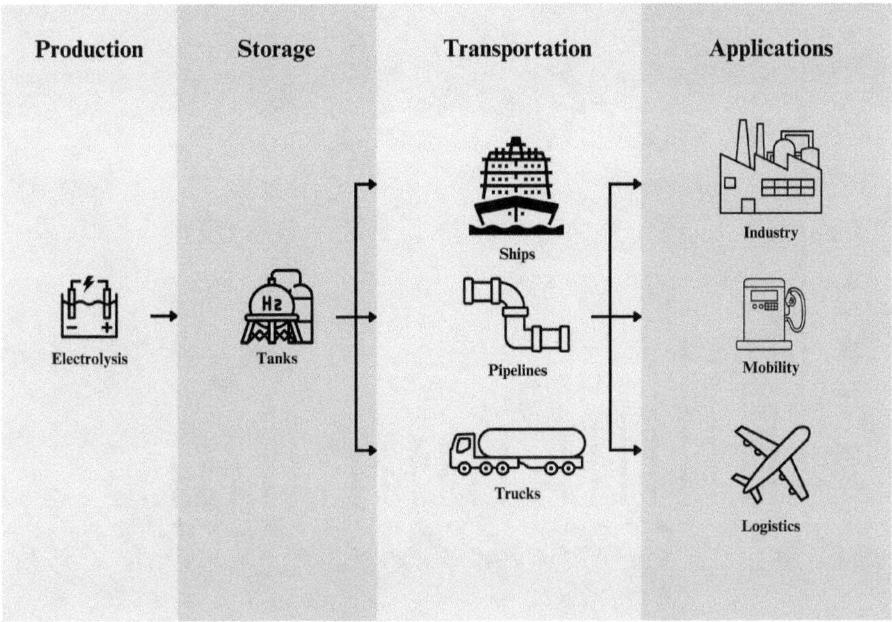

FIGURE 6.1 The components of the green hydrogen value chain.

Developing the hydrogen infrastructure is challenging due to the low energy density per volume unit and low boiling point of about −253°C. Generally, the whole hydrogen delivery operation faces key obstacles affecting efficiency and reliability. Examples of these obstacles are capital cost minimization, hydrogen leakage reduction, and maintaining the purity factor of hydrogen during the delivery and storage process.

Hydrogen transportation over long distances requires long-distance pipelines and ships in case of pipeline-setting difficulties. Further, long-distance transportation needs proper hydrogen conversion, such as liquefaction, ammonia conversion, and LOHC.

Ammonia and methanol are common and vital hydrogen carriers that considerably help facilitate the hydrogen transportation process. The main reasons for this consideration are the following [2]:

- Ammonia and methanol have a high energy density per volume unit, facilitating the storage of large amounts of energy in smaller sizes.
- Both ammonia and methanol have low flammability properties compared with hydrogen, which makes the danger level of any possible accidents during transportation low. Further, the methanol at room temperature is liquid, which ensures that transporting methanol is safer than in other transportation states.
- Ammonia and methanol are more accessible to store than hydrogen because they can be stored in containers without high-pressure tanks.
- Ammonia and methanol can be utilized in several applications, such as chemical manufacturing, biological industries, and heating.

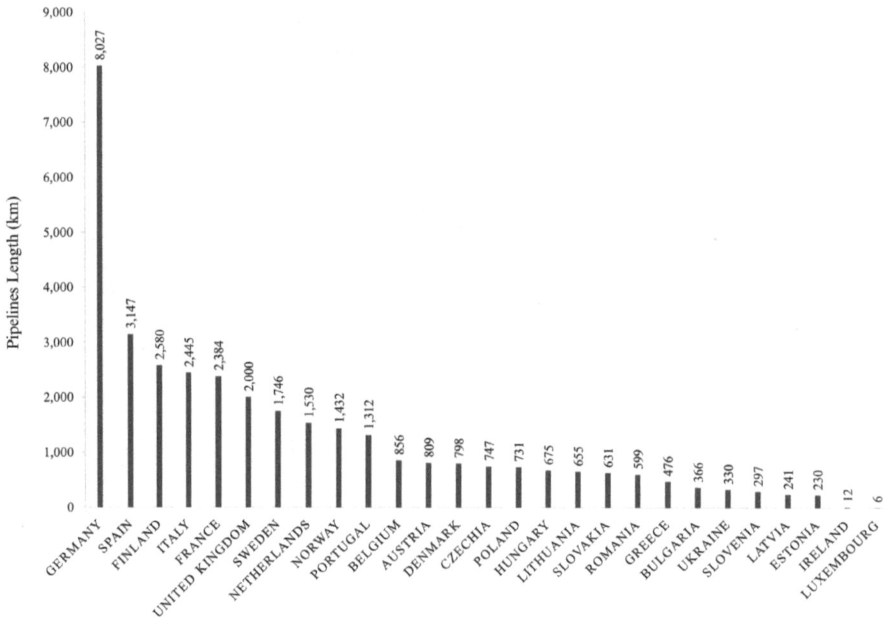

FIGURE 6.2 Hydrogen pipeline length in development in Europe 2023 by country.

6.2.1 Hydrogen Transportation by Pipelines

Pipelines are considered the most energy-efficient way of transporting hydrogen for long distances. About 5000 km of hydrogen pipelines exist, primarily in existing gray hydrogen hubs and industrial areas. Due to the expected high demand for clean hydrogen worldwide, the infrastructure deployment for hydrogen-fueled mobility is gradually growing. Therefore, several countries are currently developing to cover their future demand and achieve the net-zero goal (see Figure 6.2). Recently, investments in hydrogen infrastructure have risen to about USD 6.5 billion; 45% of these investments are in the Middle East due to the abundance of renewable energy, which will be used to produce green hydrogen [3].

The European Hydrogen Backbone (EHB) initiative presents a real case dedicated to establishing hydrogen transport infrastructure across Europe. The initiative proposes a hydrogen network of 39,700 km by 2040, with further growth expected after 2040. The European hydrogen backbone is expected to connect 21 European countries. The initiative aims to transport hydrogen over 1000 km on average €0.11–€0.21/kg of hydrogen. The estimated cost for these new infrastructures is exceptionally high due to the requirements for compression and using new pipelines in addition to the operational costs; see Figure 6.3.

Pipelines can transport gaseous hydrogen-like natural gas transport but with different considerations. Hydrogen transportation via already constructed pipelines is considered a low-cost method that can handle large amounts of hydrogen delivery. Repurposing natural gas pipelines to transmit hydrogen is one of the promising solutions to speed up the process of hydrogen delivery. This solution is considered cheap

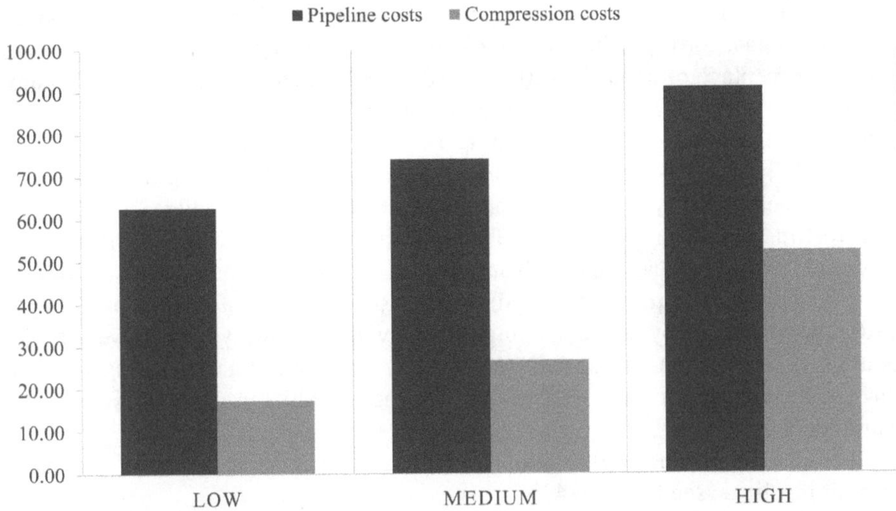

FIGURE 6.3 Projected hydrogen infrastructure costs in Europe for 2040, by range (in billion Euros) at different scenarios: low, medium, and high-cost estimations.

and fast for hydrogen usage instead of building new hydrogen transmission networks. It will decrease the investment cost by about 50–80% compared with setting up new pipelines [4]. At the same time, the construction stage cost of new pipelines may reach around $80,000/100 m [5]. Germany has initiated a joint project with several parties to specify the outlines and budget for hydrogen infrastructure. One of the plans considered through this project is to transport hydrogen using natural gas pipelines and set up hydrogen refueling terminals every 90 km on highways [6].

On the other hand, repurposing natural gas pipelines will require some considerations to adapt the installed pipelines with hydrogen properties. Examples of these considerations are as follows [4]:

- Compression method modification.
- Overall inspection of the pipeline components.
- Valves replacement.
- Gas meter replacement.

Hydrogen embrittlement occurs because of the chemical property differences between hydrogen and natural gas. This means hydrogen causes steel cracking and, consequently, pipe degradation. Several solutions to overcome the aforementioned drawbacks are proposed, such as covering the pipelines with a hydrogen barrier coating, such as a ceramic-based coating, managing the pressure swings in the pipelines, and monitoring the pipeline integrity [7].

One of the alternative methods for hydrogen delivery using pipelines is hydrogen blending. Hydrogen blending means the injection of hydrogen with natural gas steam is delivered using the installed pipelines. The blending process between hydrogen and natural gas requires specific and vital consideration due to the differences in properties between both gases. Researchers suggest different blending percentages; for example, the European Union suggests a blending percentage of about 3% of hydrogen blending with natural gas, corresponding to 1.3 million tons of hydrogen (Mt H_2) by 2030 [4]. Another percentage of about 5%–10% of hydrogen blending is accepted without any modification to the transportation infrastructure. It can be increased to reach about 20% of hydrogen blending in case of a proper change in the infrastructure to handle the properties' difference between natural gas and hydrogen [8]. Britain's gas grid network intends to blend 20% hydrogen into the gas infrastructure by 2023 [9].

Extracting pure hydrogen for specific applications, called hydrogen deblending, is considered a vital process, especially in some cases. Examples of these cases are fuel cell vehicles, which require a pure hydrogen gas of about 99.97%. On the other hand, natural gas produced by the deblending process with shallow hydrogen levels is also required in some situations, such as gas turbines and natural gas refueling stations. Further, the accepted purity level of both hydrogen and natural gas will affect the overall deblending cost.

Hydrogen pipeline networks, similar to natural gas networks, consist of four main types of pipelines; see Figure 6.4:

1. **Gathering Pipelines:** These pipelines have small diameters, collect hydrogen from production sites, and transport it to the mainline transmission grid. The diameters of gathering pipelines range from 2 to 12 inches and lower pressures; their rights-of-way (ROWs) are narrower, usually 15–30 ft wide. A pipeline right-of-way (ROW) is a safety zone, typically around 50 ft (or 25 ft on either side of the pipe), established by the operator to ensure smooth and secure pipeline operations. See Figure 6.5 to understand the right-of-way.

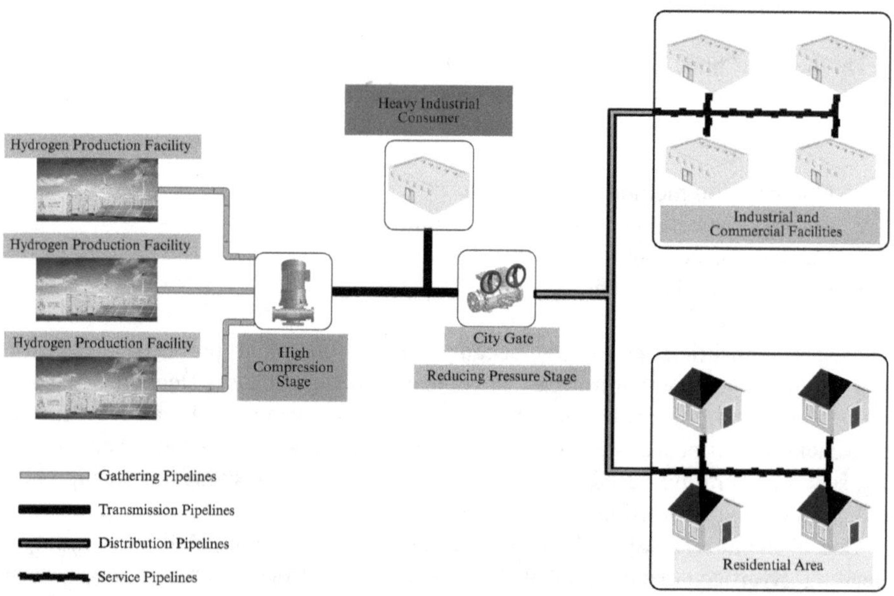

FIGURE 6.4 Types of hydrogen pipeline networks.

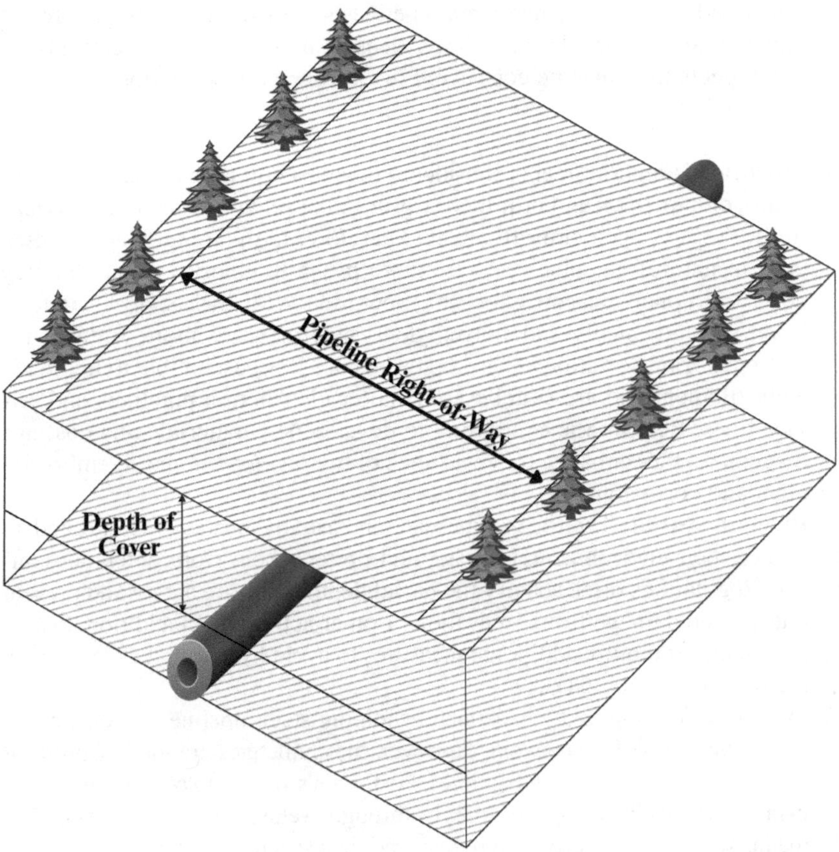

FIGURE 6.5 A pipeline right-of-way (ROW) safety zone.

2. **Transmission Pipelines:** These pipelines have large diameters to transport hydrogen over long distances, often at high pressures. Their diameters range from 24 to 48 inches, and they operate at high pressures. Therefore, compressor stations are needed to boost pressure along the transmission route. They have the widest ROWs, typically 50–150 ft. Careful planning and environmental considerations must be applied to ensure safe operation and maintenance.
3. **Distribution Pipelines:** These smaller pipelines deliver gas to local communities and industrial users. Their featured diameters range from 2 to 12 inches, and they operate at lower pressures. Their ROWs are typically narrow, ranging from 10 to 25 ft, following established infrastructure such as roads and sidewalks.
4. **Service Pipelines:** These pipelines are the smallest, delivering gas directly to residential and commercial customers.

Although the overall design of a hydrogen pipeline network is similar to that of a natural gas network, the task is challenging. Key differences in the construction, installation, and operation of hydrogen pipelines stem from their unique properties due to their smaller molecular size, lower density, and potential for embrittlement. The main aspects that must be considered in hydrogen transportation pipelines can be summarized as follows:

1. **Pipeline Material:** Steel is considered the most common material used for hydrogen pipelines and requires careful grade selection due to its susceptibility to hydrogen embrittlement. Therefore, high-strength steels, often used in natural gas pipelines, are more prone to embrittlement and may need additional measures for safe operation with hydrogen. Therefore, lower-strength steels and polyethylene pipelines are more suitable for hydrogen transportation. These materials are less susceptible to hydrogen embrittlement and are suitable for smaller distribution pipelines and service lines. Research is continuing to develop new materials and coatings that can further enhance the resistance of hydrogen pipelines to embrittlement and corrosion.
2. **Pipeline Burial:** Hydrogen pipelines are usually buried underground to reduce their environmental impact and protect them from potential damage. Regulations often specify the burial depth to ensure safety and prevent unexpected damage from excavation or other activities. In addition, hydrogen pipelines require additional measures to avoid hydrogen leakage during the implementation phase.
3. **Welding:** Welding is the backbone of joining steel pipeline sections, creating a sturdy, leak-free network. However, the challenges are more significant when it comes to hydrogen pipelines. Hydrogen's small molecular size makes it far more likely to escape, mainly through welds, than natural gas. This means welding for hydrogen pipelines requires stricter standards, specialized techniques, and thorough inspections to ensure everything holds tight.

   Operating and maintaining hydrogen pipelines present unique challenges compared with natural gas pipelines. These challenges stem from the different properties of hydrogen and its potential impact on pipeline materials and equipment.
4. **Safety:** Hydrogen pipeline safety requires addressing its wide flammability range (4%–75% in air) and low ignition energy, which elevates combustion risks compared with natural gas. Nearly invisible flames necessitate thermal imaging or flame ionization detectors for fire identification. Advanced leak detection systems, such as infrared sensors and distributed fiber-optic networks, ensure rapid response to fugitive emissions.
5. **Gas Velocity:** Hydrogen's low volumetric energy density requires higher flow rates than natural gas to deliver equivalent energy, intensifying operational challenges. The high velocities of the hydrogen can risk pipeline erosion and mechanical wear at bends or joints, threatening structural integrity. Engineers must prioritize gas velocity optimization in design to balance energy transport efficiency with material durability [10].
6. **Valves:** Valves are considered one of the main components of hydrogen pipelines that control the flow and pressure of hydrogen through the pipeline

network. However, the hydrogen properties require specialized valves to withstand their effects and ensure safe operation.

7. **Odorization:** Hydrogen is odorless and requires the addition of an odorant to ensure safety. Research is ongoing to identify suitable hydrogen odorants that are safe, effective, and compatible with pipeline materials and equipment. The odorization process for hydrogen pipelines must be carefully designed and implemented to ensure reliable leak detection and maintain public safety.

6.2.2 Hydrogen Transportation by Trucks

In the case of road transport, trucks carry hydrogen tanks that either hold gaseous or liquid hydrogen. Trucks also carry ammonia, a type of hydrogen. This method of transport is done for short distances and does not require new infrastructure to be built.

In case of gaseous hydrogen, it is a must for it to be compressed to the maximum pressure of 180 bar. After that the gas must be stored in long cylinders, which then are kept in trailers that are towed behind trucks.

That said, trucks with the capacity to store gaseous hydrogen are often referred to as tube trailers due to their use of long-tube structures. The most common type of tube trailer is the steel tube trailer, which can support up to 380 kg. In addition, other storage vessels have been built that enable the storage of 560–900 kg of hydrogen per trailer. It should be noted that these vessels are currently used to store natural gas. Further information on hydrogen storage in compressed containers can be found in Chapter 5.

6.2.3 Hydrogen Transportation by Ships

Hydrogen trading among countries is one of the considerable opportunities for the widespread use of hydrogen technology; the pipeline solution for hydrogen delivery is the most effective and cheapest method for transporting distances of about 2500–3000 km, handling hydrogen amounts of 200 ktpa (kiloton per annum).

In contrast, marine transportation can be considered a cheap solution for hydrogen transportation over more expanded distances. Accordingly, the distance and hydrogen amount are the main aspects that specify the appropriate transportation method. For example, delivering small amounts of hydrogen for short distances of less than 2000 km using the pipeline is cheaper than shipping. Repurposing the natural gas pipelines enables the employment of these pipelines to deliver hydrogen as a cost-effective solution for an extended distance of around 8000 km [11].

6.3 HYDROGEN DERIVATIVES DELIVERY

In the green hydrogen transportation stage, there are three primary forms in which hydrogen is commonly conveyed: ammonia (NH_3), liquefied hydrogen (LH_2), and LOHCs. Each energy carrier has unique characteristics and benefits that facilitate hydrogen delivery to end users. The availability of these carriers for hydrogen transportation contributes to the enrichment and enhancement of the overall hydrogen transportation stage.

6.3.1 Ammonia

Ammonia is achievable through a combination of hydrogen and nitrogen to produce liquid ammonia, which can subsequently be stored and transported in tanks. Ammonia's energy density is ranked among the highest for any combustible substance, even greater than hydrogen when measured per unit of volume under standard conditions. Hence, this fuels the release of vast quantities of energy. Depending on the running applications, ammonia can be directly used or cracked into hydrogen and nitrogen at the end-use points.

Figure 6.6 shows the conceptual diagram for ammonia, which integrates the production of green hydrogen, storage and transportation, and direct utilization of ammonia or its decomposition to hydrogen. Many domains—for example, fertilizers—have remarkably established ammonia trading, which implies that the infrastructural groundwork of ammonia transport is relatively abundant.

In addition, ammonia is stored at an ambient temperature under pressure levels of 8–10 bar or at a low temperature of –33°C in refrigerated tanks under atmospheric pressure. So, the ammonia storage and transportation phases are highly applicable, already established, and affordable. On the other hand, ammonia has several disadvantages, such as the toxic properties of ammonia in case of leakages and the cracking process of ammonia to get hydrogen, which is considered an energy-inefficient method. Also, purifying the hydrogen after the cracking process is required for use in end applications.

6.3.2 Liquefied Hydrogen

Liquid hydrogen has a much greater volumetric density than gaseous hydrogen. Thus, the quantity of liquid hydrogen that can be transported for a given storage volume is much greater than that of gaseous hydrogen that can be transported in the same volume. Liquid hydrogen can be transported without the need for any other substance for chemical reactions. Thus, hydrogen can be used directly in end-use applications

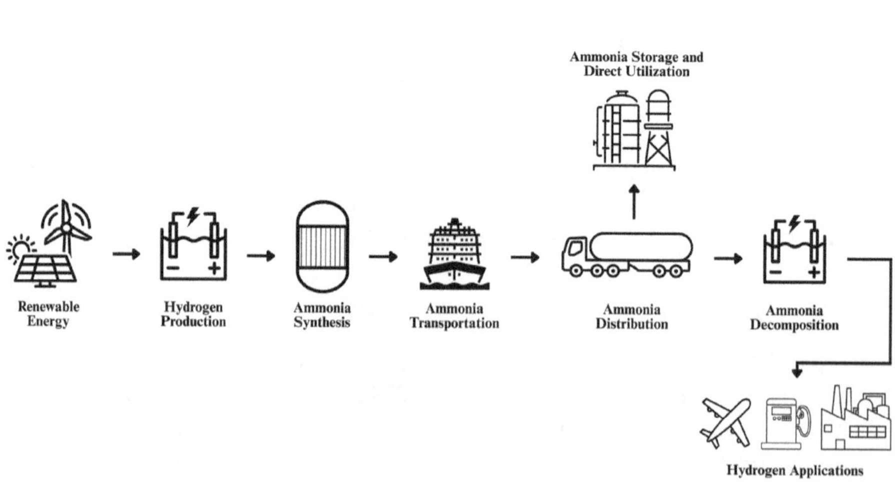

FIGURE 6.6 Conceptual diagram of the ammonia supply chain.

without further purification or reaction steps. Ammonia and LOHC, on the other hand, require specific reactions for their transport.

In the case of liquefied hydrogen, it must be stored at −253°C before loading onto fully insulated ships. In this sense, the liquefaction process wastes some hydrogen in the first phase of the trip. For some applications requiring high purity, liquefied hydrogen is good for the travel part. The worldwide hydrogen liquefaction capacity is estimated to be around 500 tons per day (tpd) as of 2022. Some small-capacity stations of 5–10 tpd have been built in the past two decades. In the last 2 years, this value has increased in the United States with new liquefaction stations of 30 tpd.

The largest station in Korea, with an operational capacity of 90 tpd [12], is expected to begin service in 2023.

An estimate of energy consumption in the hydrogen liquefaction station can be about 10 kWh/kg, equivalent to 30% of the energy preserved in hydrogen. In theory, the energy needed for the process of hydrogen liquefaction is 2.7 kWh/kg at the lowest, which is lower than those bounds in sayings (for real value) [12].

The gaseous state of hydrogen demonstrates the compressed hydrogen under a specific pressure rate, usually considered 350 bar (5000 psi) and 700 bar (10,000 psi). The density factor of compressed hydrogen is very low, which results in the need for large tanks for transportation. In addition, there are risks attached to the storage and transport of hydrogen due to its flammability and explosive nature, which necessitates a high safety factor. Compressed hydrogen has many industrial applications in medicine, food, electronics, and chemical industries. In addition, it has become an alternative fuel in the transport industry and a power source in rockets in the aerospace industry.

The American Bureau of Shipping for the Australian Provaris approved a pilot tanker for compressed hydrogen in 2021. The tanker has a subpressurized capacity of 430 t at 250 bars, while the full-scale capacity reaches 2000 t. Additionally, several Norwegian companies collaborated in making a new design for a tanker that would use hydrogen to fuel a vessel that would transfer containers filled with compressed hydrogen. The tanker is expected to incorporate around 500 containers with a 40-foot dimension, giving the vessel a total volume of 30,000 m^3.

6.3.3 Liquid Organic Hydrogen Carrier

The comparison of hydrogen storage and transport media indicates that liquefied hydrogen (LH_2), ammonia (NH_3), and LOHC differ substantially with respect to the density of hydrogen. Table 6.1 shows that ammonia has a higher volumetric hydrogen

TABLE 6.1
Comparison of the Hydrogen Density in Liquified Hydrogen, Ammonia, and Liquid Organic Hydrogen Carriers Under Specific Conditions [24]

Carrier	H_2 Density (kg-H_2/L)	Conditions
LH_2	0.071	Atmospheric pressure, −253°C
NH_3	0.121	Atmospheric pressure, −33°C
LOHC	0.047–0.15	Atmospheric temperature and pressure

TABLE 6.2
Energy Demand by Hydrogen Derivatives

Derivative	Conversion (MWh/t H_2)	Reconversion (MWh/t H_2)
Ammonia	5.57	11.2
LH_2	12	0.6
LOHC	0.5	15

density, which is important compared with LH_2. This is significant in terms of assessing the effectiveness of storage of hydrogen along with the area needed to be allocated for its use. Many organic compounds are capable of undergoing chemical reactions that result in the bonding or loss of hydrogen. LOHC is one such compound. LOHC is used for the storage and transportation of hydrogen gas. Hydrogen refueling using LOHC is achieved through a hydrogenation reaction, which occurs at elevated pressures and temperatures with the presence of a catalyst. The same process is also needed for dehydrogenation at the end-use points of hydrogen. Dibenzyl toluene is a common and good option as an example of LOHC. One cubic meter of dibenzyl toluene can store approximately 56 kg of hydrogen; thus, it can absorb hydrogen on a larger basis than it can with liquified hydrogen. Unlike liquified hydrogen, it can be transmitted without cooling. In general, dibenzyl toluene undergoes a reaction with hydrogen to form a hydrogenated dibenzyl toluene compound. The latter can easily be transported under ambient temperature and pressure.

At a later stage, hydrogen may be liberated from the brought complex by means of heating or catalytic processes. For instance, while LOHC systems have various hydrogen storage density ranges, it is important to remember that the efficiency of each carrier depends on certain thermodynamic parameters, temperature and pressure, in particular. Therefore, a suitable hydrogen carrier has to be chosen, keeping in mind the requirements of the intended application and the existing technologies for its handling and processing.

Since hydrogen has low density at room temperature, it may be transformed into various carriers, such as ammonia, liquefied hydrogen, or LOHCs. Each of these derivatives requires different amounts of energy for the intermediary conversion steps (transforming hydrogen to the carrier and back).

For easier comparison, the energy costs have been expressed in megawatt-hours per ton of hydrogen (MWh/t H_2). Table 6.2 presents energy cost values and shows how challenging each method is. In addition, the efficiency of the selected carrier illustrates how energy-intensive each method is.

6.4 HYDROGEN TRANSPORTATION COST ANALYSIS

Transportation cost mainly depends on several factors: the amount of transported hydrogen, the transportation distances, final use at the end application, and the state of hydrogen during transportation (compressed, liquefied, or converted to a chemical carrier). The effects of these factors differ based on the transportation method. The distance factor is more critical. It directly impacts pipelines' costs since the transportation cost is proportional to distance. In contrast, the transportation cost of ships mainly depends on storage and end terminals.

With large amounts of hydrogen transported for long distances, ammonia shipping is the best option from a cost-effective perspective. The most significant percentage of ammonia shipping's cost is related to the ammonia cracking process, which results in about (13%–34%) energy loss [13]. The cracking stage can be eliminated if ammonia is directly used in specific applications.

Generally, hydrogen transportation by pipeline is a cost-effective method for short distances. Utilizing the existing pipelines to transport hydrogen gives the pipelines the preference from the cost-effectiveness point of view. At the same time, establishing new pipelines leads to additional costs for the hydrogen transport process. Conversion of hydrogen to chemical carriers, cracking process at the end user to extract hydrogen, and purification stage requires specific equipment/infrastructure such as compressors, pumps, evaporators, dehydrogenation reactors, and ammonia-cracking stations. As mentioned, the need for the aforementioned equipment/infrastructure increases the overall cost [14]. Further, cracking the utilized chemical carriers has a considerable share in the overall transportation cost because the unpacking process consumes a lot of energy. The cost of the reconversion process in the ammonia case reaches more than one-third of the total cost [15].

The cost of transporting hydrogen via pipelines has a critical impact on determining the economic viability of a hydrogen economy. The levelized cost of hydrogen for pipeline transport ($LCOH_{pipe}$) includes both capital expenditures (CAPEX) and operating expenditures (OPEX).

CAPEX covers the initial costs of building the pipeline infrastructure, including materials, labor, and installation. OPEX includes the ongoing costs of operating and maintaining the pipeline, such as energy costs for compression, labor costs, and other fixed costs.

The $LCOH_{pipe}$ is calculated by considering the annualized CAPEX, OPEX, pipeline availability, and hydrogen transport capacity. The formula for calculating $LCOH_{pipe}$ is as follows [16]:

$$LCOH_{Pipe} = \frac{\text{Annualized}\ (\text{CAPEX} + \text{OPEX})}{\text{Availability} \times \text{Capacity} \times 365} \tag{6.1}$$

Where:

- Annualized CAPEX is the annual cost of capital investment, calculated using the capital recovery factor (CRF).
- OPEX is the annual operating cost, including energy, labor, and maintenance costs.
- Availability is the fraction of time during which the pipeline is operational.
- Capacity is the amount of hydrogen transported per day.

In order to accelerate the infrastructure of hydrogen transportation, the repurposing of natural gas pipelines has been recommended. The existing pipelines can significantly reduce transportation expenses, unlike the construction of new pipelines specially designed for hydrogen transport. Even though it is mentioned as the most promising of all the solutions, the highest possible mixing percentage of hydrogen and natural gas remains between 10% and 20%. Furthermore, the repurpose of natural gas pipelines also requires evaluating the condition and usability of the pipeline for hydrogen transport. Mixed hydrogen cannot be carried out through these natural gas pipelines until many changes are made, including new valves, compressors, and monitoring systems.

The liquefaction process of transporting hydrogen is considered very expensive for many reasons: the liquefaction stations use too much energy, having a long storage time necessitates a boil-off operation, and extra facilities are needed for the transport. The IRENA report (2022) predicted the possible market for hydrogen transport by 2050. This illustrates that the phases of compression and storage comprise a considerable value of the total cost of approximately \$5–\$7 billion. Figure 6.7 illustrates the estimated market potential for hydrogen transportation as for 2050.

Hydrogen's worldwide interest is expanding the hydrogen export industry/market to cover a larger area connecting the production and end-use sides. This requires export infrastructure, guidelines, and procedures that control the export process.

In 2021, several hydrogen export processes occurred in countries with different trading capacities. Table 6.3 gives examples of these hydrogen export processes

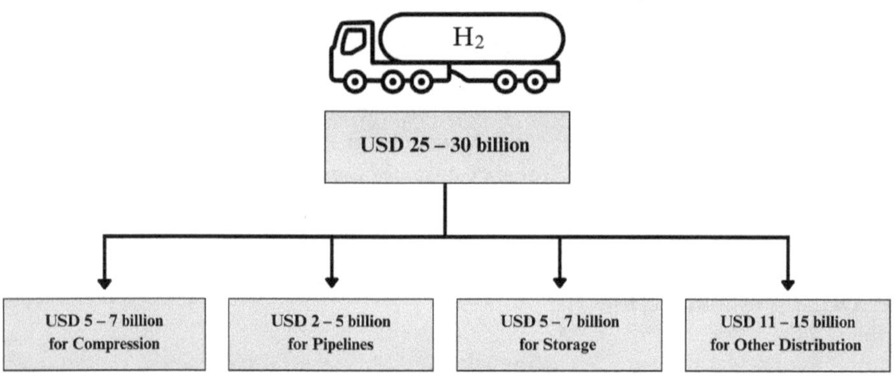

FIGURE 6.7 Estimated market potential for hydrogen transportation by 2050.

TABLE 6.3
Hydrogen Export by Country [17]

Country	Trade Amount	Trade Value \$1000
Belgium	201,338,000	63,507.36
Canada	94,088,200	57,437.78
Netherlands	90,174,100	36,237.43
Germany	13,264,000	7,140.15
Malaysia	48,544	5,583.44

classified by the exporting countries. Belgium was one of the top exporters of hydrogen in 2021, with a hydrogen trading capacity of 201,338,000 m^3 [17].

Another cost analysis example of hydrogen export to Rotterdam is conducted based on an open-source model for cost calculation. It covers the shipping costs from Australia, Saudi Arabia, the USA, Algeria, and Chile. Table 6.4 shows the shipping route data through the export process [18].

The hydrogen export requires well-specified procedures to facilitate the hydrogen transporting stage between countries. Accordingly, the hydrogen export cost shall be estimated before the actual trading to guarantee the economic aspect of the export processes. Figure 6.8 shows a schematic visualization of the main steps for estimating the hydrogen export costs. These steps help calculate the estimated green hydrogen export cost for countries with high renewable energy capacities. The scheme starts by defining the capacities of renewable energy resources for green hydrogen production and ends with the final export cost at the final point for hydrogen delivery.

The first task is to check which countries have the highest renewable energy capacities. The area's potential is assessed to plan the necessary scaffolding for exportation. In the second step, a complete dataset comprising the capacities as well as the time series of renewable energy resources is compiled. Subsequently, results are obtained and processed into an energy optimization model, where the infrastructure designs for the hydrogen export process are delineated. The optimization system seeks to construct export infrastructure at a minimal cost. After optimization, the local infrastructure for hydrogen export will be defined to determine the costs of the sub-regional infrastructure. In the end, the net export costs are computed and, together with the other costs, form a full-cost export scheme or a financial plan aimed at the entirety of expenses from production to export at the harbor.

Comprehending gas pipeline hydraulics helps in the design and operation of hydrogen pipelines. Factors such as flow rate, temperature, pipe diameter, and gas properties affect gas flow in a pipeline system. A pressure differential at the entrance and exit of a pipeline assists in gas movement. As gas flows through a pipeline, it suffers a drop in pressure because of friction with the pipe wall. Thus, a longer, rougher pipeline leads to more loss in pressure. The speed of the gas is another important factor. Increased speed can cause the erosion of the pipe walls as time passes. There is, however, a maximum value of gas speed known as erosion velocity, which must be accounted for when designing the pipeline system.

TABLE 6.4
Shipping Route Data [18]

Country	Nautical Miles/Rotterdam
Australia	10,850
Houston (USA)	5012
Jeddah (Saudi Arabia)	3997
Algeria	1773
Chile	7455

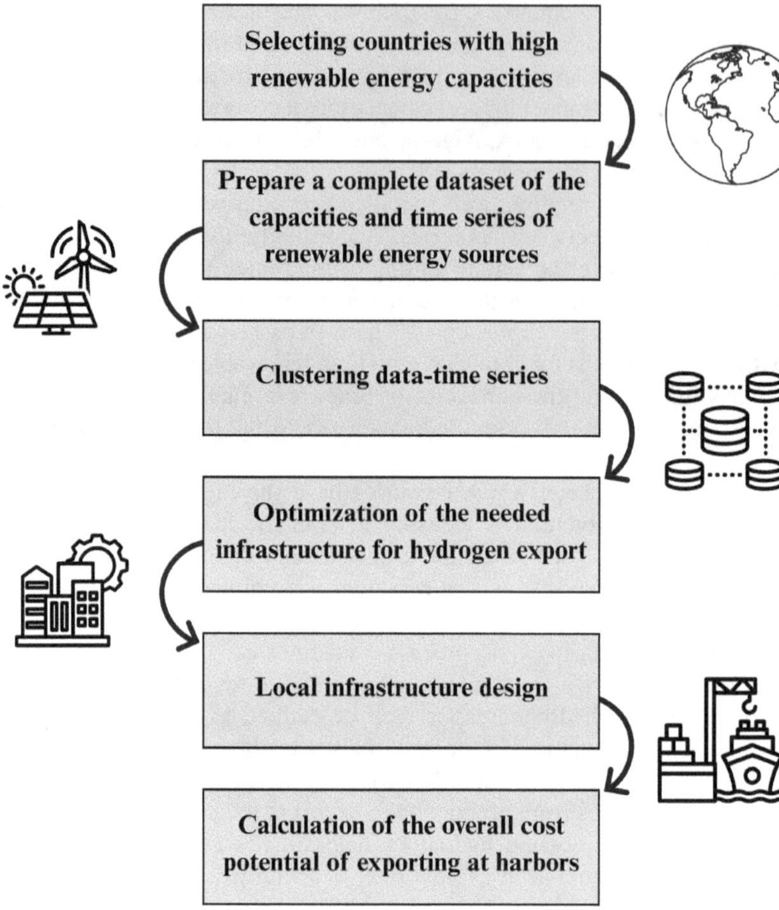

FIGURE 6.8 Schematic visualization of the main steps for estimating the hydrogen export costs.

The volumetric flow rate (Q) of hydrogen quantifies the volume of hydrogen passing a point per unit of time, which is crucial for applications such as fuel cells and chemical processing. Due to hydrogen's compressibility, temperature and pressure conditions must be specified for accurate measurements. The volumetric flow rate (Q) of the gas can be expressed as follows:

$$Q = \frac{\dot{m}}{\rho} = A \times v = \frac{\pi D^2}{4} \times v \tag{6.2}$$

As v specifies, the average fluid velocity while considering volumetric flow rate, it is evident that quantitative flow for each pipeline depends on the pipeline diameter D. With increasing diameter, the potential flow rate increases drastically. However, if the diameter is reduced, the fluid is forced to increase its speed to ensure consistent flow, which causes changes in pressure, energy, and overall efficiency. Another equally important parameter is the dimensionless Reynolds number Re. This number predicts the flow type a moving liquid: laminar or turbulent. It is determined from the equations of state of the governing parameters where fluid velocity v, density ρ, dynamic viscosity μ, and the characteristic length such as pipe diameter D are used as shown in equation 6.3. A clear-cut case of turbulent flow is when the value of Reynolds number Re is beyond 4000. At this point, the fluid in motion can be described as chaotic and irregular, with eddies and vortices. For values bordering roughly 2300, this is considered the range for laminar flow, and the fluid motion is smooth and smooth layers unidirectionally flow in parallel layers, stratified.

$$R_e = \frac{\rho \times v \times D}{\mu} \tag{6.3}$$

Another important factor to consider when addressing hydrogen transportation through pipelines is the Darcy friction factor (f). This dimensionless value considers the pressure drop and flow rate to measure resistance to flow in a pipe. It is necessary for calculating frictional losses in pipelines. The Darcy friction factor for turbulent flow in rough pipes can be computed using the Colebrook–White equation, which stands out among inadequately described formulas. The most important feature is that it reproduces the actual flow through pipes with any Reynolds number and relative roughness in detail. This is done with high head-loss accuracy, pressure loss, and other transportation fluid energy expenditures, especially through industrial pipelines where efficient and safe fluid movement is necessary. The Colebrook–White equation is expressed in equation 6.4.

$$\frac{1}{\sqrt{f}} = -2\log_{10}\left(\frac{\varepsilon/D}{3.7} + \frac{2.51}{Re\sqrt{f}}\right) \tag{6.4}$$

where f is the Darcy friction factor, which is an implicit relationship involving several key parameters: the Darcy friction factor itself, pipe roughness (ϵ), pipe diameter (D), and the Reynolds number (Re), and due to its implicit nature, it cannot be directly solved for (f). Consequently, iterative methods or numerical solvers are required to obtain the Darcy friction factor value.

Pressure drop calculations in pipelines must be carefully undertaken to design effective and safe fluid transport systems. The pressure drop (ΔP) estimation encompasses the friction loss incurred between the fluid and pipe walls and the internal fluid viscosity for a specific pipeline length (L). Due to pipe roughness and flow regime, ΔP directly depends on the Darcy friction factor (f). For estimating pressure drops, the Darcy–Weisbach equation is conventionally used and is given by equation 6.5 as follows:

$$\Delta P = f \times \left(\frac{L}{D}\right) \times \left(\frac{\rho \cdot v^2}{2}\right) \tag{6.5}$$

In order to demonstrate the application of the previously stated equations for determining volumetric flow rate, Reynolds number, Darcy friction factor, and pressure drop, the following examples will be constructed using particular assumptions.

These assumptions simplify the calculations to direct efforts toward utilizing the equations rather than computing values. Initially, it is assumed that flow is at a steady state. Moreover, hydrogen is assumed to be kept at a constant temperature of 25°C (77°F). Some average values of hydrogen properties such as density and viscosity will be used based on an assumed operating pressure. For simplicity, these rounds of elevation changes are ignored within the pipeline. An iterative solver found online or within some engineering software, can solve the implicit Colebrook–White equation. Lastly, for these example calculations, it is important to note that the impact of hydrogen embrittlement will not be included.

Example 6.1: Calculating Volumetric Flow Rate and Pipeline Diameter

A hydrogen pipeline must transport a mass flow rate ($\dot{m}$) of 5 kg/s of hydrogen. The average operating pressure is 50 bar (5 MPa); at this pressure and 25°C, hydrogen's density (ρ) is approximately 3.8 kg/m³. The desired average velocity (v) of hydrogen in the pipeline is 15 m/s.

Calculate:

a. The volumetric flow rate (Q) of hydrogen.
b. The required internal diameter (D) of the pipeline.

Solution:

a. Given that $\dot{m} = 5$ kg/s, $\rho = 3.8$ kg/m³ and using equation 6.2, the volumetric flow rate (Q) can be calculated as follows:

$$Q = \frac{\dot{m}}{\rho} = \frac{5}{3.8} = 1.316 \text{ m}^3/\text{s}$$

b. Using equation 6.2, given that Q = 1.316 m³/s (part a), v = 15 m/s, and $\pi \approx 3.1416$. The required internal diameter (D) can be calculated as follows:

$$Q = \frac{\pi D^2}{4} \times v$$

$$D = \sqrt{\frac{4Q}{\pi v}} = \sqrt{\frac{4 \times 1.316}{3.1416 \times 15}} = 0.334 \text{ m}$$

Example 6.2: Calculating Pressure Drop in the Pipeline

Using the pipeline diameter calculated in Example 6.1 (D = 0.334 m), calculate the pressure drop (ΔP) over a pipeline length (L) of 100 km (100,000 m). Assume the pipeline is made of steel with an absolute roughness (ε) of 0.046 mm (0.000046 m). The dynamic viscosity (μ) of hydrogen at the operating conditions is approximately 8.9×10^{-6} Pa·s.

Calculate:

a. The Reynolds number (Re).
b. The Darcy friction factor (f) using the Colebrook–White equation.
c. The pressure drop (ΔP).

Solution:

a. Given that $\rho = 3.8$ kg/m³, $v = 15$ m/s, $D = 0.334$ m, $\mu = 8.9 \times 10^{-6}$ Pa·s. Then, the Reynolds number (Re) can be calculated using equation 6.3 as follows:

$$R_e = \frac{\rho \times v \times D}{\mu} = \frac{3.8 \times 15 \times 0.334}{8.9 \times 10^{-6}} = 2{,}139{,}101$$

Since R_e is greater than 4000, then the flow is turbulent.

b. Given that $\varepsilon = 0.000046$ m, $D = 0.334$ m, and $Re = 2{,}139{,}101$ (part a), the Darcy friction factor has been calculated using equation 6.4 as follows:

$$\frac{1}{\sqrt{f}} = -2\log_{10}\left(\frac{\varepsilon/D}{3.7} + \frac{2.51}{Re\sqrt{f}}\right)$$

$$\frac{1}{\sqrt{f}} = -2\log_{10}\left(\frac{0.000046/0.334}{3.7} + \frac{2.51}{2{,}139{,}101\sqrt{f}}\right)$$

Solving numerically, we get:

$$f = 0.0134$$

c. Given that $f = 0.0134$, $L = 100{,}000$ m, $D = 0.334$ m, $\rho = 3.8$ kg/m³, and $v = 15$ m/s. Therefore, the pressure drop (ΔP) over 100 km can be calculated using equation 6.5 as follows:

$$\Delta P = f \times \left(\frac{L}{D}\right) \times \left(\frac{\rho \cdot v^2}{2}\right) = 0.0134 \times \frac{100{,}000}{0.334} \times \frac{3.8 \times 15^2}{2} = 1{,}715{,}119 \text{ Pa} = 1.71 \text{ MPa}$$

While choosing pipelines for hydrogen transport, it is crucial to define the thickness so that no rupture occurs from internal pressure. This essential primary safety verification is prominent in the primitive design stages. Nonetheless, choosing thickness for the pipes considers compounded issues with external forces such as loads, temperature changes, or corrosion, which are essential to a practically useful pipeline design. Hence, there is a need for an initial gap estimation for the pipeline thickness, which can be done using Simplified Barlow's Formula, which can be given as follows.

$$t = \frac{P \times D}{2 \times S \times E \times F} \tag{6.6}$$

where P represents the internal pressure the pipe must withstand, and D is the outside diameter of the pipe. S signifies the pipe material's specified minimum yield strength (SMYS), indicating its resistance to permanent deformation. E is the longitudinal joint factor, accounting for the welded joints' efficiency along the pipe's length. Finally, F is the design factor, a safety factor that ensures the pipeline can handle pressures beyond its yield strength, providing an additional margin of safety.

Example 6.3: Calculating Pipeline Wall Thickness

The maximum allowable operating pressure (MAOP) for a pipeline is 50 bar (5 MPa). The pipeline is made of API 5L X52 steel with a specified minimum yield strength (SMYS) of 360 MPa. Assume a seam joint factor (E) of 1.0 and a design factor (F) of 0.72.

Determine the minimum wall thickness (t) required for the pipeline.

Solution:

Given that P = 5 MPa = 5,000,000 Pa, D = 0.334 m (outside diameter, assuming it is close to the inside diameter for this initial calculation), S = 360 MPa = 360,000,000 Pa, E = 1.0, and F = 0.72. The minimum wall thickness (t) can be calculated using equation 6.6 as follows:

$$t = \frac{P \times D}{2 \times S \times E \times F} = \frac{5{,}000{,}000 \times 0.334}{2 \times 360{,}000{,}000 \times 1 \times 0.72} = 3.22 \times 10^{-3}\ \text{m} = 3.22\ \text{mm}$$

Therefore, a commercially available wall thickness of more than 3.22 mm shall be selected for this pipeline.

Hydrogen can be distributed from production plants to refueling stations through several methods, such as trucks or pipelines. It is necessary to calculate the key parameters for each transport method to determine the appropriate approach. This includes estimating payload, transportation cost, flow rates, and daily operating costs to determine the most cost-effective solution for a specific demand. Therefore, understanding the numerical aspects of hydrogen transportation is crucial for informed decision-making in the expanding hydrogen economy. To that end, two detailed examples will be presented. The first example provides a foundational understanding, while the second models a more realistic, large-scale hydrogen hub, demonstrating the complexities and cost implications of various delivery methods.

In order to estimate the payload, equation 6.7 can be used, and it is expressed as follows:

$$Pl = V \times \rho \tag{6.7}$$

where Pl represents payload in (kg), V is the transported volume in (m^3), and ρ is hydrogen transportation density (kg/m^3). Therefore, the transportation cost of hydrogen per kilogram (C_{kg}) can be calculated from the total cost of the total transported quantity (C_T) and the payload in kg using equation 6.8.

$$C_{kg} = \frac{C_T}{Pl} \tag{6.8}$$

For the transportation option using trucks, the number of required trips (N_T) to transfer the total quantity can be expressed as follows:

$$N_T = \frac{T_A}{T_R} \tag{6.9}$$

where T_A is the available driving time, and T_R is the round-trip time by the truck. To calculate the cumulative amount of mass (MT) transported in (kg), equation 6.10 will be used where t represents the duration over which the mass is transported.

$$M_T = \dot{m} \times t \tag{6.10}$$

Finally, the total cost (C) associated with transporting a certain amount of hydrogen mass is calculated as follows:

$$C = M_T \times C_{kg} \tag{6.11}$$

Example 6.4: Truck and Pipeline Hydrogen Transportation Calculations

A hydrogen production plant must deliver hydrogen to a refueling station 300 km away, considering truck and pipeline transport.

Truck Transportation:

- Liquid hydrogen is used.
- The truck has a tank volume of 40 m^3.
- Liquid hydrogen density is 70.8 kg/m^3.
- The truck's round-trip travel time is 6 hours.
- The total cost of a single round-trip is $1200 (including fuel, driver, and maintenance).
- The available driving time per day is 12 hours.
- The refueling station requires 5000 kg of hydrogen/day.

Pipeline Transportation:

- Gaseous hydrogen is used.
- The pipeline has a diameter of 0.6 m.
- The gaseous hydrogen density is 2.5 kg/m^3.
- The hydrogen flow velocity is 8 m/s.
- The pipeline operation cost is $0.1/kg of hydrogen transported.
    a. What is the payload of the liquid hydrogen truck?
    b. What is the transportation cost per kilogram of hydrogen delivered by truck?
    c. How many truck trips can be made per day?
    d. What is the mass flow rate of hydrogen through the pipeline?

e. How many trucks are required to meet the daily demand of the refueling station?
f. How much hydrogen (in kg) is transported by the pipeline per day?
g. What is the daily operating cost of the pipeline?
h. Which method (truck or pipeline) is more cost-effective for delivering the required 5000 kg of hydrogen daily?

Solution:

a. The truck payload can be calculated using equation 6.7 as follows:

$$Pl = V \times \rho = 40 \times 70.8 = 2832 \text{ kg}$$

b. Given that the total cost of a single round-trip is \$1200 and the truck payload is 2832 kg (part a), the cost per kilogram is approximately can be calculated using equation 6.8 as follows:

$$C_{kg} = \frac{C_T}{Pl} = \frac{1200}{2832} = \$0.42/\text{kg}$$

c. Given that the available driving time per day is 12 hours and the truck's round-trip travel time is 6 hours, the number of trips per day can be calculated using equation 6.9 as follows:

$$N_T = \frac{T_A}{T_R} = \frac{12}{6} = 2 \text{ trips/day}$$

d. Given that pipeline has a diameter of 0.6 m, the hydrogen flow velocity is 8 m/s, and the density of gaseous hydrogen is 2.5 kg/m³. The pipeline mass flow rate can be calculated using equation 6.2 as follows:

$$\frac{\dot{m}}{\rho} = A \times v$$

$$\dot{m} = A \times v \times \rho = \frac{\pi D^2}{4} \times v \times \rho = \frac{\pi (0.6)^2}{4} \times 8 \times 2.5 = 5.65 \text{ kg/s}$$

e. Since the total daily delivery from one truck = 5664 kg (two trips with 2832 kg/trip), then one truck is sufficient to cover the refueling station's demand of 5000 kg.
f. Considering that the pipeline mass flow rate is 5.65 kg/s, then the amount of hydrogen that can be transported per day (assuming 24-hour operation) equals:

$$\text{Amount of hydrogen transported per day} = 5.65 \text{ kg/s} \times 3600 \text{ seconds/h} \times 24 \text{ hours/day}$$

$$= 488{,}160 \text{ kg/day}$$

g. The daily operating cost of the pipeline (assuming 24-hour operation) can be calculated by multiplying the total daily transported amount (part f) by the operational cost per kilogram which equals \$0.1/kg as follows:

$$\text{The daily operating cost of the pipeline} = \$0.1/\text{kg} \times 488{,}160\ \text{kg/day} = \$48{,}816/\text{day}$$

h. To determine which transportation method is cost-effective, a comparison can be made as follows:
   - **Truck Cost:** One truck can deliver the 5000 kg required through two round trips per day. Each trip costs a total of \$1200. Hence, the total cost for hydrogen transportation via trucks is \$2400 per day. This can be translated into USD per kg as follows:

$$\text{Specific cost of hydrogen transport via trucks (\$/kg)} = \frac{\$2400}{5000\ \text{kg}} = \$0.48/\text{kg}$$

   - **Pipeline Cost:** As given in the example, the specific cost of hydrogen transport via pipelines is \$0.1/kg. Therefore, hydrogen transport via pipelines is way more cost-effective than hydrogen transport via trucks under these circumstances.

Example 6.5: Hydrogen Delivery Optimization: Pipeline and Truck Cost Analysis

A regional hydrogen hub is being developed to supply hydrogen to a network of refueling stations and industrial users. The hub features a large-scale electrolysis plant producing gaseous hydrogen, which is compressed for pipeline transport or liquefied for truck delivery.

Production and Demand:

- Electrolysis plant production capacity: 100,000 kg/day of gaseous hydrogen at 30 bar.
- Refueling station network demand: 60,000 kg/day.
- Industrial user demand: 30,000 kg/day.
- The remaining 10,000 kg/day is stored for backup.

Transportation Options:

- Pipeline:
  - 300 km pipeline to a central refueling station and industrial site.
  - Pipeline diameter: 0.5 m.
  - Operating pressure: 50 bar.
  - Operating temperature: 298 K.
  - Isothermal flow is assumed for simplicity.
  - The friction factor is 0.015.
  - Compressor efficiency: 85%.
  - The pipeline operation cost is \$0.1/kg of hydrogen transported.

- Truck:
  - Liquid hydrogen transport to remote refueling stations.
  - Truck tank capacity: 45 m^3.
  - Liquid hydrogen density: 70.8 kg/m^3.
  - Average round-trip distance: 400 km.
  - Average round-trip time: 8 hours.
  - Liquefaction energy consumption: 10 kWh/kg.
  - Diesel fuel cost: $1.5/L.
  - Truck fuel consumption: 0.4 L/km.
  - Driver and maintenance cost per trip: $500.
  - Electricity cost: $0.10/kWh.
    a. Calculate the mass flow rate and flow velocity of hydrogen in the pipeline.
    b. Estimate the pressure drop in the pipeline using the Darcy-Weisbach equation.
    c. Calculate the compressor power required to maintain the pipeline pressure.
    d. Determine the payload of the liquid hydrogen truck.
    e. Calculate the number of truck trips required to deliver 10,000 kg of liquid hydrogen.
    f. Calculate the total energy consumption and cost for liquefying and transporting 10,000 kg of liquid hydrogen.
    g. Compare the cost of delivering hydrogen via pipeline versus truck for the given demands.

Solution:

a. Given that the refueling station network demand is 60,000 kg/day and the industrial user demand is 30,000 kg/day. Then, the required hydrogen mass flow rate can be calculated as follows:

$$\text{Total hydrogen demand (kg/day)} = 60{,}000 \text{ kg/day} + 30{,}000 \text{ kg/day} = 90{,}000 \text{ kg/day}$$

$$\text{Total hydrogen demand (kg/s)} = 90{,}000 \text{ kg/day} \div 3600 \text{ hours/day} \div 24 \text{ hours/day} = 1.04 \text{ kg/s}$$

Given that the pipeline diameter is 0.5 m and considering that hydrogen's density at 50 bar is 40 kg/m^3, the hydrogen flow velocity can be calculated using equation 6.2 as follows:

$$\frac{\dot{m}}{\rho} = A \times v$$

$$v = \frac{\dot{m}}{A \times \rho} = \frac{\dot{m}}{(\pi D^2/4) \times \rho} = \frac{1.04}{(\pi(0.5)^2/4) \times 40} = 0.13 \text{ m/s}$$

b. Given that the friction factor is 0.015, the pipeline is 300 km long, the diameter of the pipeline is 0.5 m, the density of hydrogen at 50 bar is 40 kg/m^3, and the flow velocity is 0.13 m/s, the pressure drop can be estimated using the Darcy–Weisbach formula in equation 6.5 as follows:

$$\Delta P = f \times \left(\frac{L}{D}\right) \times \left(\frac{\rho \cdot v^2}{2}\right) = 0.015 \times \frac{300{,}000}{0.5} \times \frac{40 \times 0.13^2}{2} = 3042 \text{ Pa}$$

c. To find the required compressor power, the isothermal compression power is given by:

$$P = \frac{\dot{m} \times R \times T}{\eta \times M} \times \ln\left(\frac{P_2}{P_1}\right)$$

Where $\dot{m}$ is the mass flow rate (= 1.04 kg/s), R is the ideal gas constant (= 8.314 J/mol K), T is the temperature in (= 298 K), η is the compression efficiency (= 0.85), M is hydrogen's molar mass (= 0.002016 kg/mol), P_2 is the outlet pressure (= 50 bar), and P_1 is the inlet pressure (= 30 bar).

Therefore,

$$P = \frac{1.04 \times 8.314 \times 298}{0.85 \times 2.016} \times \ln\left(\frac{50}{30}\right) = 768{,}108 \text{ W} = 768.1 \text{ kW}$$

d. Given that the truck transportation capacity is 45 m³ and hydrogen's transportation density is 70.8 kg/m³, the truck payload can be calculated using equation 6.7 as follows:

$$Pl = V \times \rho = 45 \times 70.8 = 3186 \text{ kg/trip}$$

e. Considering a payload of 3186 kg/trip, the number of trips required to deliver 10,000 kg of liquid hydrogen can simply be calculated as follows:

$$\text{Number of trips} = \frac{\text{Transport demand}}{\text{Payload}} = \frac{10{,}000}{3186} = 3.13 \text{ trips}$$

Therefore, four trips are needed to deliver the required amount of hydrogen.

f. **Liquefaction and Transportation Energy and Cost:** Given that the specific energy consumption for hydrogen liquification is 10 kWh/kg, the total energy consumption for a truck delivery of 10,000 kg of liquid hydrogen is as follows:

$$\text{Transport energy consumption (kWh)} = 10 \text{ kWh/kg} \times 10{,}000 \text{ kg} = 100{,}000 \text{ kWh}$$

Considering an electricity rate of \$0.10/kWh, the cost of energy can be calculated as follows:

$$\text{Cost of energy } (\$) = 100{,}000 \text{ kWh} \times \$0.10/\text{kWh} = \$10{,}000$$

Since four truck trips are required (part e) with a trip distance of 400 km, the total traveled distance can be calculated as follows:

$$\text{Total traveled distance (km)} = 400 \text{ km/trip} \times 4 \text{ trips} = 1600 \text{ km}$$

Considering a fuel consumption rate of 0.4 L/km, the total fuel consumption can be calculated as follows:

$$\text{Total fuel consumption (L)} = 1600 \text{ km} \times 0.4 \text{ L/km} = 640 \text{ L of diesel}$$

Since the specific cost of diesel is assumed to be $1.5/L, the total cost of fuel can be calculated as follows:

$$\text{Total fuel cost } (\$) = 640 \text{ L} \times \$1.5/\text{L} = \$960$$

Considering that the driver and maintenance cost is $500/trip, the total cost for driver and maintenance can be calculated as follows:

$$\text{Driver and maintenance cost } (\$) = \$500/\text{trip} \times 4 \text{ trips} = \$2000$$

Therefore, the total cost of delivering 10,000 kg of liquid hydrogen via trucks equals:

$$\text{Total cost of transport } (\$) = \$10{,}000 + \$960 + \$2000 = \$12{,}960$$

Lastly, the specific total cost of transport can be calculated as follows:

$$\text{Specific total cost of transport (\$/kg)} = \frac{\$12{,}960}{10{,}000 \text{ kg}} = \$1.29/\text{kg}$$

g. **Cost Comparison:** Assuming that the compression system remains operation on a 24-hour basis, the total energy consumption can be calculated as follows:

$$\text{Daily energy consumption (kWh)} = 768.1 \text{ kW} \times 24 \text{ hours/day} = 18{,}434.4 \text{ kWh/day}$$

Considering an electricity rate of $0.10/kWh, the cost of energy can be calculated as follows:

$$\text{Cost of energy } (\$) = 18{,}434.4 \text{ kWh } \times \$0.10/\text{kWh} = \$1843.4$$

Since the specific operational cost of the pipeline system is $0.1/kg, and a total of 90,000 kg of hydrogen is to be transported per day, the specific cost of transport can be calculated as follows:

$$\text{Specific total cost of transport (\$/kg)} = \frac{\$1843.4}{90{,}000 \text{ kg}} + \$0.1/\text{kg}$$

$$= \$0.0205/\text{kg} + \$0.1/\text{kg} = \$0.1205/\text{kg}$$

Therefore, under these circumstances, the cost of hydrogen transport via pipelines is much lower than that of hydrogen's transport via trucks for the daily amount of hydrogen being moved.

6.5 MENA TO EUROPE HYDROGEN TRANSPORTATION

The need for hydrogen is expected to be high in Europe in the future. However, it cannot meet the required renewable energy output for green hydrogen production. At the same time, there is significant green hydrogen potential in the Middle East and North Africa (MENA) region, which can facilitate international and local transportation. This B enables various industries, such as industrial plants, electricity producers, or vehicle fueling stations, to utilize it. Regarding long-distance hydrogen transport, hydrogen can be supplied to European countries through existing pipelines and other means of transport, which makes collaboration between Europe and MENA beneficial.

The Europe MENA Hydrogen pipeline project is set to meet the energy requirements of various nations in the Western Sahara. Many criteria must be considered to assess which countries export hydrogen to Europe. These criteria comprise available facilities, reasonable distance for transportation, significant output, and vicinity. From this perspective, Morocco, Tunisia, Algeria, Libya, Egypt, and Saudi Arabia are primary candidates to export 6 Mt of hydrogen and 4 Mt of ammonia from MENA to Europe via pipelines [19]. Figure 6.9 illustrates the potential export capacity of MENA countries connected to Europe with pipelines [19].

Apart from abundant renewable resources in the MENA region, Gulf countries also enjoy the privilege of developing hydrocarbon infrastructure. Three Gulf Cooperation Council (GCC) members have already embarked on an ambitious hydrogen economy, and these countries are Oman, UAE, and Saudi Arabia. Oman established a hydrogen alliance referred to as Hy-Fly in August 2021. Hy-Fly comprises key energy stakeholders under the leadership of the Ministry of Energy and Minerals, which was formed to promote green hydrogen production, storage, and

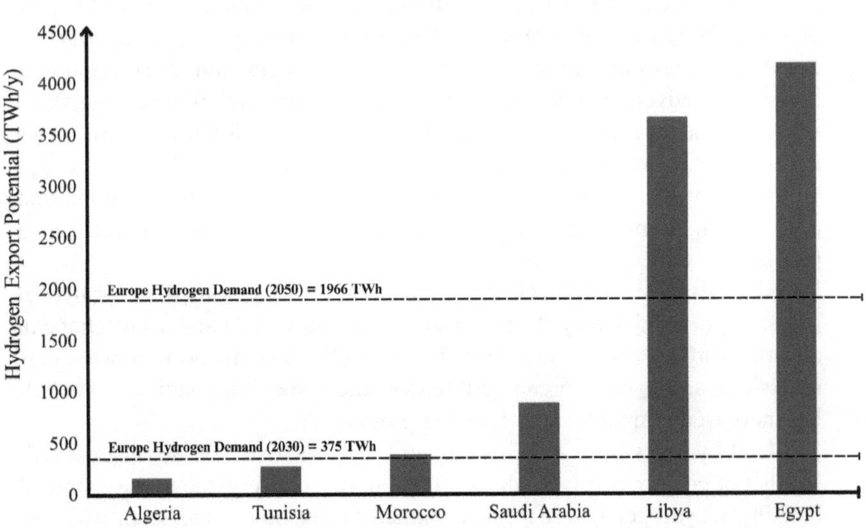

FIGURE 6.9 Hydrogen export potential via a pipeline of selected MENA countries.

export. An Oman governmental collaboration exists between OQ Gas Networks (OQGN) and Hydrogen Oman (Hydrom), which will develop the hydrogen transport infrastructure. This partnership will aid in developing an efficient green hydrogen transport network.

Saudi Arabia has an immense hydrogen production capability of nearly 160 Mtpa and an offshore pipeline delivery cost of €2.31/kg H_2. Hence, Saudi Arabia must be considered one of the foremost countries that export liquid hydrogen to Europe. If pipelines are built from the Middle East, through Saudi Arabia, and into Europe, then Europe can utilize these pipes to meet the demand exceeding 92 Mtpa [20]. Various companies have initiated several hydrogen projects in the UAE.

TAQA Utility in Abu Dhabi is developing the green hydrogen to ammonia export project [21]. Khalifa Industrial Zone will host the ammonia export terminal. Liquid ammonia will be supplied to ships bunkering ammonia as fuel. In addition, green ammonia will be stored at Khalifa Port, where it can be exported to international markets, including Europe and East Asia.

6.6 PROGRESS ON HYDROGEN DELIVERY

Researchers are developing new materials or methods to improve the delivery of hydrogen. Some of these are noted below.

- **Cryo-Compression:** Alongside other techniques, cryo-compression is one of the emerging methods that has served as a promising technique for added storage and the delivery of hydrogen [22]. It is anticipated to fill the gap between the energy-demanding liquefaction process and the low density associated with compressed gas. The process includes cooling hydrogen to near–cryogenic temperatures without complete liquefaction while compressing the gas to high pressures. The result is hydrogen storage, which is denser than gas but requires less energy than liquefaction. Furthermore, it has several advantages for heavy-duty vehicles and aviation uses, such as reduced time for refueling. Conversely, cryo-compression has several issues relating to optimizing the process and creating effective and strong storage tanks. Additionally, expectations regarding the cost of cryo-compression are very high, and decreasing the overall implementation costs remains a problem.
- **Metal Hydrides:** Metal hydrides are another promising method for hydrogen storage and delivery. They consist of certain metals and alloys that can absorb hydrogen atoms into their structure [23]. The absorption process is reversible, and hydrogen can be released under specific conditions, including increased temperature or reduced pressure.

  Metal hydrides have a high volumetric hydrogen storage density, exceeding that of compressed gas, which offers improved safety due to the chemical bonding of hydrogen within the metal lattice. On the other hand, metal hydrides still face significant challenges due to their weight, which reduces the overall energy density per unit weight and the cost of some expensive materials.

One drawback of using metal hydrides is the slow process of hydrogen and the released kinetic energy, which can limit the refueling speeds. Also, effective heat management during hydrogen absorption and release is crucial for practical applications.

- **Carbonaceous Materials for Hydrogen Storage:** Chapter 5 addresses several promising hydrogen storage methods still under research and development. These methods depend on carbonaceous materials for hydrogen storage, which can be used later for hydrogen delivery.

6.7 CHAPTER SUMMARY

This chapter provided a detailed analysis of various hydrogen transport methodologies, encompassing pipelines, ships, and trucks. Each method's limitations and suitability for hydrogen transport are thoroughly examined. The chapter extended its scope to explore alternative hydrogen carriers, specifically ammonia (NH_3), LOHC, and liquefied hydrogen (LH_2), assessing their potential to facilitate efficient, long-distance hydrogen transport. The discussion included a comprehensive evaluation of the factors that significantly influence hydrogen transport costs. These factors include the volume of hydrogen being transported, the distance of transportation, the intended end use of the hydrogen, and its physical state during transport, whether compressed, liquefied, or chemically converted. Furthermore, the chapter used case studies to illustrate the ongoing development of hydrogen transport regulations. These case studies serve to highlight the existing challenges and emerging opportunities associated with the establishment of a robust and comprehensive green hydrogen infrastructure. The chapter emphasized the growing recognition of green hydrogen's importance in supporting global decarbonization efforts. It also underlined the significant efforts dedicated to studying hydrogen's complete value chain, from its production, through storage and delivery, to its final end-use applications. The chapter pointed out that green hydrogen production relies on wind, water, and PV sources, which are not evenly distributed geographically. This necessitates the advancement of storage and energy carriers to enable efficient power transport from generation sources to end-users.

Moreover, the chapter addressed the anticipated surge in green hydrogen supply within the European market, which has the potential to outstrip internal production capacities. This anticipation underscores the critical need for establishing robust monitoring frameworks to govern intercountry hydrogen transportation and trade. The chapter also compared hydrogen's energy density with that of other fuels, noting that while hydrogen possesses a high energy density on a mass basis, its volumetric energy density is relatively low. To mitigate this, the chapter explained that hydrogen's energy density can be enhanced through compression or liquefaction, thereby enabling the storage and transportation of practical hydrogen quantities.

Finally, the chapter delineated the concept of hydrogen mobility, which encompasses the transportation of hydrogen from production sites to usage locations. The infrastructure necessary for hydrogen transportation is discussed, including pipelines, vessels, liquefaction plants, compressors, vehicles, and warehouses.

SELF-ASSESSMENT QUESTIONS

1. What factors determine the most suitable method for transporting green hydrogen?
2. How does the energy density of hydrogen compare to other fuels?
3. What are the three primary methods for transporting hydrogen?
4. Why is the development of hydrogen infrastructure considered challenging?
5. What are the main obstacles affecting the efficiency and reliability of hydrogen delivery?
6. Why are ammonia and methanol considered vital hydrogen carriers?
7. What is the most energy-efficient way of transporting hydrogen for long distances?
8. What is the approximate length of existing hydrogen pipelines globally?
9. How much have investments in hydrogen infrastructure risen recently?
10. What is the main reason for the high investment in hydrogen infrastructure in the Middle East?
11. Why is repurposing natural gas pipelines for hydrogen transport considered a promising solution?
12. What considerations are required for repurposing natural gas pipelines for hydrogen transport?
13. What is hydrogen embrittlement, and how does it affect pipelines?
14. What is hydrogen blending, and what are its potential benefits?
15. What are the different types of hydrogen pipeline networks and their functions?
16. What are the key aspects to consider in hydrogen transportation pipelines?
17. What is the most common material used for hydrogen pipelines, and what are its limitations?
18. Why are hydrogen pipelines usually buried underground?
19. What are the unique challenges in welding hydrogen pipelines?
20. What safety concerns are associated with hydrogen pipelines?
21. How does hydrogen's low volumetric energy density affect pipeline operation?
22. Why do hydrogen pipelines require specialized valves?
23. Why is odorization important for hydrogen pipelines?
24. How are trucks used for hydrogen transportation?
25. What are the three primary ways hydrogen is transported in the green hydrogen transportation stage?
26. What are the advantages and disadvantages of using ammonia for hydrogen transportation?
27. What are the advantages and disadvantages of using liquefied hydrogen for transportation?
28. What is a liquid organic hydrogen carrier (LOHC), and how is it used for hydrogen transportation?
29. What are the main factors affecting the cost of hydrogen transportation?
30. What are the challenges associated with repurposing natural gas pipelines for hydrogen transport?

31. Calculate the volumetric flow rate (Q) of hydrogen in a pipeline with a mass flow rate of 10 kg/s, given that the hydrogen density (ρ) at the operating conditions is 4.2 kg/m^3.
32. Determine the required internal diameter (D) of a hydrogen pipeline with a volumetric flow rate (Q) of 2 m^3/s and a desired average velocity (v) of 20 m/s.
33. Calculate the pressure drop (ΔP) in a hydrogen pipeline over a distance of 50 km, assuming a pipeline diameter (D) of 0.4 m, an absolute roughness (ε) of 0.05 mm, a hydrogen density (ρ) of 4 kg/m^3, an average velocity (v) of 18 m/s, and a dynamic viscosity (μ) of 9×10^{-6} Pa·s.
34. Calculate the minimum wall thickness (t) required for a hydrogen pipeline with a maximum allowable operating pressure (MAOP) of 60 bar, an outside diameter (D) of 0.4 m, a specified minimum yield strength (SMYS) of 380 MPa, a seam joint factor (E) of 1.0, and a design factor (F) of 0.72.
35. Assuming a hydrogen pipeline needs to transport a mass flow rate ($\dot{m}$) of 8 kg/s with an average operating pressure of 60 bar and a hydrogen density (ρ) of approximately 4.5 kg/m^3 at 25°C, and the desired average velocity (v) in the pipeline is 12 m/s, calculate the volumetric flow rate (Q) and the required internal diameter (D) of the pipeline.
36. Given a hydrogen pipeline with a diameter (D) of 0.3 m transporting hydrogen at a mass flow rate ($\dot{m}$) of 6 kg/s, an average operating pressure of 55 bar, and a hydrogen density (ρ) of approximately 4.1 kg/m^3 at 25°C, calculate the average velocity (v) of hydrogen in the pipeline.
37. A company is designing a hydrogen pipeline to transport gaseous hydrogen from a production plant to a storage facility 250 km away. The pipeline needs to deliver 15 kg/s of hydrogen. The average operating pressure is 65 bar, and the temperature is 25°C. The pipeline is made of steel with an absolute roughness of 0.04 mm.
    a. Calculate the volumetric flow rate of hydrogen. (Assume hydrogen density at 65 bar and 25°C is 4.0 kg/m^3)
    b. If the desired hydrogen velocity in the pipeline is 18 m/s, determine the required internal diameter of the pipeline.
    c. Calculate the Reynolds number for the flow. (Assume the dynamic viscosity of hydrogen at the operating conditions is 9.0×10^{-6} Pa·s)
    d. Calculate the Darcy friction factor using the Colebrook–White equation.
    e. Determine the pressure drop over the 250 km pipeline.
    f. Calculate the minimum wall thickness required if the pipeline's outside diameter is 0.45 m. (Assume a Specified Minimum Yield Strength (SMYS) of 400 MPa, a seam joint factor of 1.0, and a design factor of 0.72)
38. A hydrogen production plant is considering two options for transporting hydrogen to a refueling station 150 km away:

    Option A: Compressed Gaseous Hydrogen Truck Transport
    - Truck tank capacity: 500 kg of hydrogen
    - Truck round-trip time: 3 hours
    - Cost per round trip: $500

- Available driving time per day: 10 hours

Option B: Pipeline Transport

- Pipeline hydrogen flow rate: 0.5 kg/s
- Pipeline operating cost: \$0.25/kg of hydrogen transported
- The refueling station requires 2000 kg of hydrogen/day.
  a. For Option A, calculate the daily truck trips required.
  b. For Option A, calculate the daily transportation cost.
  c. For Option B, calculate the amount of hydrogen transported daily.
  d. For Option B, calculate the daily operating cost.
  e. Which option is more cost-effective for transporting the required amount of hydrogen?

39. A company is evaluating two methods for long-distance hydrogen transport: liquid organic hydrogen carriers (LOHC) and liquefied hydrogen (LH_2).
    - LOHC (Dibenzyltoluene): hydrogen storage capacity: 56 kg of hydrogen/1 m^3 of LOHC, energy required for dehydrogenation: 15 MWh/t of hydrogen
    - LH_2: Hydrogen density: 71 kg/m^3, energy required for liquefaction: 12 MWh/t of hydrogen

    The company needs to transport 500 t of hydrogen.

    a. Calculate the volume of LOHC required to transport 500 t of hydrogen.
    b. Calculate the energy required for dehydrogenation of the LOHC.
    c. Calculate the volume of LH_2 required to transport 500 t of hydrogen.
    d. Calculate the energy required for the liquefaction of hydrogen.
    e. Discuss the key factors in deciding between LOHC and LH_2 for hydrogen transport, considering energy consumption, storage volume, and other relevant factors.

REFERENCES

1. Wappler, Mona, Dilek Unguder, Xing Lu, et al. 2022. "Building the green hydrogen market–Current state and outlook on green hydrogen demand and electrolyzer manufacturing." *International Journal of Hydrogen Energy* 47 (79): 33551–33570. https://doi.org/10.1016/j.ijhydene.2022.07.253.
2. Cui, Jinyue and Muhammad Aziz. 2023. "Techno-economic analysis of Hydrogen transportation infrastructure using ammonia and methanol." *International Journal of Hydrogen Energy* 48 (42): 15737–15747. https://doi.org/10.1016/j.ijhydene.2023.01.096.
3. EHB. 2025. "Projected Hydrogen Infrastructure Costs in Europe for 2040, by Range (in Billion Euros)." Statista. Available at: https://www.statista.com/statistics/1396503/european-hydrogen-backbone-costs-and-investments-forecast/.
4. International Energy Agency. 2022. "Global Hydrogen Review 2022." Paris. Available at: https://iea.blob.core.windows.net/assets/c5bc75b1-9e4d-460d-9056-6e8e626a11c4/GlobalHydrogenReview2022.pdf.
5. Kim, Changjong, Sang Hoon Cho, Sung Min Cho, et al. 2023. "Review of Hydrogen infrastructure: The current status and roll-out strategy." *International Journal of Hydrogen Energy* 48 (5): 1701–1716. https://doi.org/10.1016/j.ijhydene.2022.10.053.

6. Federal Government. 2023. “The National Hydrogen Strategy.” Available at: https://www.bmwk.de/Redaktion/EN/Publikationen/Energie/national-hydrogen-strategy-update.pdf?__blob=publicationFile&v=2.
7. Wetegrove, Marcel, Maria Jazmin Duarte, Klaus Taube, et al. 2023. “Preventing hydrogen embrittlement: The role of barrier coatings for the hydrogen economy.” *Hydrogen* 4 (2): 307–322. https://doi.org/10.3390/hydrogen4020022.
8. Erdener, Burcin Cakir, Brian Sergi, Omar J. Guerra, et al. 2023. “A review of technical and regulatory limits for hydrogen blending in natural gas pipelines.” *International Journal of Hydrogen Energy* 48 (14): 5595–5617. https://doi.org/10.1016/j.ijhydene.2022.10.254.
9. Topolski, Kevin, Evan P. Reznicek, Burcin Cakir Erdener, et al. 2022. *Hydrogen Blending into Natural Gas Pipeline Infrastructure: Review of the State of Technology.* NREL. https://doi.org/10.2172/1893355.
10. Yue, Meiling, Hugo Lambert, Elodie Pahon, et al. 2021. “Hydrogen energy systems: A critical review of technologies, applications, trends and challenges.” *Renewable and Sustainable Energy Reviews* 146: 111180. https://doi.org/10.1016/j.rser.2021.111180.
11. IRENA. 2022. “Geopolitics of the Energy Transformation: The Hydrogen Factor.” International Renewable Energy Agency, Abu Dhabi. Available at: https://www.irena.org/-/media/Files/IRENA/Agency/Publication/2022/Jan/IRENA_Geopolitics_Hydrogen_2022.pdf.
12. Global Energy Monitor. 2024. “Countries with Europe’s Longest Proposed Hydrogen Transmission Pipeline Network in 2023 (in Kilometers).” Statista. Available at: https://www.statista.com/statistics/1389658/proposed-hydrogen-pipeline-infrastructure-length-in-europe-by-country/.
13. Wang, Changlong, Stuart D.C. Walsh, Thomas Longden, et al. 2023. “Optimising renewable generation configurations of off-grid green ammonia production systems considering Haber-Bosch flexibility.” *Energy Conversion and Management* 280: 116790. https://doi.org/10.1016/j.enconman.2023.116790.
14. Preuster, Patrick, Christian Papp, and Peter Wasserscheid. 2017. “Liquid organic hydrogen carriers (LOHCs): Toward a hydrogen-free hydrogen economy.” *Accounts of Chemical Research* 50 (1): 74–85. https://doi.org/10.1021/acs.accounts.6b00474.
15. Berger, Roland. 2021. “Hydrogen Transportation. The Key to Unlocking the Clean Hydrogen Economy.” Roland Berger GMBH, Munich, Germany. Available at: https://www.rolandberger.com/publications/publication_pdf/roland_berger_hydrogen_transport.pdf.
16. Khan, Mohd Adnan, Cameron Young, and David B. Layzell. 2021. “The techno-economics of hydrogen pipelines.” *Transition Accelerator Technical Briefs* 1 (2): 1–40. Available at: https://transitionaccelerator.ca/wp-content/uploads/2023/06/The-Techno-Economics-of-Hydrogen-Pipelines-v2.pdf.
17. World Integrated Trade Solution. 2021. “Hydrogen Exports by Country in 2021.” Available at: https://wits.worldbank.org/trade/comtrade/en/country/ALL/year/2021/tradeflow/Exports/partner/WLD/product/280410.
18. Johnston, Charles, Muhammad Haider Ali Khan, Rose Amal, et al. 2022. “Shipping the sunshine: An open-source model for costing renewable Hydrogen transport from Australia”. *International Journal of Hydrogen Energy* 47 (47): 20362–20377. https://doi.org/10.1016/j.ijhydene.2022.04.156.
19. Braun, Jan Frederik, Ad Van Wijk, and Kirsten Westphal. 2024. “Europe’s hydrogen pathways: Toward a balanced partnership with Saudi Arabia and the Gulf.” In *The Clean Hydrogen Economy and Saudi Arabia*, pp. 208–258. Routledge. https://doi.org/10.4324/9781003294290-10.

20. Jalbout, Eddy, Lucien Genge, and Felix Muesgens. 2022. "H2Europe: An analysis of long-term Hydrogen import-potentials from the MENA region." In *18th International Conference on the European Energy Market (EEM)*, Ljubljana, Slovenia, pp. 1–7. IEEE. https://doi.org/10.1109/EEM54602.2022.9921055.
21. Koch, Natalie. 2022. "Greening oil money: The geopolitics of energy finance going green." *Energy Research & Social Science* 93: 102833. https://doi.org/10.1016/j.erss.2022.102833.
22. Nivedhitha, K. S., T. Beena, Nagaraj Banapurmath, et al. 2024. "Advances in hydrogen storage with metal hydrides: Mechanisms, materials, and challenges." *International Journal of Hydrogen Energy* 61: 1259–1273. https://doi.org/10.1016/j.ijhydene.2024.02.335.
23. Bosu, Subrajit and Natarajan Rajamohan. 2024. "Recent advancements in hydrogen storage—Comparative review on methods, operating conditions and challenges." *International Journal of Hydrogen Energy* 52: 352–370. https://doi.org/10.1016/j.ijhydene.2023.01.344.
24. Spatolisano, Elvira, Federica Restelli, Laura A. Pellegrini, et al. 2024. "Liquefied hydrogen, ammonia and liquid organic hydrogen carriers for harbour-to-harbour hydrogen transport: A sensitivity study." *International Journal of Hydrogen Energy* 80: 1424–1431. https://doi.org/10.1016/j.ijhydene.2024.07.241.

7 Hydrogen Applications

7.1 INTRODUCTION

Green hydrogen holds immense potential to play a transformative role across diverse sectors worldwide due to its unique properties, including its high gravimetric energy density of 143 megajoules per kilogram (MJ/kg), its ability to store and transport renewable energy, and its capacity to partially or fully replace fossil fuels in hard-to-abate industries [1]. Its applications span a wide range of sectors including energy, industrial, chemical, transportation, and built environment sectors [2]. Specifically, green hydrogen can be utilized for electricity generation, renewable energy storage, chemical production, fuel synthetization, steel production, oil refinery, cement production, water and space heating, cooking, agriculture, and many more key applications [3–6].

Green hydrogen produces zero emissions during utilization. Hence, the incorporation of green hydrogen technologies across different operations contributes to a significant reduction in carbon emissions, accelerating the global energy transition and promoting environmental sustainability. According to the Hydrogen Council initiative [7], green hydrogen has the potential to avoid 80 gigatons (Gt) of global CO_2 emissions by 2050 while covering 22% of the final energy demand worldwide [7]. The global capacity of hydrogen is expected to reach 501.7 megatons (Mt) by 2050, which will be distributed as shown in Figure 7.1 [8].

According to BloombergNEF [8], a significant portion of the global hydrogen demand is expected to be utilized for new use cases that will start to rely on hydrogen-based technologies between 2030 and 2040. The demand for these use cases will continue to rapidly grow until 2050, making around 50.5% of the global hydrogen demand by mid-century. These use cases will include steel, aluminum, cement, residential buildings, commercial buildings, power, and other industries. On the other hand, many existing use cases, including various applications that already rely on hydrogen (e.g., oil refinery, ammonia production, methanol production) will steadily increase their reliance on hydrogen throughout the period between 2030 and 2050, corresponding to 32.5% of the global hydrogen demand by 2050. The remaining global hydrogen demand is expected to be required to decarbonize several transportation modes including aviation, shipping, and road transport with a total percentage of around 17% [8].

Altogether, green hydrogen has the significant environmental and technical potential to decarbonize many sectors worldwide by altering their operations to cope with global sustainability goals and prospects. Therefore, developing a sufficient understanding of hydrogen's integration within various sectors has become essential for engineers, researchers, and policymakers alike. In the following sections, this chapter fully identifies and examines the main applications of green hydrogen technologies for electricity generation, transportation, energy storage, built environments, and heavy industries.

DOI: 10.1201/9781003639190-7

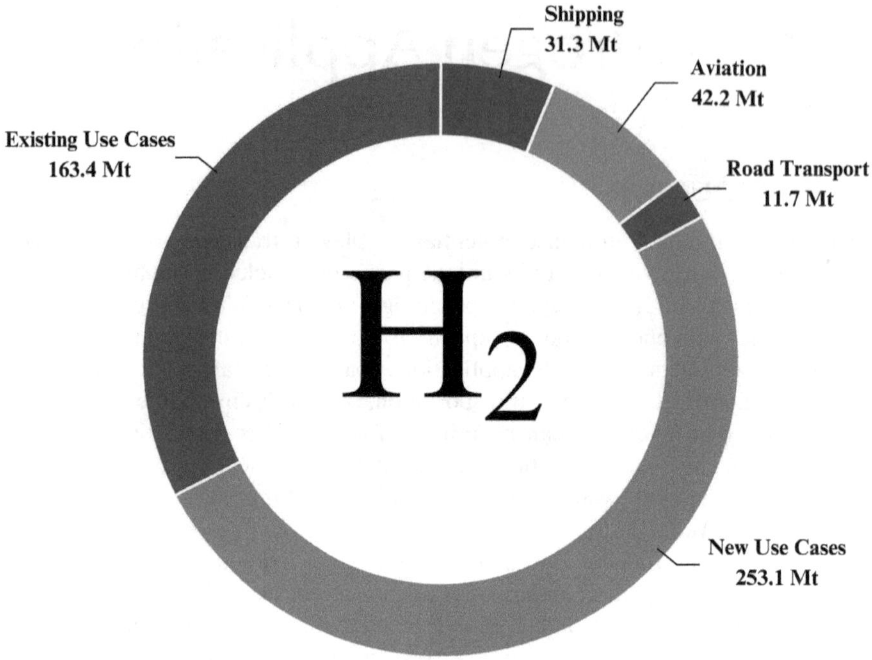

FIGURE 7.1 Global hydrogen demand on a sectoral basis by 2050. (From Bloomberg NEF [8].)

7.2 ELECTRICITY GENERATION

Electricity generation is among the largest contributors to global greenhouse gas (GHG) emissions due to the heavy reliance on fossil fuels such as coal, oil, and natural gas. In 2022, electricity generation resulted in around 14.6 Gt of GHG emissions, accounting for 27% of total global emissions throughout the year [9,10]. This trend has persisted for many years which has been prompting nations worldwide to transition toward renewable energy sources for electricity generation in efforts to mitigate the catastrophic impacts of climate change. Between the years 2014 and 2023, the global installed renewable energy capacity has increased from 1700 to 3870 gigawatt (GW), marking a 227% increase for this period with an average annual increase of around 9.58% [11]. Despite this substantial increase in renewable energy adoption worldwide, the International Energy Agency (IEA) emphasize that achieving net-zero emissions by 2050 requires increasing the global installed capacity of renewable energy systems to 11 GW by 2030, which cannot be fulfilled if the annual increase in renewable energy capacity remained at the current trend [12]. Therefore, the IEA has emphasized the crucial role of innovative concepts such as carbon capture, utilization, and storage along with green hydrogen technologies in supporting the transition toward net zero by 2050 through the decarbonization of numerous sectors worldwide [13].

Green hydrogen is considered a key player in clean electricity generation due to its ability to be fully produced from renewable energy resources. Further, hydrogen's versatility and flexibility enable seamless integration into various electricity generation technologies including fuel cells (FCs) and gas turbines (GTs), allowing for reduced carbon emissions through the provision of a reliable alternative to conventional fossil fuel-based power generation.

7.2.1 Electricity Generation via Fuel Cells

FCs are electrochemical devices that have the ability to convert chemical energy into electrical energy through a sequence of electrochemical reactions. These devices are considered as efficient and environmentally friendly sources of electricity and heat. A typical FC system consists of an FC stack integrated with several components known as balance of plant (BoP) components. These components encompass heat exchangers, compressors, power converters, hydrogen tanks, cooling systems, and other components, as shown in Figure 7.2 [14].

The hydrogen FC system efficiently converts the chemical energy of hydrogen and oxygen into electricity while producing water and heat due to the exothermic nature of the chemical reaction. As shown in Figure 7.2, hydrogen from the H_2 tank is regulated and sent into the FC stack to react with oxygen. On the other hand, oxygen is fed from the air via a compressor and then humidified in a humidifier to improve the efficiency of the electrochemical process. The FC stack is the main device responsible for converting the chemical energy of hydrogen and oxygen into useful electrical energy without resulting in any harmful emissions. This process produces water at an elevated temperature—in the form of steam in high-temperature FCs—which

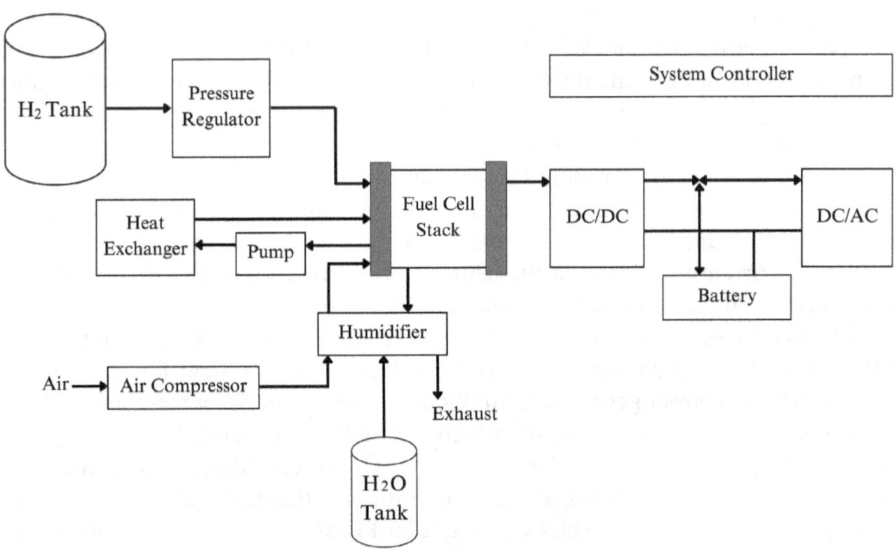

FIGURE 7.2 Fuel cell system with balance of plant components. (Adapted from Vichard Loïc et al. [14].)

can be utilized for heating purposes for many applications. This can be ensured through the integration of a heat exchanger connected to the output of the FC stack. After utilizing its high thermal energy, water can be sent back to the FC stack to act as a coolant, preventing the stack from overheating in some system arrangements. FCs produce direct current (DC) electricity, which can be directly stored in batteries after passing through DC/DC converters for stabilization. If coupled with an external application that requires an alternating current (AC) power source, a DC/AC inverter can be equipped to convert the DC electricity into AC, ensuring compatibility with grid-connected systems or other AC-powered devices. System controllers can be utilized to ensure that each process is implemented within optimal conditions for an enhanced overall efficiency [14].

The FC stack comprises several FCs connected in series to provide sufficient voltages based on the relevant application. FCs are mainly made of an anode, cathode, and an electrolyte membrane. The anode is a negative electrode where hydrogen gas is supplied, while the cathode is a positively charged electrode that is supplied with oxygen. Both the anode and the cathode are connected via an external electrical circuit, allowing electrons to pass through it to energize the load. Additionally, the electrolyte is a membrane that separates the anode and the cathode electrodes. It is crucial to prevent any direct contact between the hydrogen at the anode and the oxidant at the cathode while allowing the transfer of ions from one side to another to fulfil the electrochemical reaction [15].

FCs have been categorized into several types based on different factors such as the type of electrolyte, utilized fuel, operating temperature, efficiency, and applications, as detailed in Table 7.1 [16,17].

FCs are mainly categorized based on the integrated electrolyte, as shown in Table 7.1, which determines their operating temperature, efficiency, and fuel flexibility. In the case of polymer electrolyte fuel cell (PEMFCs), a polymeric ion exchange membrane is utilized as an electrolyte which allows for the rapid proton conduction from the anode to the cathode side while preventing electrons and reactant gases from crossing through. The flow of gases, protons, and electrons in the PEMFC system is illustrated in Figure 7.3 [18].

At the anode side, hydrogen gas (H_2) is split into protons (H^+) and electrons (e^-), typically through a platinum catalyst. The protons pass through the polymeric ion exchange membrane, while the electrons travel through an external circuit to deliver an electric current to a load. At the cathode side, oxygen gas (O_2) reacts with the protons (H^+) and electrons (e^-)—produced at the anode—to form water (H_2O) [18].

This electrochemical process occurs at a relatively low operating temperature (60°C–100°C) which enables fast startup, making PEMFCs ideal for applications that require rapid power generation. Additionally, the thin polymer membrane allows for compact and lightweight designs which makes PEMFCs suitable for transportation and mobile applications. Moreover, PEMFCs are considered among the most environmentally friendly types of FCs due to the fact that they only produce water as a byproduct. Further, the solid electrolyte in PEMFCs eliminates various issues that occur in FCs with liquid electrolytes. These include electrolyte loss, evaporation, and circulation losses which substantially reduce the overall efficiency of FC systems. Also, PEMFCs perform efficiently under variable or partial loading conditions,

TABLE 7.1
Characteristics of Several Types of Fuel Cells

Fuel Cell Type	Polymer Electrolyte Fuel Cell (PEMFC)	Alkaline Fuel Cell (AFC)	Phosphoric Acid Fuel Cell (PAFC)	Molten Carbonate Fuel Cell (MCFC)	Solid Oxide Fuel Cell (SOFC)	Direct Methanol Fuel Cell (DMFC)
Electrolyte	Polymeric ion exchange membrane	Alkaline electrolyte potassium hydroxide (KOH)	Liquid phosphoric acid (H_3PO_4) electrolyte	Molten carbonate salt mixture	Metallic oxide solid ceramic electrolyte	Solid polymer membrane
Anode Reaction	$H_2 \rightarrow 2H^+ + 2e^-$	$2H_2 + 4OH^- \rightarrow 4H_2O + 4e^-$	$2H_2 \rightarrow 4H^+ + 4e^-$	$H_2 + CO_3^{2-} \rightarrow H_2O + CO_2 + 2e^-$	$\frac{1}{2}O_2 + 2e^- \rightarrow O^{2-}$	$CH_3OH + H_2O \rightarrow CO_2 + 6H^+ + 6e^-$
Cathode Reaction	$\frac{1}{2}O_2 + 2H^+ + 2e^- \rightarrow H_2O$	$O_2 + 2H_2O + 4e^- \rightarrow 4OH^-$	$O_2 + 4H^+ + 4e^- \rightarrow 2H_2O$	$\frac{1}{2}O_2 + CO_2 + 2e^- \rightarrow CO_3^{2-}$	$O^{2-} + H_2 \rightarrow H_2O + 2e^-$	$\frac{3}{2}O_2 + 6H^+ + 6e^- \rightarrow 3H_2O$
Fuel Utilized	Pure H_2	Pure H_2	Pure H_2	H_2, CO, CH_4, other	H_2, CO, CH_4, other	CH_3OH
Operating Temperature (°C)	60–100	90–100	150–250	500–700	700–1000	60–200
Electrical Efficiency (%)	53–58	60	55	45–47	35–43	40
Applications	• Backup power. • Portable power. • Small distributed generation. • Specialty vehicle. • Transportation.	Military Space	Distributed generation	• Electric utility. • Large distributed generation.	• Auxiliary power. • Electric utility. • Large distributed generation.	Replace batteries in mobiles; computers and other portable devices
Advantages	Quick start-up and response	Has the highest electrical efficiency of all fuel cells	Cogeneration capability	Fuel flexibility	Cogeneration capability	Quick start-up time
Disadvantages	Sensitivity to contaminants	Works properly only with very pure gases	Slower start-up time	System complexity	High operating temperature	Lower efficiency

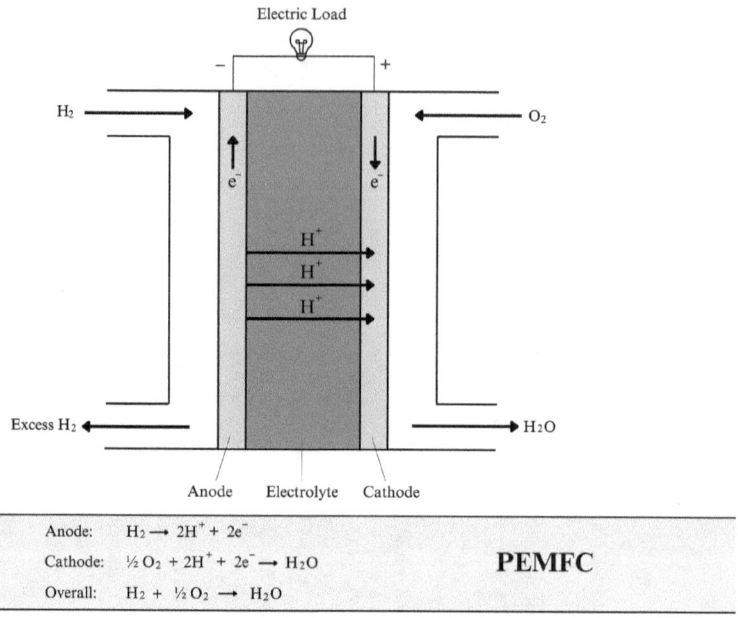

FIGURE 7.3 Standard operation of a PEMFC. (Adapted from Rezk et al. [18].)

supporting their position as a key solution for decarbonizing mobile applications which typically involve extreme variations in load demand over time [19–21].

However, PEMFCs still face some crucial challenges such as polymer membrane drying and degradation as well as the high platinum catalyst costs. Hence, effective water management strategies are being widely researched to overcome membrane dehydration in PEMFCs. Also, reinforced polymer membranes are being developed and enhanced to achieve better durability of PEMFCs. Moreover, platinum alloys and nonplatinum catalysts are being examined by PEMFC systems manufacturers with the aim to reduce the production costs of such highly efficient systems [22–24].

In alkaline fuel cells (AFCs), an alkaline electrolyte—typically a concentrated solution of potassium hydroxide (KOH) in water—is utilized to facilitate ion conduction from the cathode to the anode while preventing electrons and reactant gases from passing through. The flow of gases, hydroxide ions (OH^-), and electrons in the AFC system is illustrated in Figure 7.4 [25].

At the cathode side, oxygen gas (O_2) reacts with water (H_2O) and electrons (e^-) to form hydroxide ions (OH^-). These ions migrate through the alkaline electrolyte to the anode side where they react with hydrogen gas (H_2) to generate water (H_2O) and electrons (e^-). The released electrons travel through an external circuit to deliver electric power to a load [25].

AFCs operate at relatively low temperatures (60°C–90°C), which allows for high electrical efficiency and fast startup, making them suitable for applications that require immediate power output, similar to PEMFCs. The alkaline electrolyte enables AFCs to achieve high reaction kinetics while reducing the need for expensive noble metal

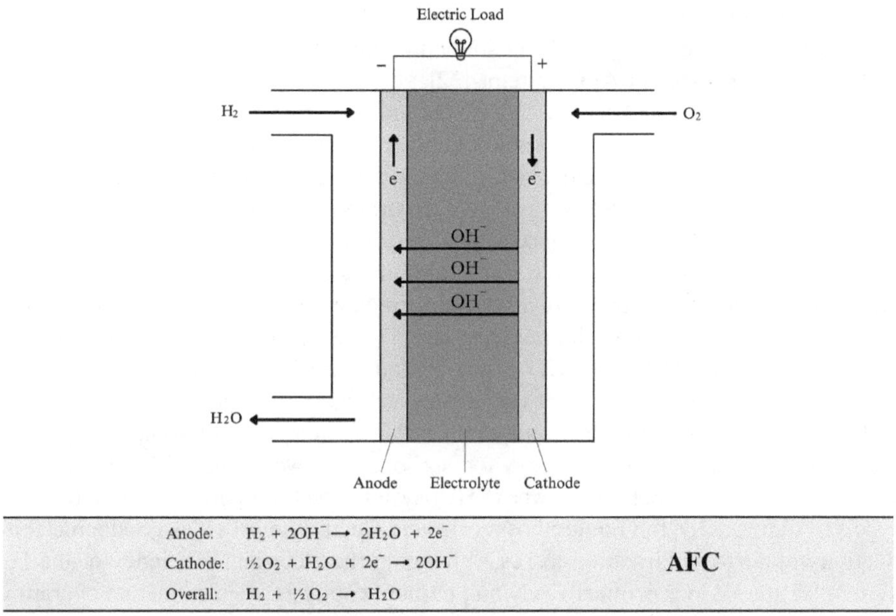

FIGURE 7.4 Standard operation of an AFC. (Adapted from Ferriday et al. [25].)

catalysts such as platinum. This crucial advantage makes AFCs an attractive option for certain stationery and aerospace applications. Moreover, AFCs are considered as the most efficient type of FCs due to their high electrochemical reaction rates and minimal polarization losses [26,27].

Despite these advantages, AFCs still face significant challenges that can hinder their performance. These challenges include the extreme sensitivity of the alkaline electrolyte to carbon dioxide as well as the potential water flooding or electrolyte dry-out. The alkaline electrolyte (KOH) in AFCs reacts with carbon dioxide (CO_2) from air or from impure hydrogen fuel intake, which leads to the formation of carbonate ions (CO_3^{2-}) deposits that accumulate in the electrolyte and can hugely reduce the ionic conductivity and degrade the performance of the AFC system over time. Further, AFCs are vulnerable to water flooding and electrolyte dry-out, both of which are related to improper water management. Water flooding occurs when excess water accumulates in the AFC system channels, reducing gas diffusion rates and accelerating electrode degradation. On the other hand, electrolyte dry-out occurs when too much water is lost from the electrolyte solution, leaving an unbalanced solution that decreases hydroxide ions (OH^-) conductivity and lowers the overall performance of the AFC system [28,29].

To address these challenges, researchers and engineers are thoroughly examining possible techniques to overcome the adverse impacts of carbonate deposition by exploring carbon-resistant alkaline electrolytes such as quaternary ammonium-based electrolytes as well as developing advanced CO_2 scrubbers to ensure high-purity oxygen and hydrogen supply to AFC systems [30,31]. Further, precise water management techniques are being carefully studied to limit the impacts of water flooding and electrolyte dry-out in AFC systems [32].

In phosphoric acid fuel cells (PAFCs), a liquid phosphoric acid (H_3PO_4) is utilized as the electrolyte, which facilitates ion conduction from the cathode to the anode while preventing electrons and reactant gases from crossing through. The flow of gases and electrons in the PAFC system is illustrated in Figure 7.5 [33].

At the anode side, hydrogen gas (H_2) is split into protons (H^+) and electrons (e^-) through a platinum-based catalyst. The protons pass through the phosphoric acid electrolyte while the electrons travel through an external circuit to generate an electric current. At the cathode side, oxygen gas (O_2) reacts with the protons and electrons to form water (H_2O). PAFCs operate at moderate temperatures (150°C–200°C), which improves tolerance to fuel impurities such as carbon monoxide (CO), especially when compared with low-temperature FCs such as PEMFCs and AFCs. This makes PAFCs particularly suitable for stationary power generation applications including combined heat and power (CHP) systems where operational reliability is crucial. Additionally, the liquid phosphoric acid electrolyte provides good ionic conductivity and low polarization losses, enhancing the electrical efficiency of the FC system. While PAFCs primarily rely on platinum catalysts, their moderate operating temperature allows for the potential development of alternative non-precious metal catalysts. PAFCs have lower power density than PEMFCs, which limits their utilization in transport applications which require light and compact FC systems design. However, PAFCs remain a viable option for large-scale stationary energy applications due to their relatively high efficiency and ability to utilize waste heat [34].

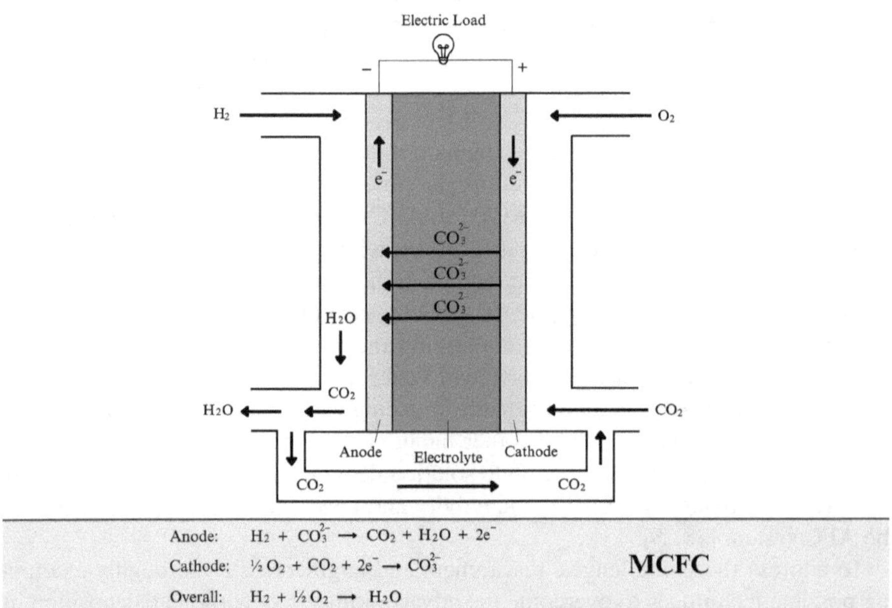

FIGURE 7.5 Standard operation of a PAFC. (Adapted from Abderezzak [33].)

PAFCs face some technical challenges that can impact their performance. One major challenge is the corrosiveness of the liquid phosphoric acid electrolyte, which can lead to material degradation over time. This requires the utilization of highly durable FC system components to ensure the longevity of PAFC systems. Further, the lower power density compared with other types of FCs limits the applications of PAFCs to stationary applications only. Moreover, PAFCs mainly rely on platinum-based catalysts, which increases their initial and operational costs due to degradation over extended use. Also, PAFCs operate at moderate temperatures, which makes their startup not as rapid as PEMFCs and AFCs while requiring higher cooling requirements to ensure the efficiency of the system. Last but not least, the liquid electrolyte utilized in PAFCs is vulnerable to being evaporated over long periods of time, requiring extra efforts to maintain their long-term reliability [35].

These challenges are being addressed by researchers and engineers to enhance the performance of PAFCs. For instance, several acid-resistant materials are being developed and explored to mitigate the corrosion-related issues in PAFCs. Further, efforts are being put to enhance the porosity of electrodes in PAFCs to improve the power density of these systems, enabling their use in nonstationary applications. Similar to the case of PEMFCs, extensive research is being carried out to enhance the economic competitiveness of such types of FCs through the development of non-precious metal and alloy catalysts to replace platinum-based catalysts. Furthermore, advanced thermal management systems and preheating mechanisms are being introduced to PAFCs to improve startup times while ensuring a high cooling efficiency. Additionally, acid retention materials, electrolyte recirculation systems, and hydrophilic coatings are being used to minimize phosphoric acid evaporation, ensuring sustained performance over longer periods of time [36–38].

Molten carbonate fuel cells (MCFCs) utilize a liquid molten carbonate salt mixture as the electrolyte, which is typically made of lithium carbonate (Li_2CO_3), potassium carbonate (K_2CO_3), or sodium carbonate (Na_2CO_3). The molten carbonate catalyst facilitates ion conduction from the cathode to the anode while preventing electrons and reactant gases from directly crossing through. The flow of gases and electrons in the MCFC system is illustrated in Figure 7.6 [39].

At the anode side, hydrogen gas (H_2) reacts with carbonate ions (CO_3^{2-}) from the electrolyte to form water (H_2O), carbon dioxide (CO_2), and electrons (e^-). These electrons travel through an external circuit to generate an electric current while carbon dioxide is recirculated to the cathode. At the cathode side, oxygen gas (O_2) from air reacts with carbon dioxide and electrons to regenerate carbonate ions (CO_3^{2-}). These ions migrate back through the electrolyte to the anode, completing the electrochemical process. MCFCs operate at high temperatures (600°C–700°C), allowing for internal fuel reforming and eliminating the need for expensive platinum catalysts. Instead, MCFCs utilize nickel-based electrodes, which have much lower costs than platinum-based ones. Further, the high operating temperature enhances reaction kinetics which makes MCFCs highly efficient and suitable for stationary large-scale power generation. Moreover, MCFCs can be efficiently utilized for applications in which both electricity and heat are required, also known as, CHP systems. Additionally, MCFCs can directly use hydrocarbon fuels such as natural gas and biogas, reducing the need for complex external reforming systems to provide pure hydrogen at the anode side [39].

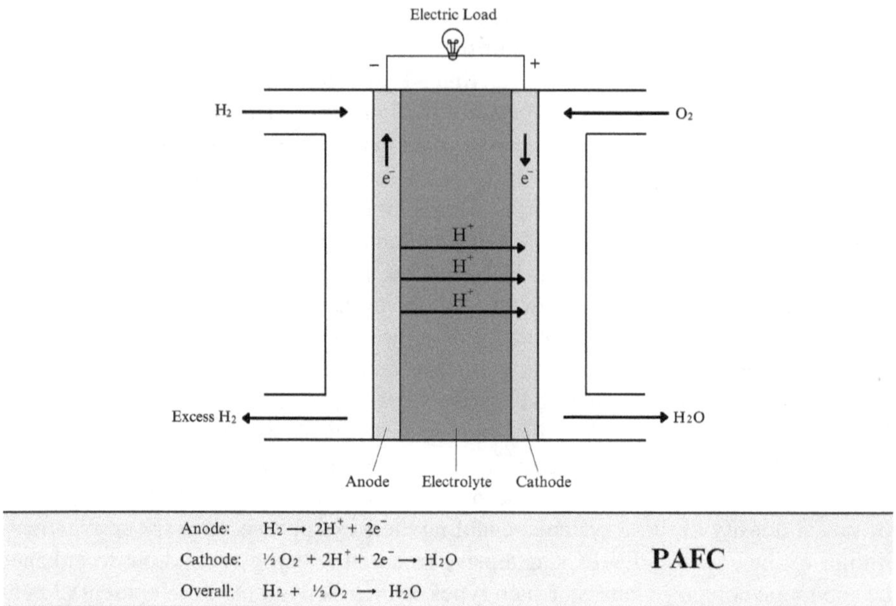

FIGURE 7.6 Standard operation of an MCFC. (Adapted from Barckholtz et al. [39].)

However, the high operating temperature of MCFCs necessitates advanced thermal management systems, which lead to some material durability challenges and require highly corrosion-resistant materials for extended operational lifespan. Further, the liquid molten carbonate electrolyte is susceptible to gradual evaporation and electrolyte redistribution, which can impact long-term stability. Moreover, MCFCs rely on nickel-based catalysts, which are prone to corrosion and sintering under prolonged high-temperature exposure. Another challenge is the recirculation of CO_2 at the cathode, which is necessary for continuous operation but can lead to complex system designs and increased operational costs. Last but not least, the high-temperature operation of MCFCs correlates to longer startup times, making them unsuitable for applications that require rapid power generation. This, along with the fact that MCFCs do not offer compact and light designs, prevents such systems from being utilized for mobile applications [40–42].

Researchers and engineers are actively working to address the challenges facing the performance and reliability of MCFC systems. For instance, temperature-resistant materials including advanced nickel alloys and ceramic coatings are being developed to improve the durability and longevity of MCFCs. Further, to mitigate electrolyte evaporation, electrolyte retention techniques such as porous electrolyte matrices and alternative carbonate compositions are being investigated for use in MCFCs. Additionally, research into optimized CO_2 recirculation mechanisms is helping to simplify MCFCs system designs and improving their operational efficiency. Moreover, advanced thermal management systems and startup optimization strategies are also being integrated into MCFC designs to shorten startup times and improve their flexibility to a wider range of applications [43–45].

Solid oxide fuel cells (SOFCs) utilize an electrolyte that is composed of a solid ceramic material which conducts oxygen ions (O^{2-}) at high temperatures (600°C–1000°C). This electrolyte is typically made of yttria-stabilized zirconia (YSZ), gadolinium-doped ceria (GDC), samarium-doped ceria (SDC), or perovskite-structured materials. YSZ is the most common electrolyte utilized in SOFCs due to its superior ionic conductivity at elevated temperatures [46]. Figure 7.7 illustrates the flow of gases and electrons in SOFCs [47].

At the cathode side, oxygen gas (O_2) from the air gains electrons and is converted into oxygen ions (O^{2-}). These ions migrate through the solid electrolyte to the anode in where they react with the fuel (H_2, CO, or hydrocarbons) to produce water (H_2O), carbon dioxide (CO_2), and electrons (e^-). Electrons travel through an external circuit to generate electricity while the oxygen ions continuously migrate through the electrolyte, completing the electrochemical cycle [47].

SOFCs operate at high temperatures (600°C–1000°C), which enhances reaction kinetics and enables internal fuel reforming, allowing for the direct use of hydrocarbon fuels such as natural gas, biogas, and syngas without an external reformer, similar to the case of MCFCs. Unlike PEMFCs and PAFCs, SOFCs do not require expensive platinum catalysts and instead utilize ceramic-based and nickel-based electrodes. Their high-temperature operation results in high electrical efficiency (50%–65%), which can be increased to 85% in CHP applications. Altogether, the relatively high efficiency along with fuel flexibility promotes SOFCs as a very suitable candidate for large-scale stationary power generation and industrial applications [48].

However, SOFCs face several technical challenges that can impact their performance and commercial viability. For example, their high operating temperature

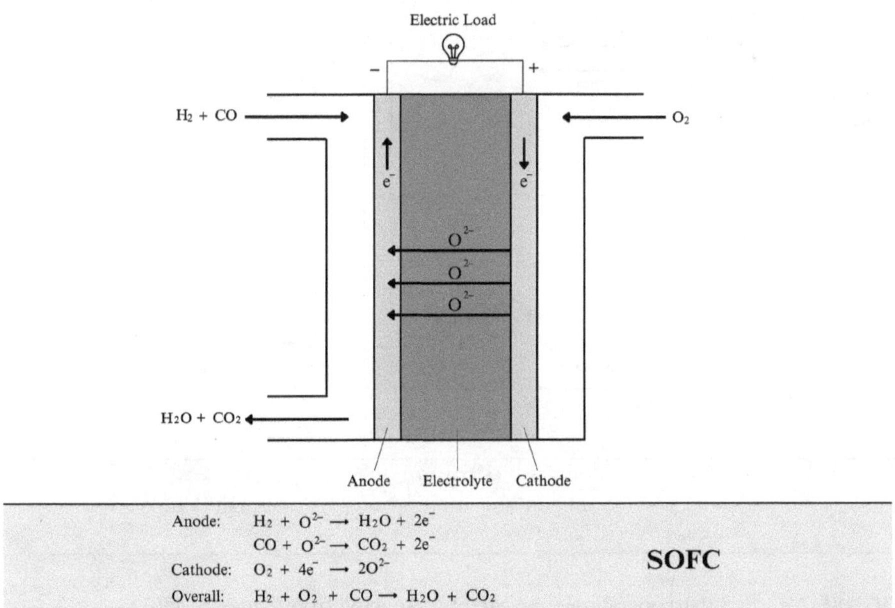

FIGURE 7.7 Standard operation of a SOFC. (Adapted from Orlando Corigliano et al. [47].)

necessitates advanced thermal management and temperature-resistant materials, leading to long startup times and mechanical stress-related durability challenges. Further, unlike MCFCs, SOFCs use a solid ceramic electrolyte which must maintain high ionic conductivity while preventing cracks and mechanical failures due to long-term thermal cycling. Additionally, nickel-based anodes used in SOFCs are susceptible to carbon deposition when exposed to hydrocarbon fuels, which can potentially reduce the overall efficiency [49,50].

To address these challenges, efforts are being made to develop alternative anode materials such as perovskite-based electrodes to prevent carbon deposition and improve durability of SOFCs. Further, graded material designs and flexible interconnects are being explored to enhance the structural integrity of SOFCs under long-term thermal cycling. Additionally, advanced thermal management strategies and optimized startup procedures are being implemented to reduce startup time and improve SOFCs system responsiveness [51–53].

Direct methanol fuel cells (DMFCs) are a type of PEMFCs that can utilize methanol (CH_3OH) as a direct fuel source instead of pure hydrogen. These FCs operate at relatively low temperatures (50°C–120°C) and do not require high-pressure hydrogen storage as liquid methanol is easier to transport and handle [54]. These characteristics make DMFCs highly attractive for applications that require fast and lightweight power generation systems, such as military equipment, portable electronics, and mobile back-up generators [55]. The typical operation of DMFCs is shown in Figure 7.8 [56].

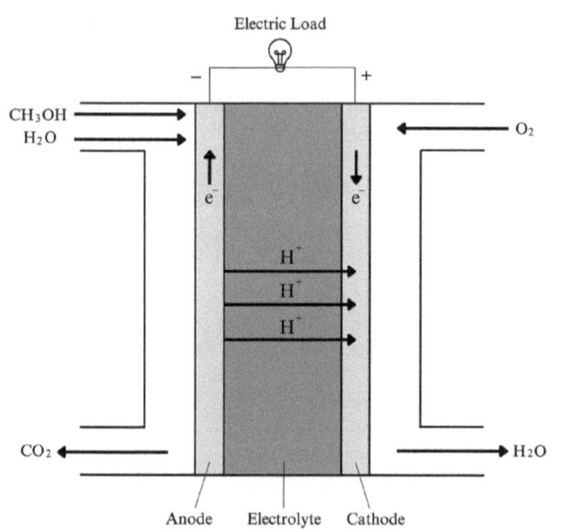

Anode: $CH_3OH + H_2O \rightarrow CO_2 + 6H^+ + 6e^-$

Cathode: $3(\frac{1}{2}O_2) + 6H^+ + 6e^- \rightarrow 3H_2O$

Overall: $CH_3OH + H_2O + 3(\frac{1}{2}O_2) \rightarrow CO_2 + 3H_2O$

DMFC

FIGURE 7.8 Standard operation of an MDFC. (Adapted from Zuo et al. [56].)

At the anode side, methanol reacts with water to produce carbon dioxide (CO_2), protons (H^+), and electrons (e^-). The protons migrate through the proton-conducting polymer electrolyte membrane toward the cathode, while the electrons travel through an external circuit to generate electricity. At the cathode, oxygen from air reacts with the protons and electrons produced at the anode to form water. The continuous transport of protons through the membrane and electrons through the external circuit maintains the electrochemical cycle in DMFCs [56].

DMFCs face several challenges that limit their efficiency and widespread adoption. One major issue is methanol crossover, where methanol permeates through the polymer electrolyte membrane to the cathode, leading to mixed potential losses and reduced overall efficiency. This phenomenon also contributes to extreme cathode degradation over time. Similar to PEMFCs and PAFCs, DMFCs require platinum-based catalysts, especially for the anode, which increases system costs and makes the FC more susceptible to catalyst poisoning from CO byproducts. Moreover, the relatively slow kinetics of methanol oxidation at the anode results in lower power densities compared with pure hydrogen FCs. Additionally, carbon dioxide generation at the anode necessitates an effective gas management system to prevent performance degradation due to CO_2 accumulation. Due to the operation of DMFCs at low temperatures, these FCs lack the ability to process alternative fuels beyond methanol, which limits their flexibility in fuel choice [57,58].

To address these challenges, researchers and engineers are working on advanced polymer membranes to be utilized in DMFCs, such as sulfonated poly ether ether ketone (SPEEK) and several composite membranes to minimize methanol crossover and improve system's efficiency. Furthermore, alternative anode catalysts such as platinum–ruthenium (Pt–Ru) alloys and other nonprecious metal catalysts are being explored to enhance reaction kinetics while reducing catalyst costs. Additionally, significant efforts are being put to improve water and gas management systems to enhance the durability and operational stability of DMFCs [59,60].

Altogether, FCs come in various types, with each distinguished by its electrolyte composition, operating temperature, efficiency, and application suitability among other differences. PEMFCs and DMFCs operate at low temperatures and offer fast startup times, making them ideal for transportation, portable electronics, and backup power applications. AFCs provide high efficiency and rapid reaction kinetics but are sensitive to CO_2 contamination, limiting their widespread use. PAFCs operate at moderate temperatures, making them well-suited for stationary power applications, particularly in CHP applications. On the other hand, high-temperature FCs, including MCFCs and SOFCs excel in large-scale power generation due to their high efficiency and ability to utilize hydrocarbon fuels directly. While each FC type has its own advantages and challenges, continuous global research is driving key innovations in enhancing the materials used in catalysts and electrolytes while also developing system designs to enhance the performance, durability, and cost-effectiveness of FCs.

Considering that a PEMFC involves the chemical reaction shown in equations 7.1 and 7.2:

Anode reaction:

$$H_2 \rightarrow 2H^+ + 2e^- \tag{7.1}$$

Cathode reaction:

$$\frac{1}{2}O_2 + 2H^+ + 2e^- \rightarrow H_2O \tag{7.2}$$

As detailed earlier, on the anode side, hydrogen (H_2) molecules are oxidized by losing electrons (e^-). The released electrons flow through the external circuit from the anode to the cathode while protons (H^+) pass through the porous electrolyte membrane. Concurrently, on the cathode side, oxygen (O_2) combines with protons and electrons to produce water molecules. Half-oxidation and reduction reactions occur at the triple phase boundary region between the gas phase, electrode, and electrolyte layers [61]. Equation 7.3 represents the overall reaction, resulting in the production of water, electricity, and heat [18].

Overall reaction:

$$H_2 + \frac{1}{2}O_2 \rightarrow H_2O + \text{Electricity} + \text{Heat} \tag{7.3}$$

The performance of any FC system indicates the effectiveness and efficiency of the system in converting chemical energy into useful electrical energy. It is typically evaluated based on several parameters, including current–voltage characteristics, efficiency, amount of fuel the cell utilizes, and the air-to-fuel ratio [62].

The operating voltage of the FC, which represents the current–voltage characteristics, is typically specified using the Nernst equation. This equation can be given as follows [63]:

$$E_{cell} = E^0_{rev} - \frac{RT}{2F}\left[\ln\left(\frac{P_{H_2O}}{P_{H_2}P^{1/2}_{O_2}}\right)\right] \tag{7.4}$$

$$E^0_{rev} = \frac{\Delta G^0_R}{nF} \tag{7.5}$$

where E^0_{rev} is standard-state reversible cell potential which equals 1.23 volts (V) and R is the universal gas constant which equals 8.314 in units of Joule per mole per Kelvin (J/mol K). T is cell temperature in Kelvin (K), F is Faraday constant which equals 96485.3 Coulombs per thousand moles of electrons (C/kmol), P_x is the partial pressure in the bulk flow, ΔG^0_R is Gibbs free energy of reaction, and n is number of moles (mol) of electronics moved during reaction.

Reaction quotient refers to the ratio of partial pressure of elements to reactants in a chemical reaction at a specific time. Accordingly, the reaction quotient of the reaction can be formed based on either the partial pressure shown in equation 7.6 or the concentration illustrated in equation 7.7.

$$Q = \frac{P_{H_2O}}{P_{H_2}P^{1/2}_{O_2}} \tag{7.6}$$

$$Q = \frac{H_2O}{[H_2][O_2]^{1/2}} \tag{7.7}$$

Hence, Nernst equation can be written as [64]:

$$E_{cell} = E_{rev}^{0} - \frac{RT}{2F}\ln(Q) \tag{7.8}$$

The anode is negatively charged due to the oxidation reaction, while the cathode is positively charged due to the reduction reaction. As long as there is no connection between electrodes, a potential difference appears between both electrodes. Therefore, the movement of ions occurs through the electrolyte until a chemical equilibrium is established between the anode and the cathode. The potential difference between electrodes represents the cell's open voltage, called "Nernst voltage," and it is the highest voltage of the FC [65].

Figure 7.9 illustrates the FC's I–V characteristics, including the over-potential losses [62]. The current density parameter controls the cell voltage. As much as the current is drawn from the cell, the voltage drops from its maximum—open circuit value—within limits. Usually, FCs have a minimum operating voltage of around (0.68–0.7 V). Any voltage below the minimum value is not preferred since it causes the anode to be oxidized by the oxygen ions flowing through the electrolyte. Therefore, to prevent anode oxidizing, the voltage should be controlled by reducing the current drawn from the FC or increasing the fuel fed into the cell.

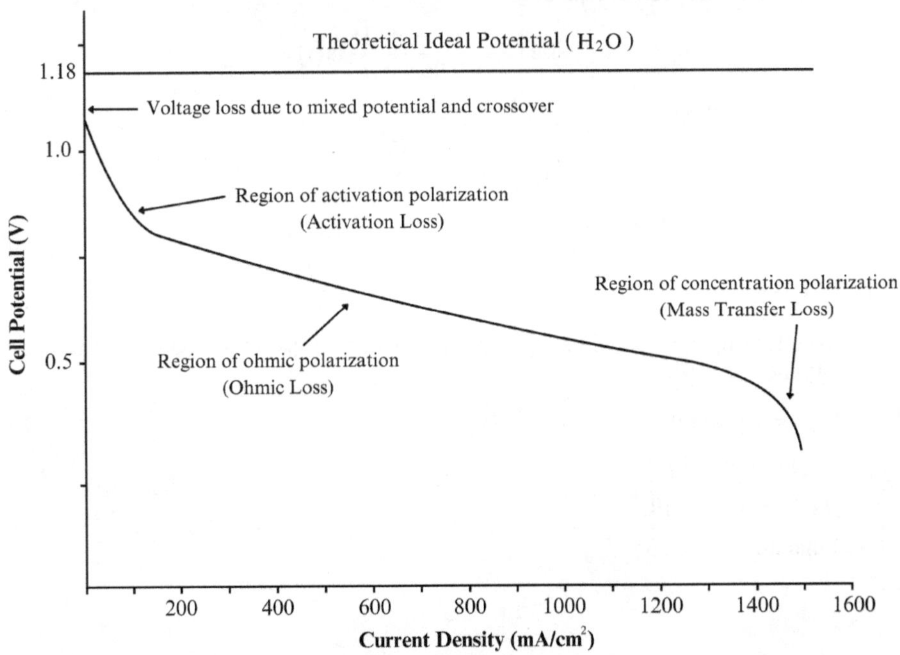

FIGURE 7.9 Voltage–current characteristic of FCs. (Adapted from Derbeli et al. [62].)

Example 7.1: Cell Potential for Hydrogen FC

Calculate the cell potential (E_{cell}) for hydrogen FC at 25°C with concentrations of H_2 and O_2 of 1 mol/L.

Solution:

E_{cell} can be calculated using the Nernst equation as follows:

$$E_{cell} = E_{rev}^0 - \frac{RT}{2F}\ln(Q)$$

The standard cell potential of this reaction is 1.23 V.

The reaction quotient of this reaction: $H_2 + \frac{1}{2}O_2 \rightarrow H_2O$ can be determined as follows:

$$Q = \frac{P_{H_2O}}{P_{H_2}P_{O_2}^{1/2}}$$

$$Q = \frac{1}{(1)(1)^{1/2}}$$

The temperature in Kelvin:

$$25 + 273.15 = 298.15\ \text{K}$$

Substituting in the Nernst Equation:

$$E_{cell} = 1.23 - \frac{8.314 \times 298.15}{2 \times 96{,}485.3}\ln(1)$$

Since ln(1) = 0, then:

$$E_{cell} = E_{rev}^0 = 1.23\ \text{V}$$

EXERCISE 7.1

Resolve Example 7.1 considering that the concentrations of H_2 and O_2 and H_2O as follows:

$(H_2) = 1.0$ mol/L

$(O_2) = 0.5$ mol/L

$(H_2O) = 0.1$ mol/L

Final answer: 1.255 V

Altogether, hydrogen can play a crucial role in decarbonizing electricity generation through FC technologies. Green hydrogen is a clean and efficient energy carrier that can be utilized as the primary fuel source in various FCs. In FCs, hydrogen undergoes an electrochemical reaction with oxygen to produce electricity while emitting only water and heat as byproducts. This makes hydrogen FCs a highly attractive solution for sustainable energy applications, ranging from portable electronics and transportation to large-scale stationary power systems. Unlike conventional combustion-based power generation, FCs offer higher efficiency, quieter operation, and significantly reduced environmental impact. As research continues to address challenges such as hydrogen storage, infrastructure development, and cost reduction, hydrogen FCs are expected to play a pivotal role in achieving a low-carbon and energy-efficient electricity generation worldwide.

7.2.2 Electricity Generation via Gas Turbines

GT are commonly employed to convert fuel energy into mechanical energy that can be utilized to produce electrical energy. GTs are currently considered as the heart of modern power plants due to their widespread adoption and flexibility to be integrated within various power generation systems. The principal operation of GT systems relies on Brayton thermodynamic cycle. Brayton cycle, like any other thermodynamic cycle, is a theoretical representation of the energy conversion process in a GT system. This cycle describes an idealized sequence of thermodynamic transformations that aim to ultimately convert heat energy into mechanical work. It provides a framework for analyzing the performance, efficiency, and limitations of real GTs by modeling the compression, heat addition, expansion, and heat rejection processes under idealized conditions. The Brayton cycle is composed of four main stages, including isentropic compression, isobaric heat addition, isentropic expansion, and isobaric heat rejection—where the term isentropic refers to constant entropy and the term isobaric refers to constant pressure—[66]. Figures 7.10 and 7.11 illustrate a standard Brayton cycle through a pressure–volume (P–V) diagram and a temperature–entropy (T–S) diagram, respectively [66].

In practical applications, deviations from the ideal Brayton cycle occur due to factors such as irreversibility, pressure losses, heat dissipation, and nonideal gas behavior. These deviations impact the actual efficiency and performance of GTs, which is limited to around (35%–45%) in simple GT systems and (55%–65%) in combined cycles [67]. Currently, the most efficient GT system worldwide is located in Japan's Chubu Electric Nishi-Nagoya power plant Block-1. The power plant utilizes the General Electric (GE) 7HA GT, which holds the highest efficiency among all GTs ever produced. Due to the utilization of GE 7HA and other key technologies for enhancing the overall efficiency of the power plant, the Nishi-Nagoya power plant has achieved an overall efficiency of 63.08%, making it the most efficient GT-based power plant worldwide [68].

The operation of actual GTs consists of four main nonideal stages: compression, combustion, expansion, and exhaust. In the first stage, the compressors compress the incoming air at high pressure. After that, the compressed air is mixed with the fuel in a combustion chamber. The resulting high pressure and temperature gases pass through turbines, resulting in the production of mechanical energy used to turn

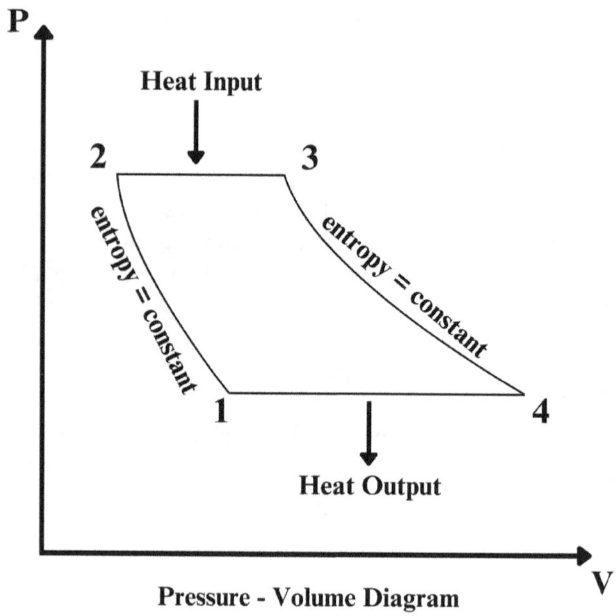

FIGURE 7.10 P–V diagram for a standard Brayton cycle. (Adapted from Yang et al. [66].)

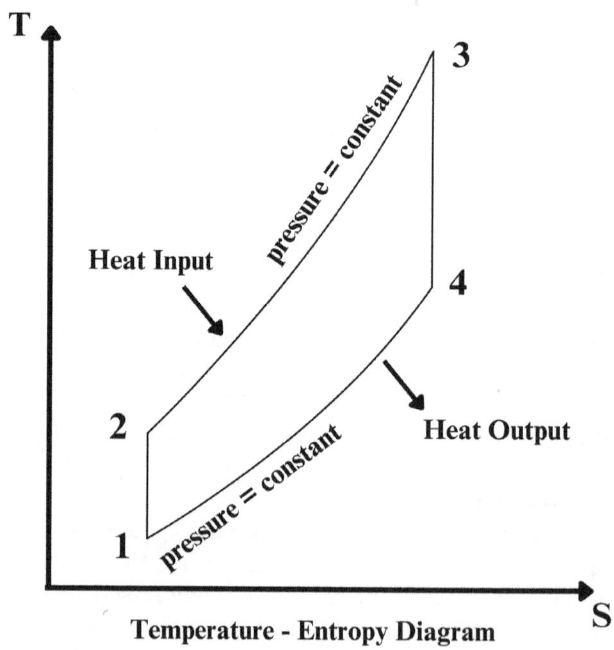

FIGURE 7.11 T–S diagram for a standard Brayton cycle. (Adapted from Yang et al. [66].)

generators and produce electrical energy. The exhaust gases from the turbine are then driven out into the atmosphere [69]. Figure 7.12 illustrates a standard GT power plant.

Conventional GTs rely on fossil fuels such as natural gas, which leads to the emission of significant amounts of carbon dioxide during the combustion stage. Recently, GTs have proven their capability to partially or fully utilize green hydrogen as a fuel, which can significantly reduce CO_2 emissions. Emissions reduction via green hydrogen utilization in GTs can be achieved through two major technologies, which are hydrogen gas turbines (HGTs) and hydrogen-enriched natural gas turbines (HENGTs) [70,71].

HGTs are advanced GTs specifically designed or retrofitted to operate using pure hydrogen as a fuel source. Unlike conventional GTs that fully rely on natural gas, HGTs combust pure hydrogen to generate power which results in a substantial reduction in CO_2 emissions. Although hydrogen utilization in GTs brought huge advantages, especially from the environmental perspective, various challenges have to be considered through the comparison between hydrogen and other hydrocarbon fuels. These challenges include the heating value factor. Hydrogen's lower heating value (LHV) is defined as 10.8 Megajoules per normal cubic meters (MJ/Nm^3) and 120 Megajoules per kilogram (MJ/kg), while the LHV of methane is 35.8 MJ/Nm^3 and 50 MJ/kg. Hence, in terms of mass, hydrogen exhibits twice the energy density of methane. However, hydrogen has an energy density that is approximately one-third that of methane when measured by volume. Accordingly, a three-time increase in hydrogen flow is needed to deliver an equivalent energy input compared with methane.

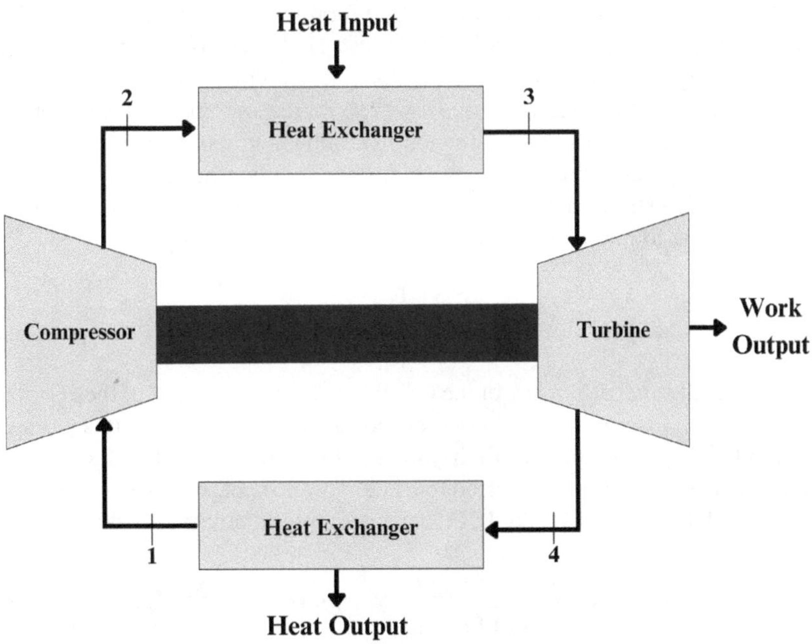

FIGURE 7.12 Standard GT power plant. (Adapted from Hai et al. [69].)

Therefore, the operation of GTs with 100% hydrogen requires specific configurations in the system to accommodate the flow rates of hydrogen [72].

The flame speed is another challenge for HGTs, which can be defined as the rate at which the unburned gases spread within the flame. Hydrogen exhibits a flame speed that is significantly faster than other hydrocarbon fuels. In the context of GTs, flame speed is critical in assessing the potential of flame spreading between the combustion zone into the premixing zone. Further, hydrogen flames have low luminosity features, which is a serious challenge to visually detecting these flames. Hence, it is necessary to adjust the flame detection system to consider this feature. Moreover, hydrogen can diffuse through seals that are considered airtight to other gases. Furthermore, hydrogen has a flammability limit of 4% (in air), while methane has a flammability limit of 5%, making hydrogen more flammable than methane. Generally, the challenges of the incorporation of hydrogen into HGTs can be controlled through various modifications in the GT systems. The proper modifications will facilitate the integration of hydrogen as fuel in GTs, which supports sustainable solutions and contributes to the transition towards green energy systems [72].

HENGTs represent a transitional approach to reducing CO_2 emissions in power generation by blending hydrogen with natural gas in existing GTs. Unlike HGTs, which operate solely on hydrogen, HENGTs use a mixture of natural gas and hydrogen. The hydrogen content in natural gas varies based on the specific turbine capabilities and the emissions control requirements. Typically, conventional GTs can withstand hydrogen blending within natural gas of around 15%, but ongoing research shows a potential for developing GTs that can withstand hydrogen blending of more than 50%. The incorporation of hydrogen into the fuel mixture in GTs enhances the combustion efficiency and reduces the carbon footprint of GT systems, as hydrogen does not contribute to CO_2 emissions. Moreover, HENGTs offer a practical solution for integrating hydrogen into the energy sector without requiring major modifications to existing natural gas infrastructure. However, increasing the hydrogen content in HENGTs necessitates careful adjustments in combustion parameters due to hydrogen's high flame speed, low luminosity, diffusion, and flammability [73].

In order to determine the efficiency, the power output, and the required power input of a GT, the following equation applies [74]:

$$\eta = \frac{P_{out}}{\dot{m}_{fuel} \times LHV_{fuel}} \tag{7.9}$$

Where η is the thermal efficiency of the GT, P_{out} is the power output of the GT in units of watts (W), $\dot{m}_{fuel}$ is the mass flow rate of the fuel in units of kilograms per second (kg/s), and LHV_{fuel} is the fuel's LHV in joules per kg (J/kg). Note that the denominator in the previous equation is referred to as the power input. In cases where a mixed fuel is utilized to power a GT, the LHV can be simply calculated as follows [74]:

$$LHV_{blend} = LHV_{fuel\ 1} \times \frac{\text{Mass of fuel 1}}{\text{Mass of fuel mix}}\% + LHV_{fuel\ 2} \times \frac{\text{Mass of fuel 2}}{\text{Mass of fuel mix}}\% \tag{7.10}$$

The LHV is the net amount of energy available from combustion, not including the latent heat of vaporization of water formed from hydrogen in the fuel. This value is a crucial characteristic to determine the viability of a fuel when used in combustion cycles. For hydrogen, the LHV equals around 120 MJ/kg, while for natural gas it equals 50 MJ/kg, considering the case of pure methane [74].

Typically, combustion processes are controlled in terms of volumetric flow rate rather than mass flow rate. Hence, it is crucial to be able to calculate the volumetric flow rate of a fuel given its mass flow rate. The volumetric flow rate of any fuel can be calculated when given its mass flow rate and density as follows [74]:

$$\dot{V}_{fuel} = \frac{\dot{m}_{fuel}}{\rho_{fuel}} \tag{7.11}$$

where $\dot{V}_{fuel}$ is the flow rate in terms of cubic meters per second (m^3/s) and ρ_{fuel} is the density of the fuel in kilograms per cubic meters (kg/m^3). It is worth mentioning that hydrogen's density at normal pressure and temperature (NTP)—at a temperature of 0°C and a pressure of 1 atm—equals 0.0899 kg/m^3, while methane has a density of 0.716 kg/m^3 at NTP.

Example 7.2: Hydrogen Utilization in Gas Turbines

Assuming a 50 MW GT with a fixed thermal efficiency of 38% is to be fed with pure methane, pure hydrogen, and a mix of 15% hydrogen and 85% methane by mass. Calculate the mass flow rate and the volumetric flow rate of each fuel mix to achieve the desired output. Recall that the LHV of hydrogen is 120 and 50 MJ/kg for methane, while the density of hydrogen is 0.0899 kg/m^3 and the density of methane is 0.716 kg/m^3.

Solution:

Since the efficiency and the power output of the GT are assumed to be fixed under all types of fuels, the required power input can be directly calculated from equation 7.9 as follows:

$$\eta = \frac{P_{out}}{\dot{m}_{fuel} \times LHV_{fuel}}$$

Required power output:

$$\dot{m}_{fuel} \times LHV_{fuel} = \frac{P_{out}}{\eta}$$

$$\dot{m}_{fuel} \times LHV_{fuel} = \frac{50}{0.38} = 131.57 \text{ MW}$$

The required mass flow rate of pure methane can then be calculated as follows:

$$\dot{m}_{methane} = \frac{131.57}{LHV_{methane}}$$

$$\dot{m}_{methane} = \frac{131.57}{50} = 2.63 \text{ kg/s}$$

Similarly, the required mass flow rate of pure hydrogen can then be calculated as follows:

$$\dot{m}_{hydrogen} = \frac{131.57}{LHV_{hydrogen}}$$

$$\dot{m}_{hydrogen} = \frac{131.57}{120} = 1.09 \text{ kg/s}$$

As shown in these results, the required mass flow rate of hydrogen is much smaller than that of methane due to the higher LHV of hydrogen on a mass basis.

For the mixed fuel of 85% methane and 15% hydrogen, the LHV can be calculated using equation 7.10 as follows:

$$LHV_{blend} = LHV_{methane} \times \frac{\text{Mass of methane}}{\text{Mass of fuel blend}} \%$$

$$+ LHV_{hydrogen} \times \frac{\text{Mass of hydrogen}}{\text{Mass of fuel blend}} \%$$

$$LHV_{blend} = 50 \times 85\% + 120 \times 15\% = 60.5 \text{ MJ/kg}$$

Therefore, the required mass flow rate of the mixed fuel can be calculated as follows:

$$\dot{m}_{blend} = \frac{131.57}{LHV_{blend}}$$

$$\dot{m}_{blend} = \frac{131.57}{60.5} = 2.17 \text{ kg/s}$$

Hence, it can be concluded that blending hydrogen with methane in GTs can reduce the required mass flow rate if only methane is utilized as a fuel. However, in order to grasp a better understanding of the volumetric flow rate for each fuel type, the following conversion step (equation 7.11) shall be applied:

$$\dot{V}_{methane} = \frac{\dot{m}_{methane}}{\rho_{methane}}$$

$$\dot{V}_{methane} = \frac{2.63}{0.716} = 3.67 \text{ m}^3\text{/s}$$

Also,

$$\dot{V}_{hydrogen} = \frac{\dot{m}_{hydrogen}}{\rho_{hydrogen}}$$

$$\dot{V}_{hydrogen} = \frac{1.09}{0.0899} = 12.12 \text{ m}^3\text{/s}$$

As the results imply, the required volumetric flow rate of hydrogen is roughly 3.3 times larger than that of methane to achieve the same power output for a fixed efficiency GT. Hence, comparing the viability of fuels based on the required mass flow rate can be misleading in many cases, while comparison based on the volumetric flow rate provides more realistic insights.

In order to calculate the volumetric flow rate for the blend fuel, the mass flow rate for hydrogen and methane within the blend must first be obtained.

$$\dot{m}_{hydrogen,blend} = \dot{m}_{blend} \times 0.15$$

$$\dot{m}_{hydrogen,blend} = 2.17 \times 0.15 = 0.325 \text{ kg/s}$$

$$\dot{m}_{methane,blend} = \dot{m}_{blend} \times 0.85$$

$$\dot{m}_{methane,blend} = 2.17 \times 0.85 = 1.844 \text{ kg/s}$$

The volumetric flow rate of each of hydrogen and methane within the blend can be calculated as follows:

$$\dot{V}_{methane,blend} = \frac{\dot{m}_{methane,blend}}{\rho_{methane}}$$

$$\dot{V}_{methane,blend} = \frac{1.844}{0.716} = 2.57 \text{ m}^3/\text{s}$$

$$\dot{V}_{hydrogen,blend} = \frac{\dot{m}_{hydrogen,blend}}{\rho_{hydrogen}}$$

$$\dot{V}_{hydrogen,blend} = \frac{0.325}{0.0899} = 3.61 \text{ m}^3/\text{s}$$

Therefore, the total volumetric flow rate of the blend becomes:

$$\dot{V}_{blend} = \dot{V}_{hydrogen,blend} + \dot{V}_{methane,blend}$$

$$\dot{V}_{blend} = 3.61 + 2.57 = 6.18 \text{ m}^3/\text{s}$$

EXERCISE 7.2

Resolve Example 7.2 considering a 40 MW GT with a thermal efficiency of 40% to be fed with pure hydrogen, pure methane, and a blend of 50% hydrogen and 50% methane.

Final answers:

$$\dot{m}_{hydrogen} = 0.833 \text{ kg/s}$$

$$\dot{m}_{methane} = 2.00 \text{ kg/s}$$

$$\dot{m}_{blend} = 1.176 \text{ kg/s}$$

$$\dot{V}_{hydrogen} = 9.27 \text{ m}^3/\text{s}$$

$$\dot{V}_{methane} = 2.79 \text{ m}^3/\text{s}$$

$$\dot{V}_{blend} = 7.36 \text{ m}^3/\text{s}$$

7.3 TRANSPORTATION

The transportation sector is the second-largest source of CO_2 emissions worldwide, with a contribution of around 25% of global emissions as of 2021 [75]. Reducing emissions in this sector presents a significant challenge due to the extreme reliance on fossil fuels to cover the energy demand for all mobility modes. Fossil fuels provide unmatched advantages when considering the transportation sector, including their widespread adoption, the ease of their handling and storage, their high energy density, and cost effectiveness [76]. As a result, efforts to reduce CO_2 emissions from the transportation sector have been primarily focused on three key approaches, which are advancements in combustion technologies, electrification, and utilizing low-carbon alternative fuels.

Advancements in combustion technologies refer to a wide range of sustainable solutions that aim to reduce CO_2 emissions resulting from conventional combustion engines. These mainly include using high-efficiency engine designs, adopting advanced combustion cycles, and developing exhaust gas treatment and emission control systems [77–83].

Enhancing the design of combustion engines to enhance their efficiency is a key approach to reducing CO_2 emissions in the transportation sector. Technologies such as turbocharging, supercharging, and direct fuel injection (DFI) have been widely adopted in many vehicles to improve the power output while optimizing fuel consumption. Turbocharging and supercharging force more air into the combustion chamber which enables smaller engines to deliver the same or higher performance as larger engines, reducing fuel consumption by (4%–10%) [77,78]. Similarly, DFI systems allow for precise fuel–air mixing which improves combustion efficiency and fuel economy by approximately 23% over conventional fuel injection systems [79]. Although these advancements lower CO_2 emissions and enhance fuel efficiency, they come with challenges such as increased engine complexity and higher production costs.

Innovative combustion cycles such as homogeneous charge compression ignition (HCCI) and reactivity controlled compression ignition (RCCI) aim to maximize thermal efficiency while minimizing emissions resulted from combustion engines. HCCI technology blends characteristics of both gasoline and diesel engines and can improve fuel efficiency while reducing NO_x emissions by (90%–98%) through promoting more uniform combustion and reducing heat losses in internal combustion engines [80]. On the other hand, RCCI can further enhance the efficiency of engines and reduce CO_2 emissions by around 30% compared with conventional engines through developing fuel mixing and ensuring controlled ignition timing [81]. Despite their advantages, both HCCI and RCCI technologies face significant hurdles in controlling combustion timing across varying engine loads and the need for advanced control systems to ensure stable performance.

Exhaust gas treatment systems such as selective catalytic reduction (SCR) and exhaust gas recirculation (EGR) play a crucial role in mitigating harmful emissions from combustion engines. SCR technologies depend on injecting a urea-based solution into the exhaust stream to reduce NO_x emissions by up to 90%, significantly contributing to meeting environmental regulations worldwide [82]. On the other hand, EGR systems recirculate a portion of exhaust gases back into the engine's intake which lowers combustion temperatures and significantly reduces NO_x emissions of up to 60% [83]. While these technologies effectively decrease harmful pollutants and help vehicles comply with global regulations on emissions, they introduce significant additional costs, increase systems complexity, and require regular maintenance to prevent clogging and efficiency losses.

Despite the revolutionary development in combustion technologies in recent years, all combustion-related technologies still fail to eliminate all of the harmful emissions resulting from different modes of transportation. Therefore, electrification has become the most attractive and promising solution for decarbonizing the transportation sector worldwide. The electrification of the transportation sector is a key strategy for global decarbonization aiming to replace fossil fuel-based propulsion systems with electric powertrains that rely on cleaner energy sources. By shifting from conventional combustion engines to battery electric vehicles (BEVs), hybrid electric vehicles (HEVs), plug-in HEVs (PHEVs), and hydrogen fuel cell electric vehicles (FCEVs), electrification can significantly reduce CO_2 emissions [84]. Types of electric vehicles are illustrated in Figure 7.13 [85].

It is important to understand that the electrification of the transportation sector will not contribute to environmental sustainability unless it is powered by renewable energy sources. When relying on renewable energy sources, PHEVs and BEVs can reduce the total life-cycle emissions by 73% and 89%, respectively, in comparison to conventional combustion engine vehicles [86]. According to the IEA [87], electric car sales increased by 3.5 million in 2023 compared with 2022, marking a 35% year-on-year growth. This represents a more than sixfold increase from sales in 2018, just 5 years prior [87].

However, challenges remain for electric vehicles including battery capacity limitations, charging infrastructure development, and grid capacity constraints. Electric vehicles are well-suited for passenger vehicles and urban mobility, but their limited range, long charging times, and battery degradation pose challenges for long-haul transportation such as buses and heavy-duty trucks. Unlike PHEVs and BEVs, FCEVs generate electricity onboard using hydrogen FCs which produce only water vapor as a byproduct while offering longer driving ranges and faster refueling times. Moreover, green hydrogen produced from renewable energy sources can serve as a sustainable alternative to fossil fuels, making FCEVs an essential component of the broader decarbonization strategy. By integrating FCEVs alongside BEVs, the transportation sector can overcome the limitations of battery-powered vehicles while ensuring a low-carbon, efficient, and scalable clean mobility [88].

In terms of lifecycle factor, FCEVs have a lifecycle of approximately 190,000 km, slightly less than the lifecycle of 200,000 km for traditional internal combustion

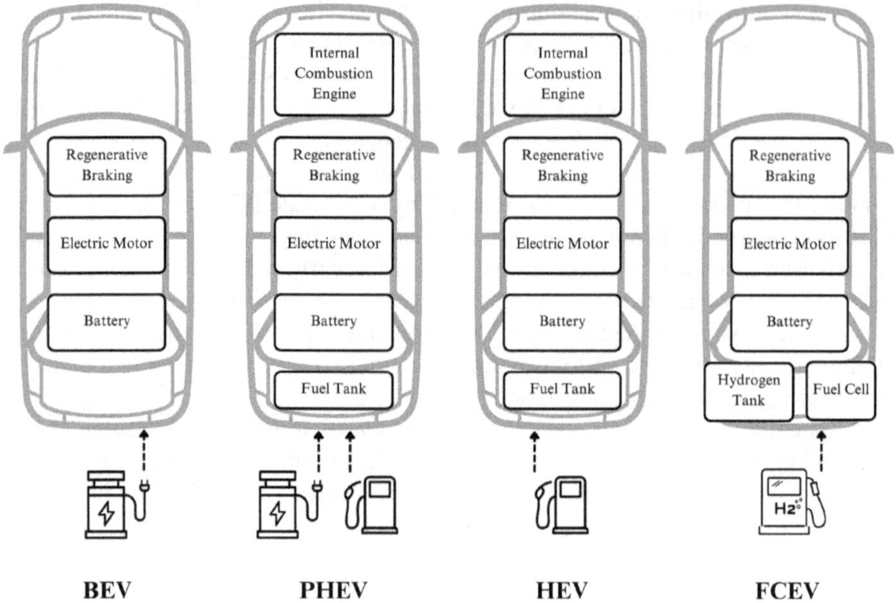

FIGURE 7.13 Types of electric vehicles. (Adapted from Foley et al. [85].)

engines. From the perspective of the CO_2 emission of 404 g/mil for the average passenger vehicle, the estimated CO_2 saving over the lifetime of one FCEV amounts to 50t. Additionally, an FCEV with a full hydrogen tank can cover up to 300 mil before requiring refueling. While a typical battery-powered vehicle with the same range on a single charge would be larger and heavier [89].

The role of hydrogen in decarbonizing the transportation sector is not limited to road transport, as key transportation modes such as aviation and shipping can also benefit from hydrogen technologies. Hydrogen presents a viable alternative to conventional jet fuel with the potential to significantly reduce emissions when produced from renewable sources. Hydrogen can be utilized in aviation through FCs or direct combustion in modified GTs. Hydrogen-powered FC aircraft generate electricity to drive electric propulsion systems, making them particularly suitable for short-haul and regional flights due to their zero-emission operation [90]. Meanwhile, hydrogen combustion in GTs can be used for longer-haul flights as it offers a cleaner alternative to conventional aviation fuels [91]. Research and development initiatives from major aircraft manufacturers, such as Airbus' ZEROe project, aim to introduce hydrogen-powered commercial aircraft by 2035, paving the way for a more sustainable aviation industry [92]. However, hydrogen storage on aircrafts presents a major challenge as it requires cryogenic tanks maintained at −253°C to remain in its liquid form [93]. These storage requirements significantly impact aircraft design, requiring innovations in aerodynamics, fuel containment, and overall aircraft efficiency. Despite these challenges, hydrogen aviation remains a promising solution for reducing the aviation industry's reliance on fossil fuels.

Furthermore, hydrogen and its derivatives, such as ammonia and synthetic fuels offer scalable and low-carbon alternatives for decarbonizing maritime transportation. One of the most promising applications of hydrogen in shipping is based on hydrogen FCs, which can enable zero-emission operations for smaller vessels, ferries, and coastal ships [94]. For long-distance maritime transport, green ammonia—produced from green hydrogen—is emerging as a leading fuel alternative due to its higher energy density, ease of storage compared with hydrogen, and compatibility with existing marine engines. Major shipping companies, including Maersk and NYK Line, are investing in ammonia-powered ships with the first commercial vessels expected to enter service before 2030 [95,96]. Additionally, other synthetic e-fuels generated via Power-to-X (PtX) technologies which utilize green hydrogen and captured CO_2 offer a drop-in fuel solution that can be used in existing ship engines, enabling a gradual transition without requiring major infrastructure changes [97]. However, hydrogen's low volumetric energy density, the need for refueling infrastructure, and the potential NO_x emissions from ammonia combustion are still major challenges for the widespread adoption of green hydrogen and its derivatives as alternative fuels for maritime transportation [98]. Despite these challenges, ongoing research and policy support are driving progress in hydrogen-based marine fuels, positioning them as a key component of the shipping industry's pathway to net-zero emissions.

7.4 ENERGY STORAGE

Energy storage (ES) mechanisms play a crucial role in providing reliability, resiliency, and efficiency features to power systems. It helps stabilize the electrical grid by addressing the imbalance between energy generation and demand. Additionally, ES enhances renewable energy sources integration to the grid by addressing the intermittent behavior of renewable energy via storing the excess energy generated at peak production times. Hence, ES significantly contributes to the reduction of GHG emissions and sustainable energy systems. Various types of ES systems can be integrated with renewable energy systems, including thermal, mechanical, electrochemical, electrical, and magnetic ES systems. Figure 7.14 provides examples of the aforementioned ES techniques [99].

Hydrogen energy storage (HydES) is a promising technique for storing large quantities of energy for long periods. They highly support the integration of renewable energy systems into the grid. HydES refers to utilizing hydrogen as a medium to store and release energy. Excess energy resulting from RES is used to produce hydrogen gas via several methods, such as electrolysis and reforming. The stored hydrogen can be converted to useful energy in the case of high demand. Figure 7.15 illustrates the schematic operation of HydES integrated with renewable energy systems [100].

Each type of ES system, including HydES, possesses its own strengths, weaknesses, and applicability features for specific cases. Different parameters can be considered to compare several ES techniques comprehensively. Table 7.2 presents a comparative analysis of different ES techniques [100]. Like other ES techniques, HydES can help provide a balanced state in the electrical grid to cover the demand

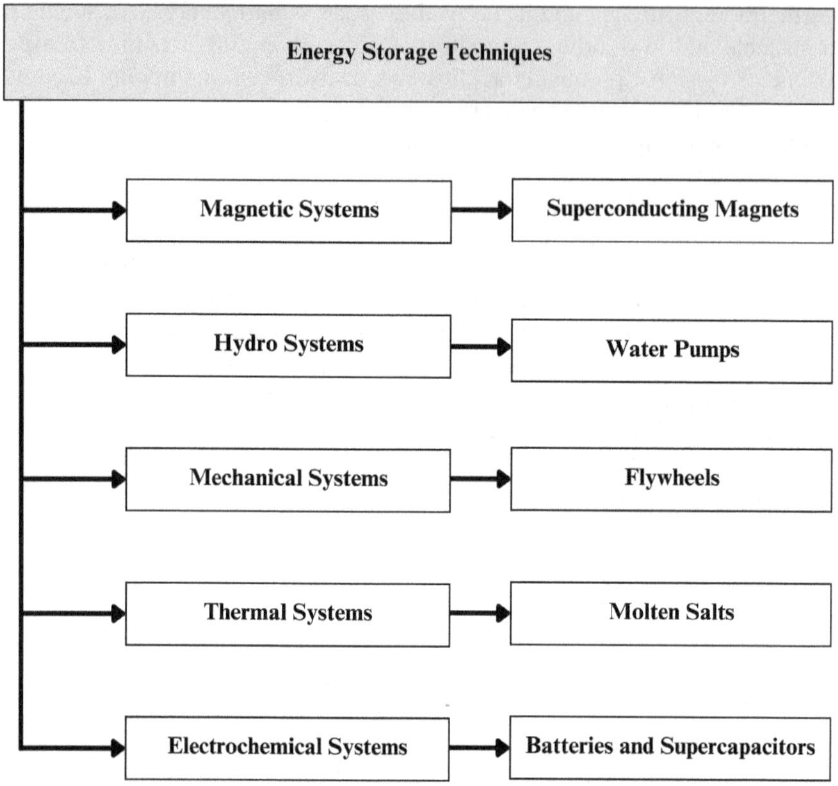

FIGURE 7.14 Main energy storage techniques. (Adapted from Olabi and Abdelkareem [99].)

in a generation shortage. Moreover, hydrogen in the HydES system can be utilized as feedstock for various industrial processes or as a fuel in the transportation sector.

7.5 HEAVY INDUSTRIES

The industrial sector is one of the most significant contributors to global GHG emissions. Heavy industries such as steel production, cement manufacturing, chemicals, and oil refining are among the biggest emitters due to their heavy reliance on fossil fuels. The steel industry alone has been emitting approximately 3.7 billion Mt of CO_2 annually since 2019, accounting for (7%–9%) of global GHG emissions [101]. Cement, chemicals, and oil and gas production accounted for 3%, 2%, and 3% of global GHG emissions in 2023, respectively [102]. Altogether, these industries must be decarbonized by all possible measures to mitigate any further deterioration of the planet and limit the catastrophic impacts of global warming. Green hydrogen can play a pivotal role in achieving decarbonization in heavy industries due to its ability to partially or fully replace existing fossil fuels.

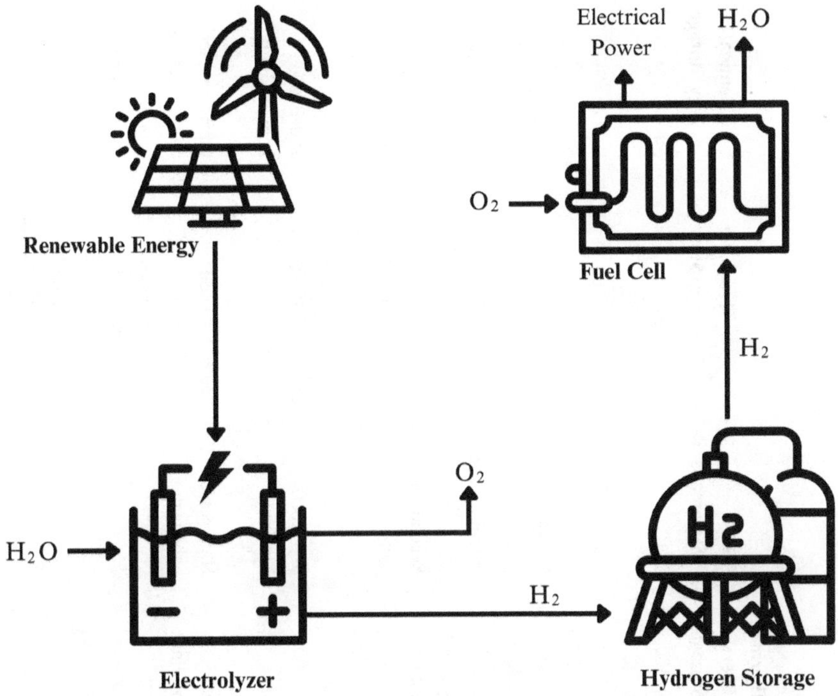

FIGURE 7.15 Integration of HydES with renewable energy systems. (Adapted from Voldsund et al. [100].)

7.5.1 Oil Refinery

The oil refinery sector produces many products that are utilized in today's demand and other industries. However, it is also a significant contributor to environmental pollution, contributing to a global emission of 40% of petroleum and gas lifecycle emissions. The worldwide Hydrogen demand in 2020 has reached 90 Mt, with approximately half of this consumption attributed to the oil refining sector. Accordingly, the oil refining sector is globally considered the largest user of hydrogen. The hydrogen utilized in the oil refinery in 2020 has been covered via gray hydrogen derived from fossil fuels, resulting in 200 Mt CO_2 emissions [103].

Hydrogen can be utilized in oil refineries through various processes, including hydrotreating and hydrocracking. Hydrotreating refers to a series of processes to remove impurities such as sulfur, nitrogen, olefins, and metals. It involves several chemical reactions that rely on adding hydrogen to the feedstock into a catalytic reactor.

An example of these reactions is the conversion of sulfur compounds into hydrogen sulfide, which can be easily separated. Accordingly, the main goals of the hydrotreating stage are the reduction of impurities in petroleum products and the enhancement of overall efficiency through reducing harmful combustion [104].

TABLE 7.2
Energy Storage Systems Characteristics

Parameter	Pumped Hydro Energy Storage	Compressed Air Energy Storage	Flywheels	Thermal Energy Storage	Supercapacitors	Hydrogen Energy Storage	Chemical Energy Storage
Energy Density (Wh/L)	0.2–2	12	20–80	80–500	2–10	600	50–500
Power Capacity (W/L)	0.1–0.2	0.2–0.6	1000–2000	N/A	40,000–120,000	0.2–20	0–400
Efficiency (%)	70–85	40–60	85–95	70–90	95–99	40–60	80–95
Lifetime (years)	50–100	25–40	>20	10–40	5	5–15	5–20
Response Time	Seconds–minutes	1–15 minutes	Millisecond–minutes	Seconds–minutes	<5 milliseconds	milliseconds–minutes	<5 milliseconds
Storage Duration	Hours–months (long term)	Hours–months (long term)	Seconds–minutes (short term)	Minutes–days (medium term)	Seconds–hours (short term)	Hours–months (long term)	Minutes–hours (short term)
Cost ($/kWh)	100–300	200–500	500–1500	50–200	500–2000	200–500	150–300
Environmental Impact	High	Medium	Very low	Low	Medium	Low	Low–medium
Applications	Ancillary services for maintaining grid stability	Back-up power supply, grid stability	Uninterruptable power supply for critical facilities	District heating and cooling systems	High power applications	Fuel cell electric vehicles, integration with renewable energy sources	Electric vehicles, residential and commercial applications

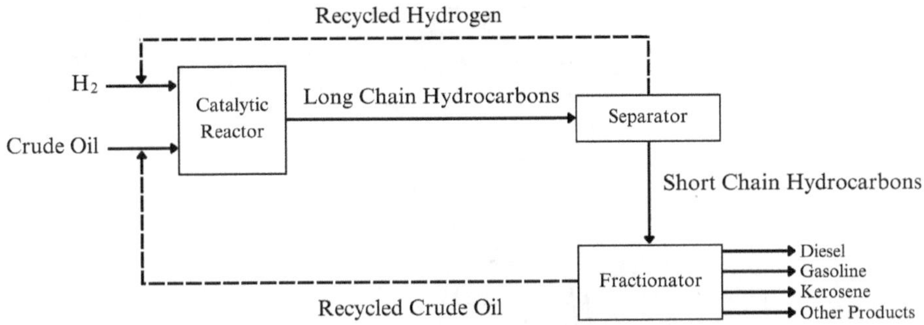

FIGURE 7.16 Hydrocracking process of crude oil. (Adapted from Song et al. [105].)

The hydrocracking stage refers to the process of breaking large hydrocarbons, such as heavy oils, into lighter and more valuable products using hydrogen and catalytic. The main components of a hydrocracker are a reactor, separator, and fractionator. Figure 7.16 illustrates the principal operation of the hydrocracking process [105].

The preheated crude oil is mixed with hydrogen within the catalytic reactor. This initiates the endothermic reaction that breaks the long-chain hydrocarbons into short-chain hydrocarbons. The short-chain hydrocarbons undergo an exothermic hydrogenation reaction involving the combination of these hydrocarbons and hydrogen to form saturated short-chain hydrocarbons. The output from the catalytic reactor is directed to a high-pressure separator, where the unreacted gaseous hydrogen is returned to the catalytic reactor, while the liquid and saturated short-chain hydrocarbons are directed to the fractionator. Within the fractionator, the useful fuels (gasoline, diesel, naphtha, and kerosene) are extracted from the fractionator's input (saturated short-chain hydrocarbons). On the other hand, the surplus crude oil is directed back to be recycled at the catalytic reactors [105].

7.5.2 Steel Production

Hydrogen utilization in steel production is considered an environmentally sustainable solution for the low-carbon emission sector. The global steel industry currently constitutes about 5% of the total energy consumption worldwide. Additionally, it contributes to (7%–9%) of global GHG emissions [101,106].

The hydrogen's utilization in the steel production field can be categorized into two main aspects, which are reducing agents in mainly the blast furnace (BF) process, gas-based direct reduction iron (DRI) process and as a fuel for heating. Figure 7.17 shows a schematic diagram of hydrogen mixed-medium injection into BF [107].

Carbon emissions resulting from the BF ironmaking process constitutes around 90% of the total value of the entire steelmaking process [108]. Consequently, there is significant pressure to minimize carbon emissions in this process. Hydrogen utilization in the BF process is considered one of the considerable solutions to overcome the aforementioned challenges. The hydrogen usage in BF relies on the injection of medium enriched with hydrogen into the BF. Some BFs have operated with an injection process of hydrogen-enriched gases such as coke over gas (COG), natural gas,

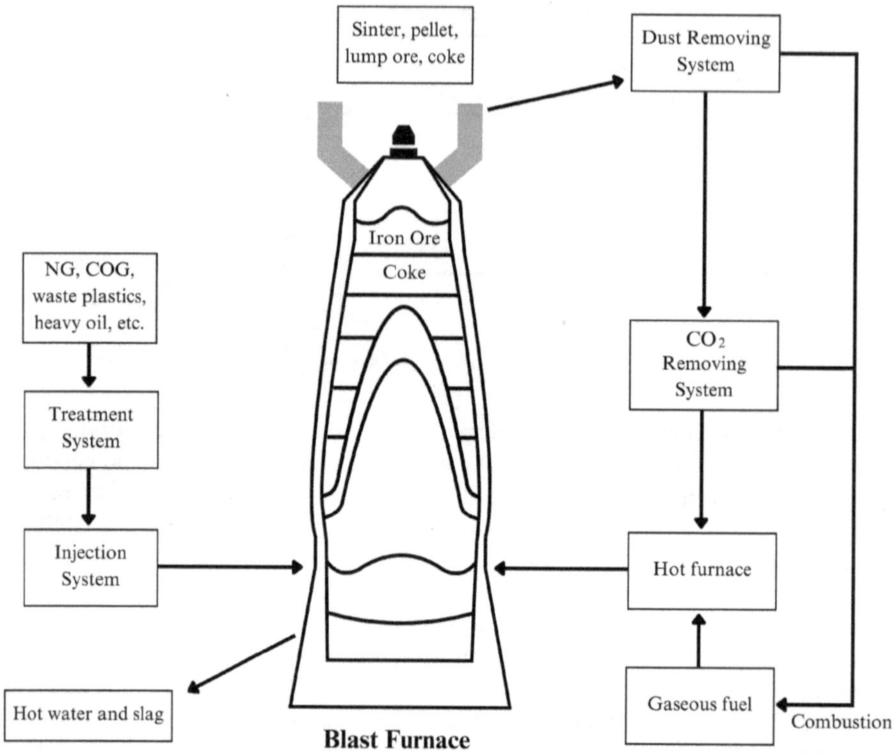

FIGURE 7.17 Schematic of Hydrogen mixed-medium injection into BF. (Adapted from Yang et al. [107].)

and waste plastic. However, carbon continues to be the primary agent for reducing iron oxides. The hydrogen mixed medium utilized for injection into the BF is comprised of NG, COG, plastic, and other materials.

According to the BF's features and the critical role of coke in BF, hydrogen cannot entirely replace the coke usage in the BF processes. However, advanced research in hydrogen technology is expected to lead to increasing the Hydrogen-to-carbon ratio, resulting in minimizing the CO_2 emissions. The DRI mechanism involves two processes: coal-based DRI reduction and gas-based DRI reduction. The Hydrogen utilization in the DRI mechanism relies on the process of gas-based DRI reduction. Throughout this process, the natural gas (NG) and output gases from steel plants are commonly utilized to produce mixed reducing gases, which have a high concentration of H_2 and a certain quantity of CO. In the pursuit of CO_2 emission reduction in steel production, the hydrogen utilization procedure or increasing the hydrogen content in mixed gases represents sustainable enhancement in the steel industry.

Generally, the main benefits of incorporating hydrogen in the steel industry include carbon emissions reduction, resource diversification, and long-term viability. Further, hydrogen features are a better reductant than CO due to its higher calorific value, faster reduction rate, and lower density.

7.5.3 Pharmaceutics

Green hydrogen utilization in medical applications has been functionally proven through several studies and experiments conducted on humans and animals. The therapeutic effects of hydrogen stem from its antioxidant features that effectively help eliminate reactive oxygen species (ROS). Diabetes, acute brain infarction, skin inflammations, side effects of radiotherapy, and skin aging are common examples of ROS diseases that have positively shown responses to hydrogen utilization [109].

Hydrogen has proven its functionality over conventional antioxidants due to several reasons which include that hydrogen has the ability to penetrate cell membranes, which enables the distribution to all organs and tissues. Also, hydrogen can effectively neutralize the most harmful and toxic hydroxyl radicals (OH) within ROS. Additionally, other experiments have demonstrated that the doses of 0.4 and 1.6 ppm have no adverse effects [110].

Hydrogen can also be utilized in the manufacturing processes in the pharmaceutical sector, including hydrogen peroxide, table sugar, and hydrochloric acid. Hydrogen peroxide, recognized for its antiseptic features, is commonly used to prevent infections that occur due to minor cuts and burns, and it can also be used as a mouth rinse. Hydrogen peroxide has demonstrated its effectiveness in the biomedical field as an oxidizing agent, surpassing other potentially harmful oxidants such as chromates and hypochlorite [111].

7.5.4 Cement Production

The cement industry is among the most carbon-emitting industries worldwide. In 2019, the cement industry resulted in 2.46 Gt of CO_2 emissions, which is 186% higher than that of 1990. This extreme increase in CO_2 emissions from the cement industry is likely attributed to the increased urbanization led by the propagation of the human population over the last years [112].

Based on global data between 2018 and 2022, producing one ton of cement corresponds to emitting 0.58 tons of CO_2, of which 40% is a result of heating and electrification processes, while the remaining 60% is related to raw material extraction and clinker production [113]. Cement clinker is an intermediate product that is utilized for cement production. Clinker is produced through heating belite, ferrite, alite, and aluminate in a kiln at high temperatures that go up to 1450°C [114]. Figure 7.18 represents the stages of cement production [115].

Initially, clay, limestone, and iron ore are mixed and ground in a grinding mill to produce raw material for clinker production. Raw materials are heated in a rotary kiln at 1450°C to produce clinker. This process is extremely energy intensive and is typically powered by conventional fossil fuels. Gypsum and limestone are then added to the clinker to produce cement [115]. CO_2 reduction from the cement industry can be carried out through energy efficiency improvements along with the utilization of clinker substitution and low-carbon fuel alternatives [116].

Green hydrogen technologies present a viable pathway for decarbonizing the cement industry by offering multiple integration opportunities. These include utilizing the oxygen and hydrogen produced via water electrolysis in clinker production

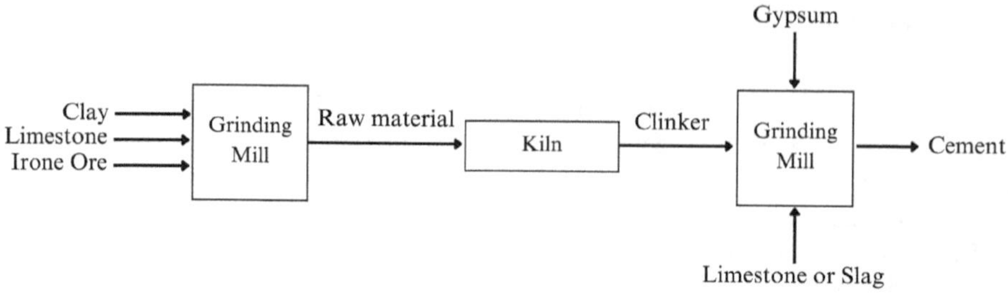

FIGURE 7.18 Stages of cement production. (Adapted from Yang et al. [115].)

and capturing CO_2 emissions for use as a feedstock in PtX applications [117]. One such approach involves oxy-combustion, which replaces traditional combustion methods by supplying a concentrated oxygen stream instead of the conventional air mixture in rotary kilns. This technique enhances clinker quality, reduces NO_x and CO_2 emissions from combustion, and produces an exhaust gas with a high CO_2 concentration, simplifying its capture and storage [118]. Additionally, hydrogen can be co-fired with natural gas in kilns to lower CO_2 emissions from combustion [119]. Moreover, green hydrogen can support electricity generation through FCs, helping to meet the energy demands of cement production [120].

Overall, the connection between green hydrogen and the cement industry is mutually beneficial, as green hydrogen contributes to decarbonizing cement production while the industry supplies the essential carbon for the PtX value chain.

7.5.5 Fertilizers Production

Agriculture is both a vital pillar of global food security and a significant contributor to GHG emissions. As for 2023, the agriculture sector contributed to 11% of global CO_2 emissions, being one of the top emitters of GHGs among many other sectors [102]. Without meaningful intervention, these emissions will continue to rise, exacerbating climate challenges and threatening food systems. Green hydrogen can play a pivotal role in decarbonizing the agriculture sector through the provision of environmentally friendly fertilizers that are based on green ammonia [121].

Ammonia plays a vital role in the production of nitrogen-based fertilizers, including di-ammonia phosphate (DAP). A percentage of 85% of the ammonia produced globally is utilized for fertilizer production. The integration of green hydrogen systems for supplying green ammonia presents a significant avenue for decarbonizing the ammonia industry. Accordingly, the potential fertilizer production from green ammonia will be highly enhanced. The production pathway of fertilizers based on both renewable resources and fossil fuels is illustrated in Figure 7.19. Green fertilizers are produced with chemical and physical composition processes that are approximately identical to fossil fuel-based fertilizers. However, the carbon footprint of green fertilizers is significantly reduced since renewable resources generate them [122].

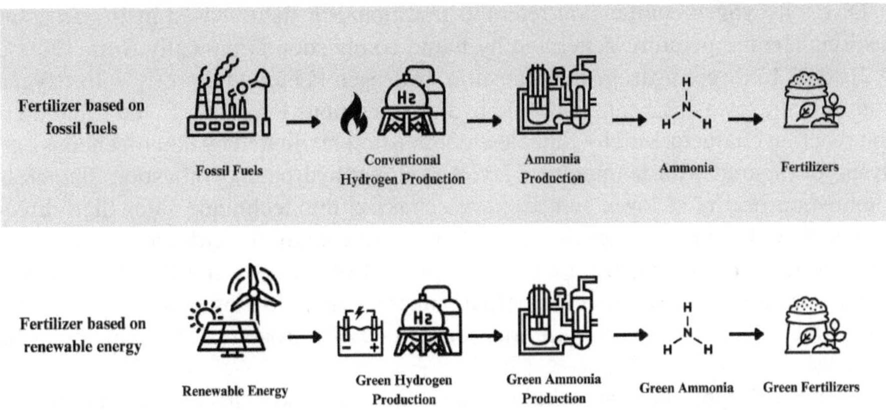

FIGURE 7.19 Fertilizer productions pathways.

Incorporation of green hydrogen in fertilizers helps the agricultural sector to enhance soil fertility by providing essential nutrients and supporting healthy crop growth. It also contributes to a sustainable and resilient food value chain.

7.6 BUILT ENVIRONMENTS

The built environment sector plays a crucial role in shaping modern society, yet it remains one of the largest contributors to global GHG emissions. Buildings and infrastructure account for nearly 6% of global energy-related CO_2 emissions as for 2023, primarily due to energy consumption in heating, cooling, lighting, and material production [102]. By implementing innovative technologies such as green hydrogen, carbon capture, and smart energy systems, the sector can significantly lower its carbon impact while enhancing resilience and efficiency.

7.6.1 Fuel for Cooking

Around three billion individuals rely on traditional carbon-based fuels for daily cooking requirements [6]. Hydrogen utilization in this sector presents an opportunity to reduce the significant dependencies on conventional fuel resources, facilitating clean and secure cooking and reducing pollutant gas emissions. It is worth noting that hydrogen utilization for clean and efficient cooking aligns with several sustainable development goals.

Several cooking technologies utilized worldwide can be classified based on energy sources such as fossil fuels, RES, utility grids, agricultural wastes, and others. These technologies include biogas, solar cooking, natural gas, solid fuel, and electrical systems. Additionally, the hydrogen cooking mechanism is one of the cooking technologies that can be used in direct hydrogen combustion cookers, catalytic hydrogen combustion cookers, or hybrid hydrogen cooking systems.

Direct hydrogen combustion refers to traditional or flame-based hydrogen combustion. The temperature generated by flame combustion is typically from 1200°C to 2100°C. In direct hydrogen combustion, hydrogen is burned directly with oxygen from the air, releasing heat. Catalytic hydrogen combustion refers to the full oxidation reaction characterized by flameless combustion facilitated by heterogeneous catalysts. Compared with temperature levels in direct hydrogen combustion, flameless combustion occurs at lower temperatures, making this technique safer than direct combustion. Temperature levels in low-temperature catalytic hydrogen combustion range between room temperature to 500°C and between 500°C and 1200°C in hybrid catalytic combustion. Another factor that makes catalytic hydrogen combustion safer than direct combustion is the minimal emissions of nitrogen oxides which can be negligible [123].

Hybrid hydrogen cooking technique refers to cooking systems powered by electrical energy produced from hydrogen combustion, or hydrogen cooking systems with thermal energy produced as waste heat. The generation of electrical power utilized for cooking involves the use of small alternators to transform mechanical energy into electrical energy. Microcombustors can be combined with thermo-PV to produce electrical energy by transforming thermal energy from hydrogen combustion into electrical energy. In contrast, FCs generate electrical energy via the electrochemical oxidation of hydrogen. An alternative hybrid hydrogen cooking system type can be cookers fueled by the heat generated from hydrogen combustion, paired with inventive waste heat recovery mechanisms. These mechanisms are utilized to produce electrical energy or store thermal energy for household electrical devices or hot water needs [6].

7.6.2 Space and Water Heating

Space heating and domestic hot water systems contribute significantly to building environmental emissions in Europe, with approximately 80% of household energy consumed [124]. The use of fossil fuels remains widespread in the heating industry across many European countries. Therefore, the continued development in decarbonizing this sector holds significant potential to contribute towards climate neutrality by 2050.

In the heating sector, hydrogen can be utilized as an energy source for various heating systems, including catalytic boilers, gas-powered heat pumps, and boilers that employ direct flame combustion. Additionally, CHP devices designed for natural gas could be adapted for use with hydrogen. Hydrogen represents a sustainable option for meeting both electrical power and heat needs in CHP systems.

A hydrogen boiler utilizing direct flame combustion operates in the same manner as gas boilers. The primary difference lies in the fuel source, as the hydrogen boilers burn hydrogen instead of natural gas. On the other hand, similar to NG boilers, the direct combustion of hydrogen generates a series of flame jets that heat water [125].

In the case of a catalytic boiler, hydrogen gas is directed over a metal catalyst with high reaction properties, resulting in an exothermic chemical reaction that generates heat for both space and hot water without a flame. This method leads to minimal nitrogen oxide emissions, and the produced heat is potentially more manageable than other conventional flame burners.

7.7 CHAPTER SUMMARY

This chapter presents green hydrogen utilization in several applications and sectors, including electricity generation, transportation, energy storage, industrial, residential, and commercial sectors. It presents hydrogen utilization in the electricity generation field via FCs and GTs. Additionally, this chapter presents the utilization of hydrogen in the transportation sector by powering FC electric vehicles or as a fuel for combustion engines. Hydrogen energy storage systems have been illustrated in this chapter as a promising technique for supporting the integration of renewable energy systems to the grid and storing large quantities of energy for long periods. Furthermore, this chapter mentions several industrial sectors that can benefit from green hydrogen in different processes during their operation. Hydrogen contributes significantly to chemical production, steel manufacturing processes, and oil refining. It can also be utilized in heating and cooking applications. The widespread adoption of green hydrogen across these sectors signifies a paradigm shift towards cleaner, more sustainable energy solutions.

SELF-ASSESSMENT QUESTIONS

1. What are the main sectors that can benefit from green hydrogen applications?
2. How does green hydrogen contribute to the decarbonization of electricity generation?
3. What are the two major technologies for reducing CO_2 emissions in gas turbines using hydrogen?
4. Explain how fuel cells generate electricity using hydrogen.
5. What are the key challenges associated with fuel cell technologies?
6. How does hydrogen play a role in the transportation sector, particularly in fuel cell electric vehicles (FCEVs)?
7. What are the advantages of hydrogen fuel cells over conventional combustion engines?
8. Describe the significance of hydrogen in aviation and maritime transportation.
9. What is the impact of hydrogen adoption in steel production?
10. How can hydrogen be used to reduce emissions in the cement industry?
11. Why is green hydrogen considered a sustainable alternative in oil refineries?
12. How does hydrogen energy storage (HydES) work, and why is it important for integrating renewable energy?
13. Compare the different types of energy storage systems with hydrogen energy storage.
14. What role does green hydrogen play in the production of fertilizers?
15. How is hydrogen used in built environments for cooking and heating?
16. What are the benefits and challenges of hydrogen-powered gas turbines?
17. Explain the role of oxy-combustion in reducing emissions in cement production.
18. What are the main barriers to large-scale hydrogen adoption in the energy sector?

19. How does hydrogen compare with natural gas in terms of energy density and efficiency?
20. What future developments are needed to enhance the viability of green hydrogen technologies?

REFERENCES

1. Tashie-Lewis, Bernard Chukwudi and Somtochukwu Godfrey Nnabuife. 2021. "Hydrogen production, distribution, storage, and power conversion in a hydrogen economy: A technology review." *Chemical Engineering Journal Advances* 8: 100172. https://doi.org/10.1016/j.ceja.2021.100172.
2. Mneimneh, Farah, Hasan Ghazzawi, Mohammad Abu Hejjeh, Matteo Manganelli, and Seeram Ramakrishna. 2023. "Roadmap to achieving sustainable development via green hydrogen." *Energies* 16 (3): 1368. https://doi.org/10.3390/en16031368.
3. Altaf, Cigdem Tuc, Orcun Demir, Tuluhan Olcayto Colak, et al. 2024. "Decarbonizing the industry with green hydrogen." In *Towards Green Hydrogen Generation*, pp. 1–48. Wiley. https://doi.org/10.1002/9781394234110.ch1.
4. Muhsen, Hani, Mohammed Al-Mahmodi, Rashed Tarawneh, Asma Alkhraibat, and Ala'aldeen Al-Halhouli. 2023. "The potential of green hydrogen and power-to-X utilization in Jordanian industries: Opportunities and future prospects." *Energies* 17 (1): 213. https://doi.org/10.3390/en17010213.
5. Oliveira, Alexandra M., Rebecca R. Beswick, and Yushan Yan. 2021. "A green hydrogen economy for a renewable energy society." *Current Opinion in Chemical Engineering* 33: 100701. https://doi.org/10.1016/j.coche.2021.100701.
6. Mukelabai, Mulako D., Upul K. G. Wijayantha, and Richard E. Blanchard. 2022. "Hydrogen for cooking: A review of cooking technologies, renewable hydrogen systems, and techno-economics." *Sustainability* 14 (24): 16964. https://doi.org/10.3390/su142416964.
7. Hydrogen Council. 2021. "Hydrogen for Net Zero." Brussels: Hydrogen Council. Available at: https://hydrogencouncil.com/wp-content/uploads/2021/11/Hydrogen-for-Net-Zero.pdf (accessed February 15, 2025).
8. Bloomberg NEF. 2024. "New Energy Outlook 2024: Executive Summary." Bloomberg Finance. Available at: https://assets.bbhub.io/professional/sites/24/847353_NEO24_ExecSum.pdf (accessed February 17, 2025).
9. International Energy Agency. 2023. "CO_2 Emissions in 2022." Paris: IEA. Available at: https://iea.blob.core.windows.net/assets/3c8fa115-35c4-4474-b237-1b00424c8844/CO2Emissionsin2022.pdf (accessed February 19, 2025).
10. Crippa, Monica, Diego Guizzardi, Federico Pagani, et al. 2023. "GHG Emissions of All World Countries." Luxembourg: Publications Office of the European Union. Available at: https://edgar.jrc.ec.europa.eu/booklet/GHG_emissions_of_all_world_countries_booklet_2023report.pdf (accessed February 19, 2025).
11. International Renewable Energy Agency. 2024. "Renewable Capacity Statistics 2024." Abu Dhabi: IRENA. Available at: https://www.irena.org/-/media/Files/IRENA/Agency/Publication/2024/Mar/IRENA_RE_Capacity_Statistics_2024.pdf (accessed February 20, 2025).
12. International Energy Agency. 2024. "Renewables 2024". Paris: IEA. Available at: https://www.iea.org/reports/renewables-2024 (accessed 20 February, 2025).
13. International Energy Agency. 2021. "Net Zero by 2050 A Roadmap for the Global Energy Sector." Paris: IEA. Available at: https://iea.blob.core.windows.net/assets/deebef5d-0c34-4539-9d0c-10b13d840027/NetZeroby2050-ARoadmapfortheGlobalEnergySector_CORR.pdf (accessed 21 February, 2025).

14. Vichard, Loïc, N. Yousfi Steiner, Noureddine Zerhouni, et al. 2021. "Hybrid fuel cell system degradation modeling methods: A comprehensive review." *Journal of Power Sources* 506: 230071. https://doi.org/10.1016/j.jpowsour.2021.230071.
15. Singla, Manish Kumar, Parag Nijhawan, and Amandeep Singh Oberoi. 2021. "Hydrogen fuel and fuel cell technology for cleaner future: A review." *Environmental Science and Pollution Research* 28 (13): 15607–15626. https://doi.org/10.1007/s11356-021-13191-3.
16. Dahiru, Ahmed Rufai, Ari Vuokila, and Mika Huuhtanen. 2022. "Recent development in Power-to-X: Part I—A review on techno-economic analysis." *Journal of Energy Storage* 56: 105861. https://doi.org/10.1016/j.est.2022.105861.
17. Mazza, Andrea, Ettore Bompard, and Gianfranco Chicco. 2018. "Applications of power to gas technologies in emerging electrical systems." *Renewable and Sustainable Energy Reviews* 92: 794–806. https://doi.org/10.1016/j.rser.2018.04.072.
18. Rezk, Hegazy, Seydali Ferahtia, Ali Djeroui, et al. 2022. "Optimal parameter estimation strategy of PEM fuel cell using gradient-based optimizer." *Energy* 239: 122096. https://doi.org/10.1016/j.energy.2021.122096.
19. Chen, Kui, Salah Laghrouche, and Abdesslem Djerdir. 2021. "Performance analysis of PEM fuel cell in mobile application under real traffic and environmental conditions." *Energy Conversion and Management* 227: 113602. https://doi.org/10.1016/j.enconman.2020.113602.
20. Fan, Liyun, Kui Xu, Zejun Jiang, et al. 2024. "Advances of membrane electrode assembly aging research of proton exchange membrane fuel cell under variable load: Degradation mechanism, aging indicators, prediction strategy, and perspectives." *Ionics* 30 (9): 5111–5140. https://doi.org/10.1007/s11581-024-05661-8.
21. Mathur, Lakshya, Yeon Namgung, Hosung Kim, et al. 2023. "Recent progress in electrolyte-supported solid oxide fuel cells: A review." *Journal of the Korean Ceramic Society* 60 (4): 614–636. https://doi.org/10.1007/s43207-023-00296-3.
22. Wang, Xiaorong, Yin Ma, Ji Gao, et al. 2021. "Review on water management methods for proton exchange membrane fuel cells." *International Journal of Hydrogen Energy* 46 (22): 12206–12229. https://doi.org/10.1016/j.ijhydene.2020.06.211.
23. Liu, Binghui, Yuting Duan, Yang Pang, et al. 2023. "Metal-organic framework-decorated PTFE reinforced membranes with immobilized caffeic acid radical scavenger towards improved performance and durability of PEMFC." *Chemical Engineering Journal* 477: 146955. https://doi.org/10.1016/j.cej.2023.146955.
24. Shroti, Nivedita and Maria K. Daletou. 2022. "The Pt–Co alloying effect on the performance and stability of high temperature PEMFC cathodes." *International Journal of Hydrogen Energy* 47 (36): 16235–16248. https://doi.org/10.1016/j.ijhydene.2022.03.109.
25. Ferriday, Thomas Benjamin and Peter Hugh Middleton. 2021. "Alkaline fuel cell technology—A review." *International Journal of Hydrogen Energy* 46 (35): 18489–18510. https://doi.org/10.1016/j.ijhydene.2021.02.203.
26. Tao, Zhengwang, Chenyi Wang, Xiaoyan Zhao, et al. 2021. "Progress in high-performance anion exchange membranes based on the design of stable cations for alkaline fuel cells." *Advanced Materials Technologies* 6 (5): 2001220. https://doi.org/10.1002/admt.202001220.
27. Krishnan, Subramani, Vinzenz Koning, Matheus Theodorus de Groot, et al. 2023. "Present and future cost of alkaline and PEM electrolyser stacks." *International Journal of Hydrogen Energy* 48 (83): 32313–32330. https://doi.org/10.1016/j.ijhydene.2023.05.031.
28. Cofell, Emiliana R., Uzoma O. Nwabara, Saket S. Bhargava, et al. 2021. "Investigation of electrolyte-dependent carbonate formation on gas diffusion electrodes for CO_2 electrolysis." *ACS Applied Materials & Interfaces* 13 (13): 15132–15142. https://doi.org/10.1021/acsami.0c21997.

29. Cheng, Chaochao, Bowen Wang, Zhi Liu, et al. 2022. "Numerical investigation of design and operating parameter effects on permeability-differentiated alkaline fuel cell with metal foam flow field." *Applied Thermal Engineering* 207: 118183. https://doi.org/10.1016/j.applthermaleng.2022.118183.
30. Patil, Smitha S., V. Madhura, Irshad Kammakakam, et al. 2022. "Quinuclidinium-piperidinium based dual hydroxide anion exchange membranes as highly conductive and stable electrolyte materials for alkaline fuel cell applications." *Electrochimica Acta* 426: 140826. https://doi.org/10.1016/j.electacta.2022.140826.
31. Paglini, Roberto, Marta Gandiglio, and Andrea Lanzini. 2022. "Technologies for deep biogas purification and use in zero-emission fuel cells systems." *Energies* 15 (10): 3551. https://doi.org/10.3390/en15103551.
32. Mahmoud, Ahmed Mohamed Ahmed, Kimio Yoshimura, and Yasunari Maekawa. 2021. "Alkaline fuel cells consisting of imidazolium-based graft-type anion exchange membranes: Optimization of fuel cell conditions to achieve high performance and durability." *Journal of Membrane Science* 620: 118844. https://doi.org/10.1016/j.memsci.2020.118844.
33. Abderezzak, Bilal. 2018. "Introduction to hydrogen technology." In *Introduction to Transfer Phenomena in PEM Fuel Cell*, pp. 1–51. Elsevier. https://doi.org/10.1016/B978-1-78548-291-5.50001-9.
34. Bhosale, Amit C., S.R. Suseendiran, Ramadoss Ramya, et al. 2022. "Phosphoric acid fuel cells." In *Comprehensive Renewable Energy*, pp. 437–458. Elsevier. https://doi.org/10.1016/B978-0-12-819727-1.00006-6.
35. Jiang, San Ping, Qingfeng Li, San Ping Jiang, et al. 2022. "Phosphoric acid fuel cells." In *Introduction to Fuel Cells: Electrochemistry and Materials*, pp. 649–671. Springer Nature. https://doi.org/10.1007/978-981-10-7626-8_14.
36. Zhang, Xiaoming, Zhao An, Zhangxun Xia, et al. 2023. "Phosphoric acid resistance PtCu/C oxygen reduction reaction electrocatalyst for HT-PEMFCs: A theoretical and experimental study." *Applied Surface Science* 619: 156663. https://doi.org/10.1016/j.apsusc.2023.156663.
37. Huang, Yicheng, Xuelian Xiao, Huifang Kang, et al. 2022. "Thermal management of polymer electrolyte membrane fuel cells: A critical review of heat transfer mechanisms, cooling approaches, and advanced cooling techniques analysis." *Energy Conversion and Management* 254: 115221. https://doi.org/10.1016/j.enconman.2022.115221.
38. Wani, Ajaz Ahmad, Norazuwana Shaari, Siti Kartom Kamarudin, et al. 2024. "Critical review on composite-based polymer electrolyte membranes toward fuel cell applications: Progress and perspectives." *Energy & Fuels* 38 (19): 18169–18193. https://doi.org/10.1021/acs.energyfuels.4c02516.
39. Barckholtz, Timothy A., Kevin M. Taylor, Sundar Narayanan, et al. 2022. "Molten carbonate fuel cells for simultaneous CO_2 capture, power generation, and H_2 generation." *Applied Energy* 313: 118553. https://doi.org/10.1016/j.apenergy.2022.118553.
40. Kim, Kab In, Han Sung Kim, and Sung Pil Yoon. 2022. "A study on the characteristics of electrolyte redistribution through the surface modification of an anode for molten carbonate fuel cells." *Journal of Alloys and Compounds* 903: 163956. https://doi.org/10.1016/j.jallcom.2022.163956.
41. Bian, Jing, Liqiang Duan, and Yongping Yang. 2023. "Simulation and economic Investigation of CO_2 separation from gas turbine exhaust gas by molten carbonate fuel cell with exhaust gas recirculation and selective exhaust gas recirculation." *Energies* 16 (8): 3511. https://doi.org/10.3390/en16083511.
42. Gao, Liangjuan, J. Robert Selman, and Philip Nash. 2023. "Materials and wetting issues in molten carbonate fuel cell technology: A review." *Journal of Materials Science* 58 (41): 15936–15972. https://doi.org/10.1007/s10853-023-08958-7.

43. De Miguel, Maria Teresa, María Isabel Lasanta, Gustavo García-Martín, et al. 2022. "Temperature effect and alloying elements impact on the corrosion behaviour of the alloys exposed to molten carbonate environments for CSP application." *Corrosion Science* 201: 110274. https://doi.org/10.1016/j.corsci.2022.110274.
44. Kandhasamy, Sathiyaraj, Geir Martin Haarberg, Signe Kjelstrup, et al. 2021. "Thermo-electrochemical cell performance and physicochemical properties of the molten carbonate electrolyte dispersed with different solid oxides." *Electrochimica Acta* 386: 138481. https://doi.org/10.1016/j.electacta.2021.138481.
45. Bosio, Barbara, Maurizio Archetti, Emilio Audasso, et al. 2023. "Process analysis of a molten carbonate fuel cell on-board application to reduce vessel CO_2 emissions." *Chemical Engineering and Processing-Process Intensification* 190 (2023): 109415. https://doi.org/10.1016/j.cep.2023.109415.
46. Priya, S. Dharani, A. Immanuel Selvakumar, and A. Samson Nesaraj. 2020. "Overview on ceramic and nanostructured materials for solid oxide fuel cells (SOFCs) working at different temperatures." *Journal of Electrochemical Science and Technology* 11 (2): 99–116. https://doi.org/10.33961/jecst.2019.00612.
47. Corigliano, Orlando, Leonardo Pagnotta, and Petronilla Fragiacomo. 2022. "On the technology of solid oxide fuel cell (SOFC) energy systems for stationary power generation: A review." *Sustainability* 14 (22): 15276. https://doi.org/10.3390/su142215276.
48. Singh, Mandeep, Dario Zappa, and Elisabetta Comini. 2021. "Solid oxide fuel cell: Decade of progress, future perspectives and challenges." *International Journal of Hydrogen Energy* 46 (54): 27643–27674. https://doi.org/10.1016/j.ijhydene.2021.06.020.
49. Xie, Jiamiao, Wenqian Hao, and Fenghui Wang. 2022. "Parametric study on interfacial crack propagation in solid oxide fuel cell based on electrode material." *International Journal of Hydrogen Energy* 47 (12): 7975–7989. https://doi.org/10.1016/j.ijhydene.2021.12.153.
50. Shabri, Hazrul Adzfar, Mohd Hafiz Dzarfan Othman, Mohamad Azuwa Mohamed, et al. 2021. "Recent progress in metal-ceramic anode of solid oxide fuel cell for direct hydrocarbon fuel utilization: A review." *Fuel Processing Technology* 212: 106626. https://doi.org/10.1016/j.fuproc.2020.106626.
51. Samreen, Ayesha, Muhammad Sudais Ali, Muhammad Huzaifa, et al. 2024. "Advancements in perovskite-based cathode materials for solid oxide fuel cells: A comprehensive review." *The Chemical Record* 24 (1): e202300247. https://doi.org/10.1002/tcr.202300247.
52. Deepi, Alagu Segar, Srinivasan Dharani Priya, Arputharaj Samson Nesaraj, et al. 2022. "Component fabrication techniques for solid oxide fuel cell (SOFC)—A comprehensive review and future prospects." *International Journal of Green Energy* 19 (14): 1600–1612. https://doi.org/10.1080/15435075.2021.2018320.
53. Yu, Fangyong, Tingting Han, Zhigang Wang, et al. 2021. "Recent progress in direct carbon solid oxide fuel cell: Advanced anode catalysts, diversified carbon fuels, and heat management." *International Journal of Hydrogen Energy* 46 (5): 4283–4300. https://doi.org/10.1016/j.ijhydene.2020.10.259.
54. Din, Muhammad Aizaz Ud, Muhammad Idrees, Sidra Jamil, et al. 2023. "Advances and challenges of methanol-tolerant oxygen reduction reaction electrocatalysts for the direct methanol fuel cell." *Journal of Energy Chemistry* 77: 499–513. https://doi.org/10.1016/j.jechem.2022.11.023.
55. Ahmed, Aser Alaa, Malik Al Labadidi, Ahmed T. Hamada, et al. 2022. "Design and utilization of a direct methanol fuel cell." *Membranes* 12 (12): 1266. https://doi.org/10.3390/membranes12121266.
56. Zuo, Yuanhui, Wenchao Sheng, Wenquan Tao, et al. 2022. "Direct methanol fuel cells system—A review of dual-role electrocatalysts for oxygen reduction and methanol oxidation." *Journal of Materials Science & Technology* 114: 29–41. https://doi.org/10.1016/j.jmst.2021.10.031.

57. Yujing, Wu, Zhou Hongxiu, Zhang Ao, et al. 2022. "Numerical simulation of methanol crossover in flowing electrolyte-direct methanol fuel cell." *Journal of Power Sources* 519: 230801. https://doi.org/10.1016/j.jpowsour.2021.230801.
58. Shi, Jingwei, Baoling Wang, and Sujuan Hu. 2024. "From photo-assisted methanol catalytic oxidation to direct methanol fuel cells: Applications of semiconductors-based electrode." *Surfaces and Interfaces* 46: 103970. https://doi.org/10.1016/j.surfin.2024.103970.
59. Wu, Juan, Shijun Nie, Hai Liu, et al. 2022. "Design and development of nucleobase modified sulfonated poly (ether ketone) membranes for high-performance direct methanol fuel cells." *Journal of Materials Chemistry A* 10 (37): 19914–19924. 10.1039/D2TA03166C.
60. Abdullah, Norulsamani, Saidur Rahman, Azran Mohd Zainoodin, et al. 2022. "Comparative study for electrochemical and single-cell performance of a novel MXene-supported platinum–ruthenium catalyst for direct methanol fuel cell application." *Journal of Electroanalytical Chemistry* 925: 116884. https://doi.org/10.1016/j.jelechem.2022.116884.
61. Gorin, Melody, Mehri Shabani, Sébastien Votat, et al. 2024. "Application of fungal-based microbial fuel cells for biodegradation of pharmaceuticals: Comparative study of individual vs. mixed contaminant solutions." *Chemosphere* 363: 142849. https://doi.org/10.1016/j.chemosphere.2024.142849.
62. Derbeli, Mohamed, Oscar Barambones, and Lassaad Sbita. 2018. "A robust maximum power point tracking control method for a PEM fuel cell power system." *Applied Sciences* 8 (12): 2449. https://doi.org/10.3390/app8122449.
63. Falcão, Daniela, and Alexandra Pinto. 2020. "A review on PEM electrolyzer modelling: Guidelines for beginners." *Journal of Cleaner Production* 261: 121184. https://doi.org/10.1016/j.jclepro.2020.121184.
64. Patel, Ravi and Dipankar Deb. 2018. "Parametrized control-oriented mathematical model and adaptive backstepping control of a single chamber single population microbial fuel cell." *Journal of Power Sources* 396: 599–605. https://doi.org/10.1016/j.jpowsour.2018.06.064.
65. Bilgili, Muhittin and Yunus Emre Gönülaçar. 2024."Investigation of the effect of operating parameters on Nernst voltage in hydrogen-oxygen fuel cells." *Isı Bilimi ve Tekniği Dergisi* 44 (1): 59–69. https://doi.org/10.47480/isibted.1494033.
66. Yang, Jingze, Zhen Yang, and Yuanyuan Duan. 2022. "A review on integrated design and off-design operation of solar power tower system with S–CO_2 Brayton cycle." *Energy* 246: 123348. https://doi.org/10.1016/j.energy.2022.123348.
67. James, Allister, and Sudhir Rajagopalan. 2014. "Gas turbines: operating conditions, components and material requirements." In *Structural Alloys for Power Plants*, pp. 3–21. Woodhead Publishing. https://doi.org/10.1533/9780857097552.1.3.
68. Stan, Cornel. "Jet engine flow instead of nuclear reaction: Energy for heat and electricity. 2024. "In *Energy Scenarios for the Future: Ways to Climate-friendly Mobility, Heating and Industry*, pp. 163–177. Berlin, Heidelberg: Springer Berlin Heidelberg. https://doi.org/10.1007/978-3-662-69687-3_8.
69. Hai, Tao, Mohammed M. Alhaider, Pooya Ghodratallah, et al. 2024. "Techno-economic-environmental study and artificial intelligence-assisted optimization of a multigeneration power plant based on a gas turbine cycle along with a hydrogen liquefaction unit." *Applied Thermal Engineering* 237: 121660. https://doi.org/10.1016/j.applthermaleng.2023.121660.
70. Zhou, Haiqin, Jiye Xue, Haobu Gao, et al. 2024. "Hydrogen-fueled gas turbines in future energy system." *International Journal of Hydrogen Energy* 64: 569–582. https://doi.org/10.1016/j.ijhydene.2024.03.327.
71. Cecere, Donato, Eugenio Giacomazzi, Antonio Di Nardo, et al. 2023. "Gas turbine combustion technologies for hydrogen blends." *Energies* 16 (19): 6829. https://doi.org/10.3390/en16196829.

72. Goldmeer, Jeffrey. 2019. "Power to Gas: Hydrogen for Power Generation Fuel-Flexible Gas Turbines as Enablers for a Low or Reduced Carbon Energy Ecosystem." *GE*. Available at: https://www.wika.com/media/Others/EN-GB/powertogas_hydrogenforpowergeneration_en_gb.pdf.
73. Ryu, Ju-Yeol, Sungho Park, Changhyeong Lee, et al. 2023. "Techno-economic analysis of hydrogen–natural gas blended fuels for 400 MW combined cycle power plants (CCPPs)." *Energies* 16 (19): 6822. https://doi.org/10.3390/en16196822.
74. Cengel, Yunus A. and Michael A. Boles. 2002. "Thermodynamics: An engineering approach." *Sea* 1000 (8862): 287–293. ISBN 10: 0071150714.
75. Hwang, Joonsik, Krisha Maharjan, and HeeJin Cho. 2023. "A review of hydrogen utilization in power generation and transportation sectors: Achievements and future challenges." *International Journal of Hydrogen Energy* 48 (74): 28629–28648. https://doi.org/10.1016/j.ijhydene.2023.04.024.
76. Neves, Sónia Almeida and António Cardoso Marques. 2021. "The substitution of fossil fuels in the US transportation energy mix: Are emissions decoupling from economic growth?." *Research in Transportation Economics* 90: 101036. https://doi.org/10.1016/j.retrec.2021.101036.
77. Nain, Ajay. 2020. "Turbocharger Optimization of Diesel Engine for Fuel Economy Improvement Using 1-D Thermodynamic Analysis." No. 2020-01-5050. *SAE Technical Paper*. https://doi.org/10.4271/2020-01-5050.
78. Petitjean, Dominique, Luciano Bernardini, Chris Middlemass, et al. 2004. "Advanced Gasoline Engine Turbocharging Technology for Fuel Economy Improvements." No. 2004-01-0988. *SAE Technical Paper*. https://doi.org/10.4271/2004-01-0988.
79. Chincholkar, Shrikrushna, and Jiwak Suryawanshi. 2016. "Gasoline direct injection: an efficient technology." *Energy Procedia* 90: 666–672. https://doi.org/10.1016/j.egypro.2016.11.235.
80. Srinivasan, Jayakrishna, Abhishek Krishna Swamy, Pradeep Madanagopalan, et al. 2021. "Performance and emission characteristics of a methane fuelled HCCI engine at various injection location and operating speed." *Materials Today: Proceedings* 46: 1022–1027. https://doi.org/10.1016/j.matpr.2021.01.216.
81. Benajes, Jesús, Antonio García, Javier Monsalve-Serrano, et al. 2018. "Potential of RCCI series hybrid vehicle architecture to meet the future CO_2 targets with low engine-out emissions." *Applied Sciences* 8 (9): 1472. https://doi.org/10.3390/app8091472.
82. Van Caneghem, Jo, Johan De Greef, Chantal Block, et al. 2016. "NO_x reduction in waste incinerators by selective catalytic reduction (SCR) instead of selective non catalytic reduction (SNCR) compared from a life cycle perspective: A case study." *Journal of Cleaner Production* 112: 4452–4460. https://doi.org/10.1016/j.jclepro.2015.08.068.
83. Bakar, Anuar Abu, Sheikh Alif Ali, Mohammed Ismail Russtam Suhrab, et al. 2022. "A review of various non-emission reduction methods in marine diesel engines." *Specialusis Ugdymas* 2 (43): 43. https://sumc.lt/index.php/se/article/view/1494.
84. Masclans Abelló, Pol. 2021. "Design and Validation of a Consumption Framework for Mixed Fleets Considering ICEV, HEV, PHEV, BEV and FCEV." Master's thesis, Universitat Politècnica de Catalunya. https://hdl.handle.net/2117/347198.
85. Foley, Bethany, Kenan Degirmenci, and Tan Yigitcanlar. 2020. "Factors affecting electric vehicle uptake: Insights from a descriptive analysis in Australia." *Urban Science* 4 (4): 57. https://doi.org/10.3390/urbansci4040057.
86. Buberger, Johannes, Anton Kersten, Manuel Kuder, et al. 2022. "Total CO_2-equivalent life-cycle emissions from commercially available passenger cars." *Renewable and Sustainable Energy Reviews* 159: 112158. https://doi.org/10.1016/j.rser.2022.112158.
87. International Energy Agency. 2024. "Global EV Outlook 2024—Moving Towards Increased Affordability". Available at: https://iea.blob.core.windows.net/assets/a9e3544b-0b12-4e15-b407-65f5c8ce1b5f/GlobalEVOutlook2024.pdf.

88. Parikh, Ahan, Manan Shah, and Mitul Prajapati. 2023. "Fuelling the sustainable future: a comparative analysis between battery electrical vehicles (BEV) and fuel cell electrical vehicles (FCEV)." *Environmental Science and Pollution Research* 30 (20): 57236–57252. https://doi.org/10.1007/s11356-023-26241-9.
89. Lamichhane, Bipin, Syed Ali Rezaei, and Mehdi Kouhi Habibi. 2022. "Economic Analysis of Hydrogen Fuel Cell Electric Vehicles (FCEV)." Tampere University. https://doi.org/10.13140/RG.2.2.31586.61123.
90. Svensson, Christian, Amir A. M. Oliveira, et al. 2024. "Hydrogen fuel cell aircraft for the Nordic market." *International Journal of Hydrogen Energy* 61: 650–663. https://doi.org/10.1016/j.ijhydene.2024.02.382.
91. Roper, Jack. 2022. "In for the long H_2aul." *Aerospace Testing International* 2022 (4): 14–20. https://doi.org/10.12968/S1478-2774(23)50280-5.
92. Retallack, Liv, Satya Shah, and Syed Hasan. 2023. "Commercial aircraft manufacturer strategic adoption of hydrogen decarbonisation technologies." In *2023 28th International Conference on Automation and Computing (ICAC)*, Birmingham, United Kingdom, pp. 01–06. IEEE. https://doi.org/10.1109/ICAC57885.2023.10275239.
93. Prewitz, Marc, Jonas Schwärzer, and Andreas Bardenhagen. 2023. "Potential analysis of hydrogen storage systems in aircraft design." *International Journal of Hydrogen Energy* 48 (65): 25538–25548. https://doi.org/10.1016/j.ijhydene.2023.03.266.
94. Shakeri, Nastaran, Mehdi Zadeh, and Jorgen Bremnes Nielsen. 2020. "Hydrogen fuel cells for ship electric propulsion: Moving toward greener ships." *IEEE Electrification Magazine* 8 (2): 27–43. https://doi.org/10.1109/MELE.2020.2985484.
95. Sánchez, Antonio, MA Martín Rengel, and Mariano Martín. 2023. "A zero CO_2 emissions large ship fuelled by an ammonia-hydrogen blend: Reaching the decarbonisation goals." *Energy Conversion and Management* 293: 117497. https://doi.org/10.1016/j.enconman.2023.117497.
96. Akgül, Ersin Fırat. 2024. "Navigating decarbonization: Examining shipping companies' fleet modernization strategies worldwide." *Dokuz Eylül Üniversitesi Denizcilik Fakültesi Dergisi* 16 (1): 1–21. https://doi.org/10.18613/deudfd.1418186.
97. Martin, Jonas, Anne Neumann, and Anders Ødegård. 2023. "Renewable hydrogen and synthetic fuels versus fossil fuels for trucking, shipping and aviation: A holistic cost model." *Renewable and Sustainable Energy Reviews* 186: 113637. https://doi.org/10.1016/j.rser.2023.113637.
98. La Corte, Daniele, Marina Maddaloni, Reza Vahidzadeh, et al. 2025. "Recovered ammonia as a sustainable energy carrier: Innovations in recovery, combustion, and fuel cells." *Energies* 18 (3): 508. https://doi.org/10.3390/en18030508.
99. Olabi, Abdul Ghani and Mohammad Abdelkareem. 2021. "Energy storage systems towards 2050." *Energy* 219: 119634. https://doi.org/10.1016/j.energy.2020.119634.
100. Voldsund, Mari, Adriana Reyes-Lúa, Chao Fu, et al. 2023. "Low carbon power generation for offshore oil and gas production." *Energy Conversion and Management:* X 17: 100347. https://doi.org/10.1016/j.ecmx.2023.100347.
101. Chen, Jing, Yi Xing, Yan Wang, et al. 2022. "Application of iron and steel slags in mitigating greenhouse gas emissions: A review." *Science of the Total Environment* 844: 157041. https://doi.org/10.1016/j.scitotenv.2022.157041.
102. United Nations Environment Programme. 2024. "No More Hot Air … Please! With a Massive Gap Between Rhetoric and Reality, Countries Draft New Climate Commitments." Available at: https://doi.org/10.59117/20.500.11822/46404.
103. International Energy Agency. 2021. "Global Hydrogen Review 2021". IEA. Available at: https://iea.blob.core.windows.net/assets/5bd46d7b-906a-4429-abda-e9c507a62341/GlobalHydrogenReview2021.pdf.
104. Lindfors, Christian, Douglas C. Elliott, Wolter Prins, et al. 2022. "Co-processing of biocrudes in oil refineries." *Energy & Fuels* 37 (2): 799–804. https://doi.org/10.1021/acs.energyfuels.2c04238.

105. Song, Wenjiang, Vladimir Mahalec, Jian Long, et al. 2020. "Modeling the hydrocracking process with deep neural networks." *Industrial & Engineering Chemistry Research* 59 (7): 3077–3090. https://doi.org/10.1021/acs.iecr.9b06295.
106. Liu, Wenguo, Haibin Zuo, Jingsong Wang, et al. 2021. "The production and application of hydrogen in steel industry." *International Journal of Hydrogen Energy* 46 (17): 10548–10569. https://doi.org/10.1016/j.ijhydene.2020.12.123.
107. Yang, Liuyimei, Qijun Zheng, and Aibing Yu. 2021." Numerical simulation of solid flow and segregation in a blast furnace by coupling granular rheology and transport equation." *Chemical Engineering Science* 242: 116741. https://doi.org/10.1016/j.ces.2021.116741.
108. Zhao, Jun, Haibin Zuo, Yajie Wang, et al. 2020. "Review of green and low-carbon ironmaking technology." *Ironmaking & Steelmaking* 47 (3): 296–306. https://doi.org/10.1080/03019233.2019.1639029.
109. Chung, Myung-Hee and Jai Youl Ro. 2021. "The medical uses of hydrogen." *Food Supplements and Biomaterials for Health* 1 (1). https://doi.org/10.52361/fsbh.2021.1.e5.
110. Hirano, Shin-ichi, Yusuke Ichikawa, Bunpei Sato, et al. 2020. "Hydrogen is promising for medical applications." *Clean Technologies* 2 (4): 529–541. https://doi.org/10.3390/cleantechnol2040033.
111. Okolie, Jude A., Biswa R. Patra, Alivia Mukherjee, et al. 2021. "Futuristic applications of hydrogen in energy, biorefining, aerospace, pharmaceuticals and metallurgy." *International Journal of Hydrogen Energy* 46 (13): 8885–8905. https://doi.org/10.1016/j.ijhydene.2021.01.014.
112. Chen, Cuihong, Ruochong Xu, Dan Tong, et al. 2022. "A striking growth of CO_2 emissions from the global cement industry driven by new facilities in emerging countries." *Environmental Research Letters* 17 (4): 044007. https://doi.org/10.1088/1748-9326/ac48b5.
113. International Energy Agency. 2023. "Tracking Clean Energy Progress 2023." IEA, Paris. Available at: https://www.iea.org/reports/tracking-clean-energy-progress-2023, License: CC BY 4.0.
114. Oguntola, Olurotimi and Steven Simske. 2023. "Continuous assessment of the environmental impact and economic viability of decarbonization improvements in cement production." *Resources* 12 (8): 95. https://doi.org/10.3390/resources12080095.
115. Yang, Zequn, Bengen Gong, Jianping Yang, et al. 2019. "Trace element partition in a coal-feed industry furnace." In *Emission and Control of Trace Elements from Coal-Derived Gas Streams*, pp. 173–226. Woodhead Publishing. https://doi.org/10.1016/B978-0-08-102591-8.00006-4.
116. Fennell, Paul S., Steven J. Davis, and Aseel Mohammed. 2021. "Decarbonizing cement production." *Joule* 5 (6): 1305–1311. 10.1016/j.joule.2021.04.011.
117. Nhuchhen, Daya R., Song P. Sit, and David B. Layzell. 2022. "Decarbonization of cement production in a hydrogen economy." *Applied Energy* 317: 119180. https://doi.org/10.1016/j.apenergy.2022.119180.
118. Carrasco-Maldonado, Francisco, Reinhold Spörl, Kristina Fleiger, et al. 2016. "Oxy-fuel combustion technology for cement production–state of the art research and technology development." *International Journal of Greenhouse Gas Control* 45: 189–199. https://doi.org/10.1016/j.ijggc.2015.12.014.
119. El-Emam, Rami S. and Kamiel S. Gabriel. 2021. "Synergizing hydrogen and cement industries for Canada's climate plan–case study." *Energy Sources, Part A: Recovery, Utilization, and Environmental Effects* 43 (23): 3151–3165. https://doi.org/10.1080/15567036.2021.1936699.
120. Thomas, John Meurig, Peter P. Edwards, Peter J. Dobson, et al. 2020. "Decarbonising energy: The developing international activity in hydrogen technologies and fuel cells." *Journal of Energy Chemistry* 51: 405–415. https://doi.org/10.1016/j.jechem.2020.03.087.

121. Alwazeer, Duried, John T. Hancock, Grace Russell, et al. 2024. "Molecular hydrogen: A sustainable strategy for agricultural and food production challenges." *Frontiers in Food Science and Technology* 4: 1448148. https://doi.org/10.3389/frfst.2024.1448148.
122. Pagani, Gianni, Yashar Hajimolana, and Canan Acar. 2024. "Green hydrogen for ammonia production—A case for the Netherlands." *international Journal of Hydrogen Energy* 52: 418–432. https://doi.org/10.1016/j.ijhydene.2023.06.309.
123. Kim, Jongho, Jianglong Yu, Soonho Lee, et al. 2021. "Advances in catalytic hydrogen combustion research: Catalysts, mechanism, kinetics, and reactor designs." *International Journal of Hydrogen Energy* 46 (80): 40073–40104. https://doi.org/10.1016/j.ijhydene.2021.09.236.
124. Famiglietti, Jacopo, Tommaso Toppi, Davide Bonalumi, et al. 2023. "Heat pumps for space heating and domestic hot water production in residential buildings, an environmental comparison in a present and future scenario." *Energy Conversion and Management* 276: 116527. https://doi.org/10.1016/j.enconman.2022.116527.
125. Dodds, Paul E., Iain Staffell, Adam D. Hawkes, et al. 2015. "Hydrogen and fuel cell technologies for heating: A review." *International Journal of Hydrogen Energy* 40 (5): 2065–2083. https://doi.org/10.1016/j.ijhydene.2014.11.059.

8 Hydrogen Safety

8.1 INTRODUCTION

The safety aspects regarding the storage and production of hydrogen facilities, along with their future uses, such as in transportation, are factors that may influence public acceptance of the technology, which, in essence, is a social concern. These concerns could be grouped into physiological, physical, and chemical dangers. Therefore, hazards and risks, together with the measures to reduce them, have to be defined for the user's protection. This means analyzing the dangers and risks that hydrogen poses and locating the zones where these risks may be present in hydrogen-handling operations.

Hydrogen gas is flammable; therefore, it can lead to accidents if it is not handled wisely. While there are certain risks associated with hydrogen use, accidents have been low, even when operating heavy machinery. One such risk involves storing the gas. Hydrogen must be kept in either cryogenic pressure, high-pressure, or cryo-compressed containers. Once these tanks or containers fail, it can lead to a leak or release of hydrogen gas. Additionally, areas where hydrogen is used must be adequately ventilated to prevent the displacement of oxygen, which could lead to asphyxiation. Hydrogen used in fuel cells in vehicles also poses a danger to the driver as well as other road users.

Fuel cell vehicles use hydrogen fuel to power pressure cells that generate electricity. In case of collision, hydrogen can leak, increasing the risk of fire. Luckily, the vehicle is well equipped to handle such situations, as it is built with pressure relief valves to avoid overloading containment barriers, helping the vehicle withstand accidents.

8.2 HISTORY OF HYDROGEN-RELATED ACCIDENTS

The German airship Hindenburg caught fire on 6 September 1937 while attempting to land at a naval station in New Jersey, United States, where it was scheduled to be stationed. Hindenburg was filled with hydrogen gas, which played a role in the disaster. The precise reason has never been definitively established. However, it is thought to have a blend of strong winds and a gas leak; there is also a chance of sabotage. The incident has had considerable media coverage due to its impact on the airship industry and business in general [1].

A recent accident occurred in Towanda, Pennsylvania, in 2022. This incident occurred at Global Tungsten & Powders Corp, where a hydrogen tank exploded. According to a company spokesperson, five workers were injured, but their injuries were not considered severe. The cause of the incident was investigated to identify the reasons, but it has not been announced yet.

DOI: 10.1201/9781003639190-8

Many accidents are recorded in various locations worldwide; some causes are identified, while many incidents have unknown reasons. In the literature, prior hydrogen accidents were returned to the following reasons.

- Hanau, Germany, 1999: The tank was supposed to be positioned horizontally but was instead placed vertically. This orientation concentrated forces on the top of the tank, causing it to rupture and explode.
- Taiwan (Formosa Plastics Group refinery): In 2015, hydrogen leaked from a pipe, causing a chemical plant explosion.
- Gangneung, South Korea 2019: Oxygen entered the tanks used for storing Hydrogen.
- Illinois (AB Specialty Silicones in Waukegan) 2019: The operator made a mistake and added an improper ingredient and hydrogen gas was released, which caused an explosion. This resulted in the death of four workers and the severe injury of a fifth worker due to an explosion.
- Changhua City 2020: An incident occurred where a hydrogen tanker truck met an accident on a freeway in Changhua City, which caused an explosion.
- South Africa (Medupi Power Station) 2021: An operator error caused an error in procedures during the purging of hydrogen from the generator.

For any hydrogen incident, the reason for occurrence can be attributed to several causes:

- Human actions (mistakes made by humans during installation or maintenance).
- Breakdown or malfunction of equipment or materials.
- Damage caused by corrosion.
- Excess pressure.
- Increased fragility of storage tanks in cold temperatures.
- The explosion is caused by liquid boiling and expanding into vapor.
- Destruction or tearing due to the impact of shock waves from nearby explosions.

These accidents are considered a source of awareness and lessons to be learned regarding the accident's reasons and damage.

8.3 HYDROGEN PROPERTIES AND THE SOURCES OF HAZARDS

Hydrogen peroxide is imperceptible; neither an odor nor a color can be associated with it, hence it cannot be sensed. Under normal atmospheric conditions, it exists in gaseous form, but hydrogen gas can be turned into liquid when a combination of increased pressure and decreased temperature is exerted. The gas turns to liquid at –252.87°C under a pressure of 0.1013 MPa (1.013 bar). Hydrogen can crystallize at even lower temperatures and can exist in solid forms, such as either alpha-hydrogen or beta-hydrogen. The relation between hydrogen temperature and pressure is illustrated in Figure 8.1.

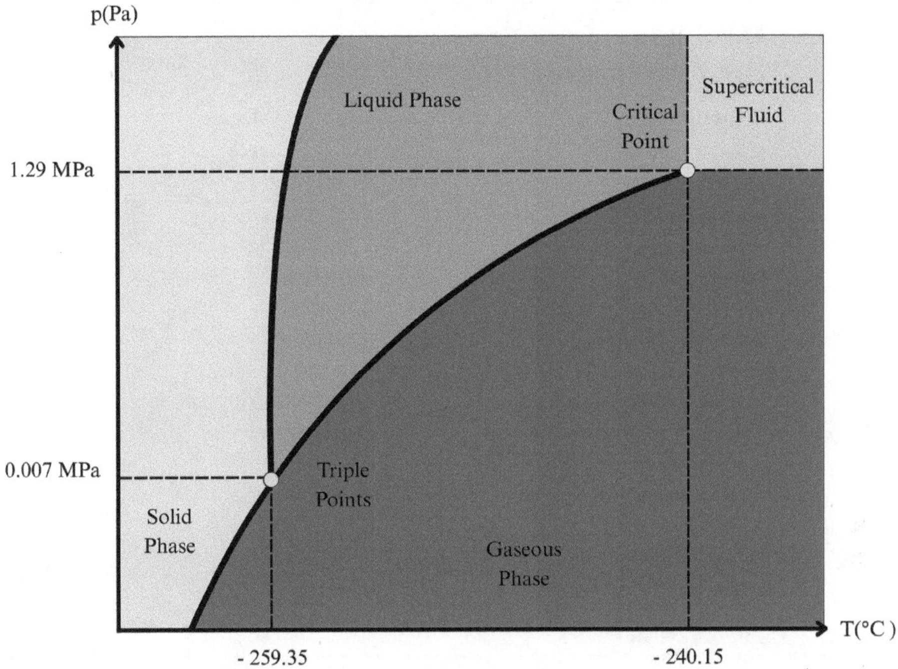

FIGURE 8.1 Hydrogen pressure vs. temperature diagram.

The triple point of hydrogen occurs at a temperature of −259.35°C and a pressure of 0.007 MPa, where all three phases of hydrogen (solid, liquid, and gas) coexist in equilibrium. Therefore, studying hydrogen phases and its triple point is essential for understanding matter's fundamental properties and developing cryogenic technologies.

Though hydrogen gas is not toxic, the oxygen surrounding us can be displaced, leading to a hydrogen-rich area. Such displacement can prove harmful, as a person can get asphyxiated if the oxygen content present is minimal. Hydrogen, in its previous states, is colorless and odorless, so an odorant may be added for detection purposes. It is also nontoxic at low concentrations but may displace oxygen in the air and cause asphyxiation at high concentrations [2]. Regarding corrosion, it may cause the corrosion of some metals at high temperatures and pressures. Other key properties of hydrogen gas are in Table 8.1.

Moreover, the flame produced by a mixture of hydrogen, air, and oxygen is almost invisible in daylight and emits mostly in the infrared and ultraviolet regions. Exposure to high levels of thermal radiation emitted from a hydrogen flame can cause severe damage (Table 8.2) [3].

Liquefied hydrogen has a low boiling point of −253°C and can cause frostbite or hypothermia if it contacts the skin or eyes. Continuous vaporization of liquid hydrogen occurs, and if it is confined within a pipe, the pressure will increase significantly as it heats up (Table 8.3).

TABLE 8.1
Hydrogen Properties [3]

Property	Value
Molecular formula	H_2
Molecular weight	2.016 g/mol
Boiling point	−252.87°C
Melting point	−259.14°C
Density at STP	0.0899 g/L
Specific gravity	0.0696
Solubility in water	0.001 g/L
Flammability range (in air) at 1 atm	4%–74% (by volume)
Detonable limits (in air) at 1 atm	18.2%–58.9% (by volume)
Autoignition temperature at 1 atm	500°C–577°C
Ignition energy	0.02 mJ
Thermal conductivity	0.1805 W/m K
Viscosity	8.7610–6 Pas
Heat of combustion	141.8 kJ/mol
Global warming potential	0 (none)

TABLE 8.2
Gaseous Hydrogen Hazard-Associated Properties

Property	Hazard Description
Colorless	It cannot be detected in any amount.
Volumetric leakage	Volumetric leakage from containers and pipelines is anticipated to be larger than gaseous methane leakage by a factor of 1.3–2.8 and four times greater than air leakage under comparable conditions.
Invisible flame	A flame produced by a mixture of hydrogen, air, and oxygen is almost invisible in daylight, emitting primarily in the infrared and ultraviolet regions.
Flam radiation	Severe damage may occur due to exposure to high levels of thermal radiation emitted. Atmospheric moisture absorbs the thermal radiation from the hydrogen flame.
Density at boiling point 1 atm	1.33 kg/m^3
Heat capacity	8.1 kJ/kg °C, at 256°C

TABLE 8.3
Liquid Hydrogen Hazard-Associated Properties

Property	Hazard Description
Boiling point	The boiling point of hydrogen is low, which is −253°C at sea level (at pressure 1 bar, 14.7 psia). Therefore, liquid hydrogen touching the skin, or the eyes can lead to frostbite or hypothermia. In addition, breathing in vapor or cold gas can cause asphyxiation.

(Continued)

TABLE 8.3 (*Continued*)
Liquid Hydrogen Hazard-Associated Properties

Property	Hazard Description
Evaporation	Liquid hydrogen continuously vaporizes, turning into a gaseous state.
Pressure increases	Liquefied hydrogen confined within a pipe between two valves will eventually be heated up and reach the temperature of its surroundings. In this case, the pressure will increase significantly.
Density at boiling point 1 atm	67.758 kg/m^3
Heat capacity	14.3 kJ/kg °C, at 25°C
Heat capacity	8.1 kJ/kg °C, at 256°C

8.4 HYDROGEN ACCIDENT RISKS

Hydrogen's accident risks can be classified into several categories: Industrial failure, ignition and fire, leakage, etc.

8.4.1 Industrial Accidents

Analyzing the accidents that have occurred, it is noticeable that industrial failures can be attributed primarily to the mechanical failures of containment vessels, piping, or auxiliary components due to factors such as brittle failure, Hydrogen embrittlement (HE), or freeze-up, the reaction of the fluid with a contaminant such as air, failure of safety devices, and operational errors.

To avoid these failure reasons/causes, safety measures such as regular inspections and maintenance, proper safety equipment and procedures, and regular safety training are essential to prevent accidents.

Moreover, the systems should be designed to ensure that a single failure does not cause a chain of failures, which could increase the potential for hazards and damage.

8.4.2 Ignition and Fire

Several components of hydrogen systems might catch fire or explode due to various ignition sources. These sources can be one but not limited to the following causes.

- **Hot Surfaces**: Materials heated to a sufficient temperature can ignite flammable materials, which include hot plates, engine parts, and overheated machinery.
- **Flames and Hot Gases**: Open flames, sparks, and hot gases from burning materials or processes (such as welding) can ignite nearby flammables.
- Mechanically generated sparks are created by friction, impact, or grinding of metals. These can be hot enough to ignite certain flammable mixtures.

- Normal operation of electrical equipment can produce sparks or hot surfaces sufficient for ignition.
- Stray electric current, cathodic corrosion protection: Unintended electrical currents, such as those from faulty wiring or cathodic corrosion protection systems, can generate heat and sparks, posing an ignition risk.
- The sudden release of accumulated static charge can create a spark, common in dry environments and with certain materials.
- Lightning is a powerful natural electrical discharge that can easily ignite flammable materials over a wide area.
- **Electromagnetic Fields (9 kHz to 300 GHz)**: Radio waves in this frequency range can induce currents in conductive materials, leading to heating and potential sparks.
- **Electromagnetic Radiation**: Radiation in specific frequency or wavelength ranges (e.g., lasers, microwaves) can cause heating and ignition if absorbed by a flammable material.
- **Ionizing Radiation**: High-energy radiation (alpha, beta, gamma, X-rays) can cause chemical reactions that generate heat and may lead to ignition in some cases.
- **Ultrasonics**: High-frequency sound waves can generate heat through friction and vibration, potentially igniting sensitive materials.
- **Adiabatic Compression and Shockwaves**: Rapid compression of gases, as in explosions or shockwaves, can generate high temperatures that ignite flammable mixtures.
- Exothermic reactions are chemical reactions that release heat. If the heat is not dissipated quickly enough, it can ignite surrounding materials.

When gaseous hydrogen is present, it quickly disperses with air turbulence, increasing the rate of gaseous hydrogen dispersion. In the case of a liquid hydrogen (LH_2) spill, evaporation can happen quickly, resulting in a flammable mixture forming. Even if there are no ignition sources at the leakage location, a fire could still occur if the flammable mixture moves and reaches an ignition source. Safely handling liquid hydrogen hinges on several critical engineering and operational factors. Material selection is paramount, and it includes tanks, vessels, and related equipment to withstand cryogenic temperatures without becoming brittle or succumbing to thermal cycling stresses. Adequate insulation is vital to minimize heat leaks and prevent hydrogen boil-off, the evaporation of liquid hydrogen due to heat input. In addition, pressure relief mechanisms are necessary to avert over-pressurization caused by boil-off or other process upsets.

Furthermore, all relief devices and valves must be designed explicitly for liquid hydrogen service, ensuring they function reliably and avoid blockages due to air solidification at cryogenic temperatures. Pressure and temperature sensors should also identify potentially dangerous conditions such as elevated pressure or excessive vaporization. Further, strong operational protocols are essential and must be considered when handling LH_2. Developing comprehensive emergency procedures to tackle possible release, fire, and explosion situations is vital.

Gas and leak detection systems and automatic controls such as isolation and shutdown valves can effectively contain and mitigate potential releases. Further, ignition sources must be identified and eliminated using appropriately rated equipment in hazardous areas. Finally, minimizing congestion and avoiding vertical confinement can promote the effective dispersion of any released hydrogen, reducing the risk of explosions.

Therefore, it is essential to understand the characteristics of hydrogen related to the ignition to avoid explosion accidents. For instance, the auto-ignition temperature is the minimum temperature required to initiate the combustion reaction of a fuel-oxidizer mixture in the absence of an external source of ignition. The standard auto-ignition temperature of hydrogen in air is around 500°C [4], which is relatively higher than many hydrocarbons. In addition, it is essential to remember that objects with a lower temperature can also cause ignition under prolonged contact, even at less than atmospheric pressure, due to the accumulative effect of the heat source. Another important parameter to consider is the minimum ignition energy, which signifies the least amount of electric energy stored in the discharge circuit, minimizing losses along the leads, that can ignite a still mixture when discharged across a spark gap at its most ignitable composition [5].

8.4.3 Leakage

The leak can happen inside or outside the system. It can be hazardous if air or contaminants leak slowly into a cold liquid hydrogen LH_2 or slush liquid hydrogen (SLH_2) system through cryo-pumping. These leaks can be caused by damaged seals, gaskets, valves, flanges, or equipment and can lead to further damage.

When liquid hydrogen (LH_2) leaks, it can vaporize and create fog due to moisture condensation, which can serve as a warning. However, if leaks go undetected, they can result in fires or explosions.

The best approach to avoid or mitigate the hydrogen leak is to keep it within the piping and use specific setups that employ gas detection systems to ensure safe operations. Others can use electronic leak monitoring instead of detection measures. In indoor piping, it is essential to monitor the pressure continuously when the system is not in use. Excess flow protection should also be in place. This can include employing excess flow valves, flow-limiting orifices, control systems designed to restrict total flow, or a combination of the previous. Since hydrogen flame is invisible to the naked eye and is odorless, hydrogen gas and flame detectors must be utilized, and they should be linked to both audible and visible alarms. Additionally, detectors may trigger the isolation or shutdown of the source valve at the hydrogen supply.

8.4.4 Embrittlement

HE causes metal degradation. This phenomenon depends on the purity of the metals, the time of exposure to hydrogen, and the conditions of the surroundings.

The more the metal is pure, the more embrittlement occurs. As shown in Figure 8.2a, metal exposure to hydrogen results in cracking, and these cavities might be filled with hydrogen molecules, creating pressure inside them, as shown in Figure 8.2b. Despite the challenges caused by embrittlement, several solutions have been recently proposed. For instance, certain metal alloys, such as austenitic

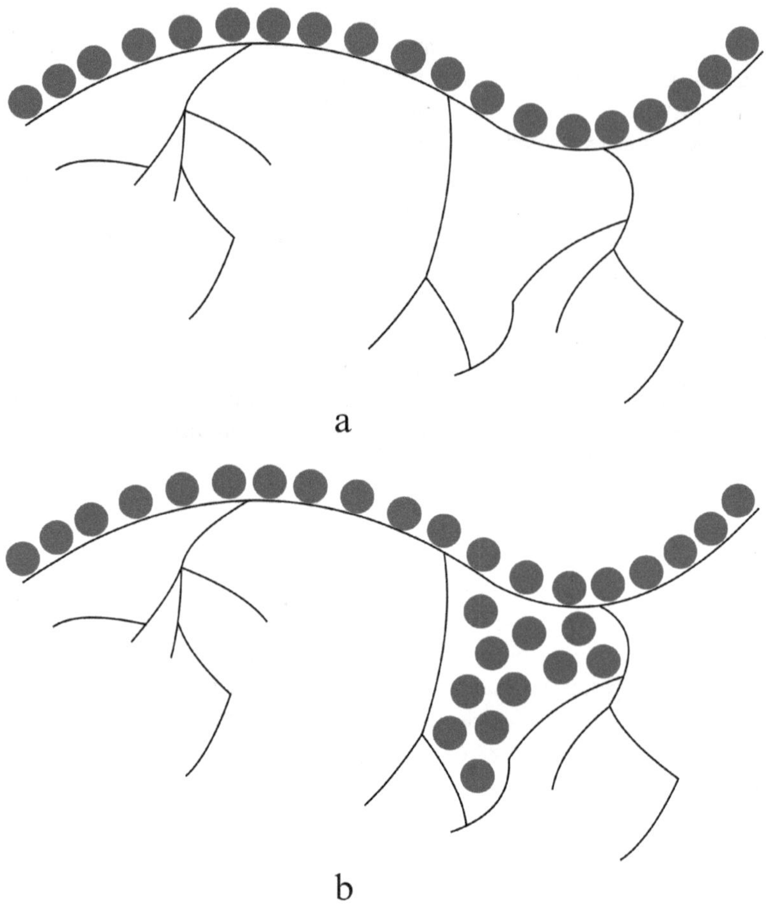

FIGURE 8.2 (a) Cracking due to exposure to hydrogen. (b) Cavities might be filled with hydrogen molecules.

stainless steels, aluminum alloys, low-alloy ferritic steels, and copper alloys, exhibit more excellent resistance to HE [6,7]. The use of austenitic stainless steel and aluminum alloys, which can withstand cryogenic temperatures, is preferred for liquid hydrogen systems. Conversely, metals such as high-strength ferritic alloys, cast iron, titanium alloys, and nickel alloys, which are more susceptible to embrittlement, should be avoided. Recently, high-entropy alloys (HEAs) have been explored as a potential solution to the HE challenge in conventional hydrogen transport and storage materials. HEAs are designed with increased ductility and phase stability to enhance resistance to HE [8,9]. They incorporate multiple elements, achieving high entropy levels by promoting microstructural modifications and improved mechanical properties. Furthermore, research into leak-resistant coatings, including ceramic-based barriers such as oxides, nitrides, and carbides and alternative pipe materials such as polymers with metallic linings or epoxy coatings, continues to advance. The ASME B21.12 standard guides repurposing pipelines, including pressure reduction, laboratory testing options, and successful examples, such as the repurposed 12 km high-pressure hydrogen pipeline in the Netherlands, which demonstrates the feasibility of mitigating HE through careful material selection and implementation of mitigation procedures.

8.5 LOWER AND UPPER EXPLOSIVE LIMITS OF HYDROGEN

Both the lower and upper explosive limits (LEL and UEL) are defining parameters regarding the combustible capability of a vapor or gas in the presence of air. The mixture is capable of achieving self-ignition and sustaining combustion when ignited at the LEL, which is the lower limit. At this level, the mixture is too lean to burn due to the insufficient fuel it possesses. On the other hand, UEL indicates the upper level of concentration, beyond which the mixture becomes too rich to ignite as there is not sufficient oxygen for combustion. In a nutshell, flammability has an upper and lower boundary, which is identified by the LEL and UEL. Additionally, if an ignition source is present, it defines the concentration of gas that has the potential to explode. Obscured by the explosion range are safety parameters that must be addressed in industries dealing with ignitable substances. It is fundamental to have an understanding of these limits for precise risk evaluation, using precautionary measures to avert explosive concentration, and providing a reliable emergency response mechanism to deal with potential leakage or spillage. All LEL and UEL values are not the same for all gases and vapors; these values are usually incorporated in their respective safety data sheets and should be carefully considered. Figure 8.3 compares the LEL and UEL of hydrogen with other gases.

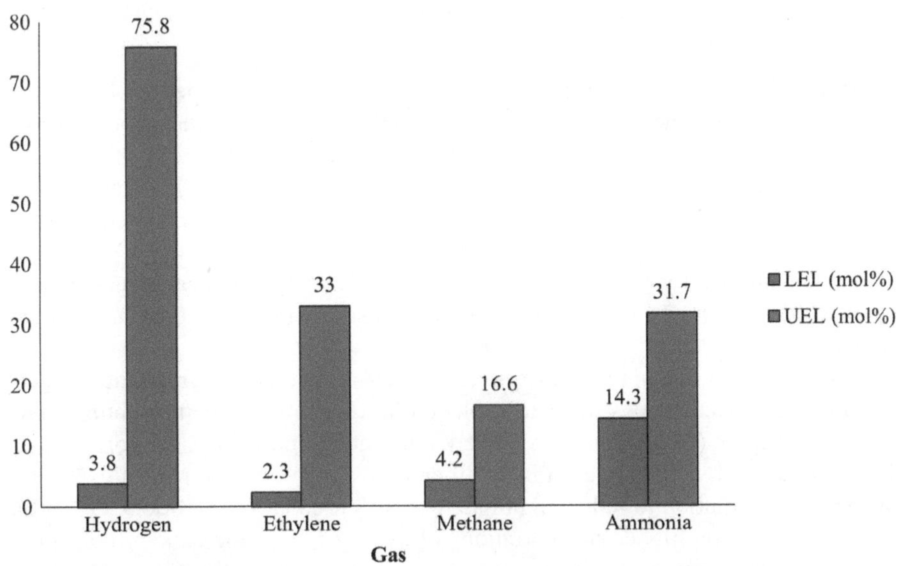

FIGURE 8.3 Explosion limits of hydrogen, ethylene, methane, and ammonia in the air according to DIN 51649-1.

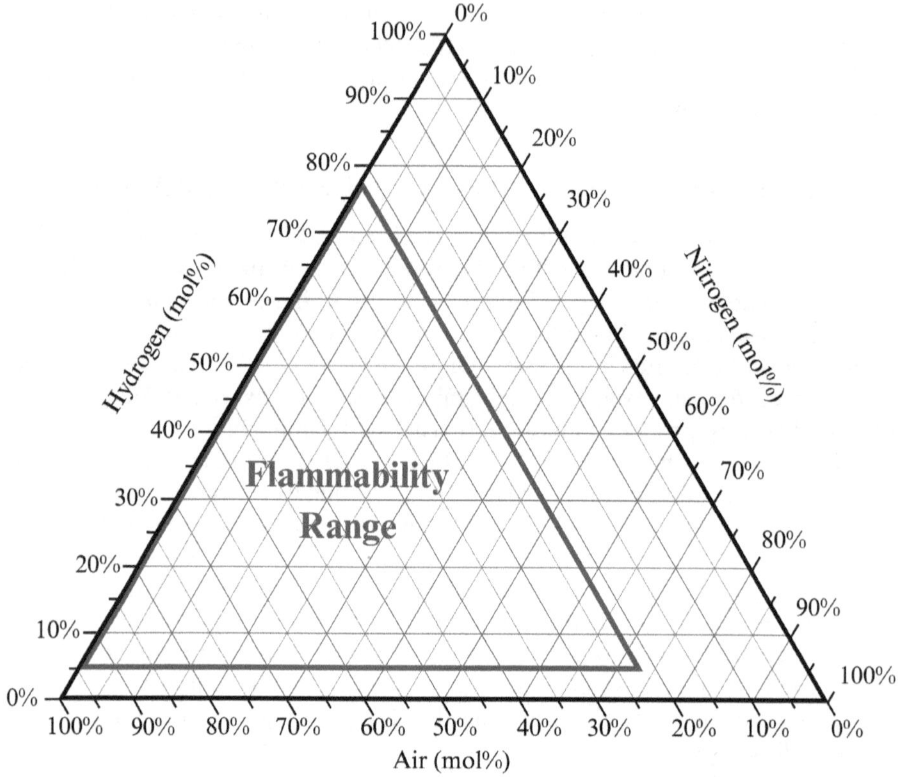

FIGURE 8.4 Maximum flammable hydrogen for which a mixture of the flammable gas /in nitrogen is not flammable in air (mol%).

Since the infrastructure to transport hydrogen through new pipelines could take years to be ready. The repurpose of the natural gas pipelines was proposed as a promising solution for hydrogen transportation at the current time. The transportation of hydrogen with natural gas impacts the explosive limit of the fuel–gas mixture. Also, it affects the minimum concentration level in the air required for combustion to occur when an ignition source is present; see Figure 8.4. Le Chatelier's Principle is a method applied to systems at equilibrium. It offers a helpful conceptual analogy for understanding how a fuel gas mixture's composition influences flammability and explosive limits.

Mixing a small amount of hydrogen (10%–50%) to methane significantly widens the range of concentrations that can explode (the UEL), while the minimum explosive concentration (LEL) remains relatively stable at around 4.5%. Therefore, mixing the hydrogen with methane, even in small amounts, increases the risk of explosion. One proposed solution to reduce flammability is to use carbon dioxide (CO_2) in the mixture. Therefore, higher concentrations of hydrogen with methane require more CO_2 to effectively prevent ignition, which is essential for industries that produce biogas, where CO_2 is readily available. In addition, the explosion region increases with increasing hydrogen fraction in nonlinear form; see Figure 8.5.

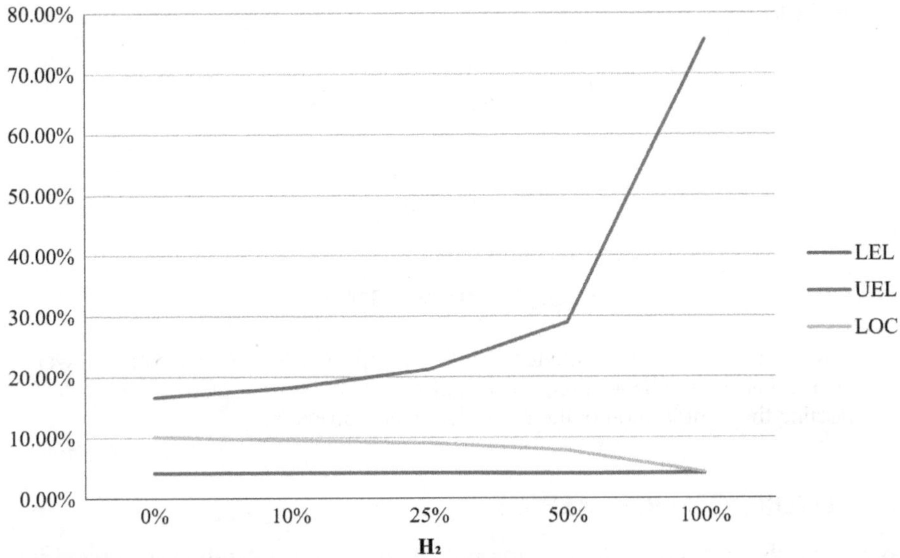

FIGURE 8.5 Hydrogen percentage mixing with methane impacts LEL, UEL, and LOC.

8.6 LE CHATELIER'S PRINCIPLE AND THE BURGESS–WHEELER EQUATION

The lower explosive limit (LEL) of a fuel gas mixture is the minimum concentration in air required for combustion to occur when an ignition source is present. Le Chatelier's Principle is a method applied to systems at equilibrium. It offers a helpful conceptual analogy for understanding how a fuel gas mixture's composition influences flammability and LEL. While Le Chatelier's Principle is a conceptual analogy here, in practical terms, empirical rules can be used to estimate the LEL of fuel gas mixtures, known as mixture rules. One of the common rules is the Burgess–Wheeler equation:

$$LEL_{mixture} = \frac{1}{\sum_{i=1}^{n} (Y_i / LEL_i)} \tag{8.1}$$

Where:

- $LEL_{mixture}$ is the LEL of the gas mixture.
- Y_i is the volume fraction (or mole fraction, approximately the same for gases at low pressures) of component (i) in the fuel gas mixture.
- LEL_i is the lower explosive limit of pure component i.

Example 8.1: Low Level of Explosion for Gas Mixture

Suppose you have a mixture of Methane (CH4): 80% by volume, LEL ≈ 5%, and Propane (C_3H_8): 20% by volume, LEL ≈ 2.1%. Find the LEL of the mixture.

Solution:

$$\text{LEL}_{\text{mixture}} = \frac{1}{(0.8/5\%) + (0.2/2.1\%)}$$

$$\text{LEL}_{\text{mixture}} = \frac{1}{25.52}$$

$$\text{LEL}_{\text{mixture}} = 0.0392 \approx 3.92\%$$

So, the estimated LEL of this mixture would be approximately 3.92%. This is lower than the LEL of methane alone (5%) and closer to the LEL of propane (2.1%), reflecting the contribution of the more flammable propane.

8.7 HYDROGEN RISK ASPECTS

As previously mentioned, the hazards regarding hydrogen can be categorized into physiological, physical, and chemical risks. Although the usage of hydrogen gas is beneficial, there are certain risks that require attention while handling it. These risks can be classified into three categories: physiological risks, physical risks, and chemical risks. The physiological risks of hydrogen include asphyxiation, burns from heat, frostbite, internal body temperature drop, and exsanguination due to pressure. These types of risks are illustrated in Figure 8.6.

Asphyxiation is the risk that takes place when the attending hydrogen concentration is so high that oxygen becomes unavailable in the air.

- Thermal and cryogenic burns are dangerous burns that occur once gaseous or liquid hydrogen comes in contact with the skin or tissues of the eyeball, resulting in frostbite, among other injuries.

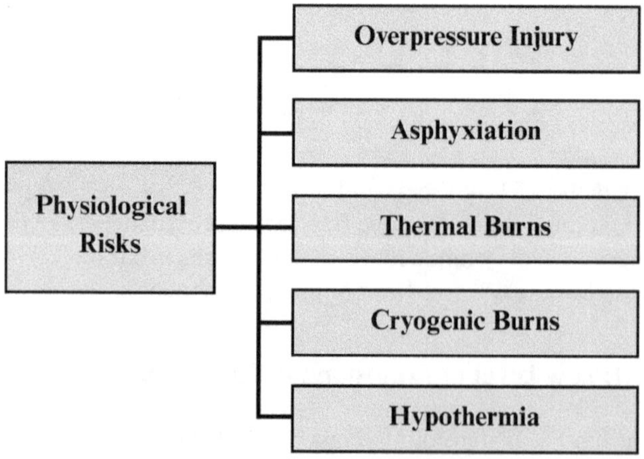

FIGURE 8.6 Physiological hazards are associated with hydrogen.

- Hypothermia is the risk that occurs when hydrogen vapor or liquid is ingested or breathed in, leading to the loss of heat by the body.
- Overpressure injury arises once hydrogen in gaseous form is stored or transported under high pressure, causing damage to lungs, ears, or other organs.

The physical hazards consist of HE and the thermal stability of certain structural materials. Recall that metals become brittle and cracked when hydrogen gas diffuses into them. This process is known as HE. In this case, the chemical hazards include the flammability of hydrogen, sources of ignition, and explosion phenomena. General considerations and accident statistics refer to the following overall risk of accidents and the history of such accidents [1].

The ignition of hydrogen is one of the prime concerns due to the presence of a spark or flame. Ignition sources, in this case, would include any heat source, electricity, and everything else that has the capability of igniting hydrogen gas.

Hydrogen is well known for exploding with the surrounding air, and this can only happen if there is sufficient oxygen and the required amount of heat. This would call forth explosion phenomena that bear the characteristics of hydrogen explosions. These would include the rate at which pressure rises, the duration of the explosion, and the extent of damage incurred.

To reduce these risks, it is crucial to understand the characteristics of hydrogen gas, the materials and equipment used in handling it, and appropriate engineering controls. This includes appropriate storage and transportation vessels, hydrogen concentration measurement in the work atmosphere, and accident prevention safety measures. These approaches will assist in safely handling the gas while protecting individuals from potential harm.

8.8 HYDROGEN STORAGE SAFETY

There are numerous problems regarding the storage and handling of hydrogen. Some of the methods regarding the storage of hydrogen include:

- Storing hydrogen gas (GH_2) under low pressure. In this case, the risk will be the hydrogen's uptake by the metal to become a metal hydride.
- Liquid hydrogen (LH_2). Liquid hydrogen is three times more energy dense than gasoline in gravimetric energy density, yet lower in volumetric energy density. Both forms of hydrogen are extremely combustible, pose cryogenic threats, and cause pressure buildups. Unlike gaseous hydrogen, river hydrogen is stored and transported in a liquid state, which adds to the risks associated with very low temperatures and rapid vaporization [4,5].

Liquid hydrogen is advantageous where high purity is required. Nevertheless, it is not energy-neutral since it suffers from boil-off losses, overpressure remedial action by stabilizing temperature and liquefaction energy.

In most cases, hydrogen is stored either in a gaseous state or in under liquid form at moderate temperatures and mild to high temperatures and pressures. The storage tank has to be built with suitable materials and has pressure-release vents together with flashback arresters.

However, there is still a significant potential hazard for the mechanical explosion of hydrogen vessels when exposed to high temperature or thermal radiation, especially if the vessel is designed to store hydrogen gas under high pressure. In case of overheating, the vessel may burst, spilling its flammable content, which can ignite and cause a fireball or jet fire.

Liquefied hydrogen vessels operate at moderate to low pressures, and their walls can be designed with lower pressure resistance than that for storage of hydrogen gas. However, overheating can still cause superheating of the liquid phase, which can result in an explosion.

Hydrogen can also be stored in a liquid state under cryogenic temperatures. However, liquid hydrogen has low viscosity and surface tension, which can cause it to leak easily. If a leak occurs, hydrogen will quickly evaporate and form a cloud of flammable gas that can ignite if exposed to a spark or flame.

Cryogenic storage of hydrogen also presents hazards related to extremely low temperatures. If inhaled, contact with liquid hydrogen can cause severe cold burns, frostbite, and damage to the eyes and respiratory system. The risk of such injuries is increased when handling liquid hydrogen in poorly ventilated areas. It is important to notice that adequate hydrogen ventilation requires careful consideration of several factors, such as the type of ventilation, forced or natural, depending on the required availability, reliability, effectiveness, and cost. For natural ventilation, inlets should be positioned near ground level and outlets at a high level, designed to prevent short-circuiting and promote crossflow across potential leak points. Mechanical ventilation rates must be sufficient to dilute hydrogen release, reduce reactivity and cloud size, and prevent accumulation. Local extraction ventilation points should be positioned around potential release points. It must be determined whether ventilation and extraction systems run continuously or are triggered by detectors; in the latter case, detector locations and set points are critical. Equipment that could act as an ignition source should be interlocked to shut down if the ventilation/extraction system fails. Finally, the level of congestion and confinement should be minimized to promote airflow and mitigate against gas accumulation.

8.9 HYDROGEN TRANSPORTATION SAFETY

Transporting hydrogen gas is done with tube trailers, which have cylinders able to carry hydrogen gas at pressures up to 2400 psi (165 bar), and some specialized types can even reach 6000 psi (413.68 bar). These steel or aluminum cylinders are encased in carbon fiber for added strength and safety. They must comply with very stringent safety requirements and have bursting disks fitted to reduce pressure build-up. If a fire occurs, the bursting disk will melt, allowing the cylinder to be vented [10].

Gaseous hydrogen refueling stations can be restocked by swapping out empty tube trailers for full ones, using cylinder banks, or replacing stationary cylinders [5,6].

Liquid hydrogen is transported in liquid semitrailers with 54,000–70,030 L capacity. They bring it to the customer's position, pouring it into stationary tanks. The liquid hydrogen tankers are constructed as stationary tanks, which are thermos bottles and are subject to the regulations for carrying hazardous cargo. The tanks are constructed from double-walled steel with vacuum insulation in between the walls. All these tanks are fitted with vents and overpressure relief devices that vent pressurized gas when it exceeds the safe level.

Overpressure relief devices and vents ensure safety and stability, and vacuum insulation maintains the low temperature of the hydrogen's liquid form. The extremely cold liquid hydrogen is transported in cryogenic trailers built like thermos bottles with double walls. Should the pressure in the tank rise because hydrogen is warming and transitioning to gas? Pressure relief devices will be activated to release the pressure.

8.10 FUEL CELL ELECTRIC VEHICLE SAFETY

Like any vehicle equipped with an internal combustion engine, fuel cell electric vehicles (FCEVs) are also safe to drive. Still, there are possible risks. Manual protocols to handle FCEVs during emergencies will become important as these vehicles grow in popularity. While the market for FCEVs remains modest, it is developing at a fast pace. Supportive public integration of FCEVs requires responsible social adoption of this innovative green technology.

FCEVs are also eco-friendly and capable of eliminating harmful carbon emissions from operating vehicles. These vehicles are designed to automatically shut off during a collision, lowering the likelihood of fires starting (in the event of a leak, hydrogen leakage detectors will stop burning more hydrogen and alert listeners) [13]. Moreover, the hydrogen fuel is stored within unifiable high-pressure tanks, which are specially designed to be shock and leak-proof. It should be noted that the responders first in line will need to take precautions when there's a hydrogen fuel leak, as hydrogen burns fiercely and might combust easily. Responders also need to be trained to manage hydrogen fuel cell vehicles during emergencies because their structure and how they operate differ from standard internal combustion engines. For optimal weight distribution, space efficiency, and easy access during hydrogen station refilling, the hydrogen tank of FCEVs is usually positioned at the rear of the vehicle under the trunk or cargo area. FCEV components are displayed in Figure 8.7.

8.11 FUELING STATIONS SAFETY

FCEV stations store and dispense hydrogen fuel, which is available at specialized stations with several safety features such as combustive leak geology, storage, delivery of hydrogen, and overpressure protection vents. The amount of stored hydrogen and pressure varies between stations, 300–700 bar [14]. True zero offers a Del Mar location with fueling options at 350 and 700 bars. True zero is a publicly traded corporation providing hydrogen fueling stations and related services. For more insights into the available hydrogen stations, the Alternative Fuels Data Center website provides a database of the available fueling stations and the available pressure [15].

Stations can store anywhere from 100 to 1000 kg of hydrogen, less than 20,000 gallons (about 75,708.2 L) to 40,000 gallons (about 151,416.4 L) in underground tanks. Hydrogen can be produced on-site through electrolysis or obtained from natural gas and delivered in either liquid or gaseous form. Some stations receive hydrogen through an underground pipeline, while others store it in double-walled steel tanks

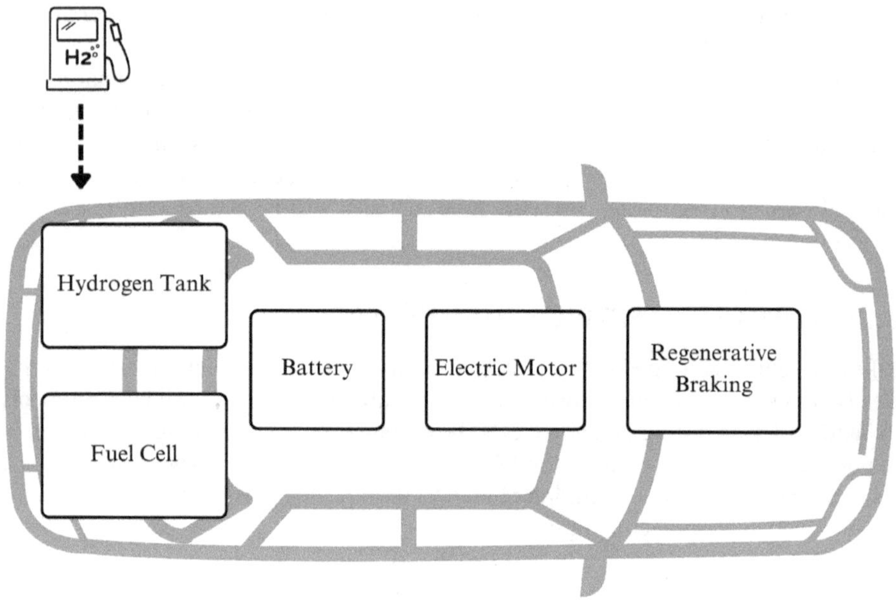

FIGURE 8.7 FCEV construction and main components.

with vacuum insulation. Regardless of the storage method, only gaseous hydrogen is dispensed to the vehicle [16].

It is important to classify hazard regions in hydrogen fueling stations for safety measures by defining different hazard levels according to the hydrogen concentration. Figure 8.8 presents a simplified, two-dimensional layout of safety zones within a hydrogen fueling station, illustrating the strategic design of safety areas about potential hydrogen release points. The plot utilizes a coordinating system with X and Y axes to represent the spatial dimensions of the station. Distinct safety zones are delineated using dashed lines, each labeled according to its associated hazard level: Zone 0 (highest hazard), Zone 1 (high hazard), Zone 2 (moderate hazard), and nonhazardous (minimal hazard). These zones are depicted as rectangular areas strategically positioned to ensure that high-risk areas are adjacent to potential hydrogen release points.

Colored circular markers represent key components of the fueling station. A blue circle signifies the hydrogen storage location, a green circle indicates the dispenser location and an orange circle denotes the control area. The arrangement of these components and the safety zones reflects fundamental hazard safety design principles, with the highest hazard zone (Zone 0) situated closest to the storage unit and the control area within the nonhazardous zone. A simulated hydrogen concentration distribution is superimposed onto the safety zone layout, employing a gradient color spectrum. Lighter shades indicate higher hydrogen gas concentrations, particularly within the storage area (Zone 0), while darker shades signify lower concentrations. This distribution pattern is based on a simplified Gaussian function, visually representing potential gas dispersion.

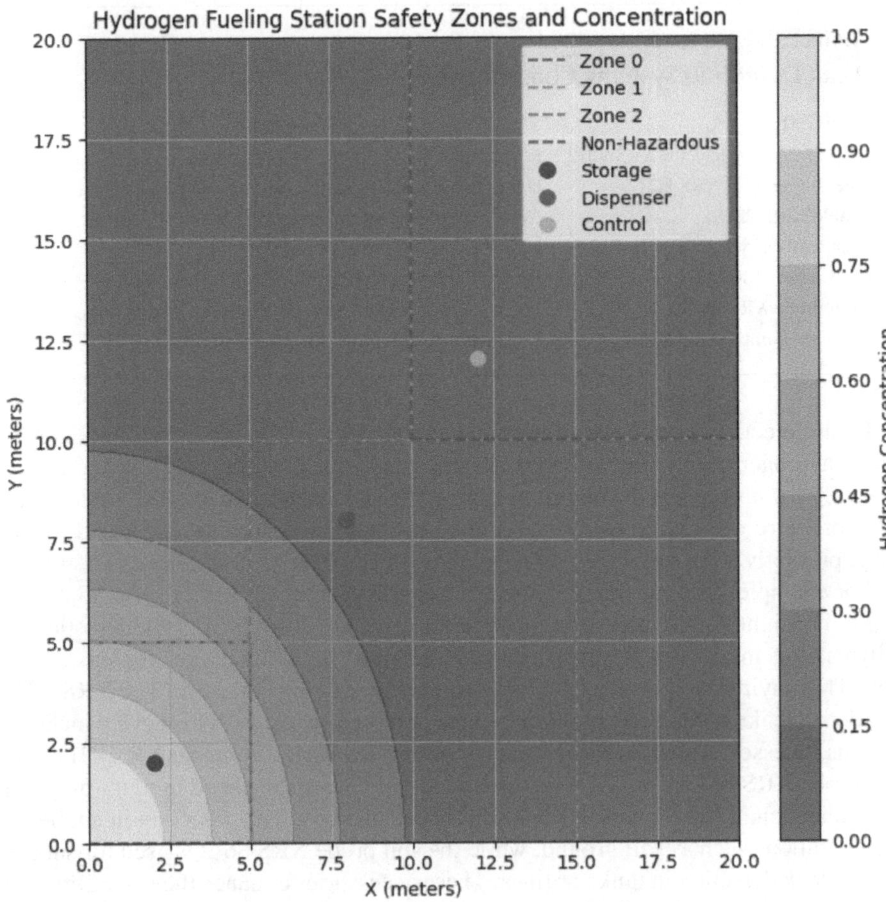

FIGURE 8.8 Simplified 2D hydrogen fueling station safety zones.

It is crucial to emphasize that the numerical values defining the zone boundaries and concentration levels in this scenario are entirely fictional and abstract, bearing no resemblance to any verified real-world situation. Consequently, the figure's purpose is purely schematic, intended to convey the conceptual framework of safety zoning principles and hazard data visualization in the context of a hydrogen fueling station. The actual determination of safety zones necessitates rigorous computational fluid dynamics (CFD) analysis, adherence to relevant codes and standards, and consideration of specific site and design conditions.

8.12 SAFETY DEVICES

Pressure relief valves prevent the over-pressurization of hydrogen storage tanks and pipelines. They are typically set to open slightly above the system's normal operating pressure to allow for small pressure fluctuations [12].

TABLE 8.4
Leak Detection Systems Components

Category	Purpose
Hydrogen leak detector XRS 9012	Main device for hydrogen leak detection.
Measuring head XRS H21	Hand-held measuring head connected to the leak detector.
Cable XRS C21S	Cable for connecting the measuring head to the detector.
Standard carry handle XRS P12	Optional handle for easier transportation.
Soil sensor XRS 8212	Sensor for leak detection in soft ground conditions.
Soil probe XRS 8612	Probe for leak detection in subterranean tanks or lines.
Vacuum pump	Used with the soil probe for gas absorption.

Leak detection systems are important safety devices for producing and storing green hydrogen. They are designed to detect leaks in pipelines and storage tanks, allowing for prompt repair and minimizing the risk of gas release. The system utilizes pressure drop monitoring and hydrogen-specific sensors to detect and locate leaks promptly.

For example, a commercial Hydrogen Leak Detector, XRS 9012, is available. It is a lightweight, handy measuring device for efficient, cost-saving leak detection. It utilizes a gas mixture of 5% hydrogen and 95% nitrogen as a trace gas for leak detection. This environmentally friendly mixture is safe and noncorrosive. The XRS 9012 can detect leaks of any size, providing quick and precise results. This device includes various accessories such as the measuring head XRS H21, soil sensor XRS 8212, and soil probe XRS 8612. The XRS H21 is a hand-held measuring head used for pinpoint leak detection. The soil sensor XRS 8212 is designed for leak detection in challenging conditions such as soft ground, while the soil probe XRS 8612 is used for subterranean leak detection in tanks or lines. These accessories enhance the versatility and effectiveness of the XRS 9012 system (Table 8.4).

- **Emergency Shutdown Systems**: These systems automatically shut down equipment in an emergency, reducing the risk of injury or damage [17].
- **MV Butterfly Valve**: The MV Butterfly Valve by IMI Orton includes emergency shutdown systems. It performs well under extreme temperatures and pressures. The valve is bidirectional and can be automated, making it suitable for emergency shutdown applications, including HIPPS. The vacuum-jacketed option makes it ideal for liquid hydrogen service.
- **C-Rex™**: The IMI Truflo C-Rex™ valve is designed for demanding applications, including emergency shutdown systems. The C-Rex™ valve is suitable for high-cycle switching applications, handling solid/slurry media, and operating at high temperatures and cryogenic conditions. Its top-entry design allows for easy in-line maintenance.
- **O-Rex™**: The IMI Truflo O-Rex™ valve is also designed for demanding applications, including emergency shutdown systems. See Table 8.5 to explore the differences between the three types of compressors.

TABLE 8.5
Examples of Emergency Shutdown Systems

Valve Model	Size Range (inches)	Pressure Range (bar)	Temperature Range (°C)
MV Butterfly	2–160	150–1500	−253 to 649
C-Rex™	1–40	150–2500	−196 to 649
O-Rex™	1–40	150–4500	−196 to 649

TABLE 8.6
Types of Fire Detectors

Type of Detector	Description
Thermal detectors	Detect fires by sensing changes in temperature or heat flux caused by the fire.
Smoke detectors	Detect fires by sensing smoke particles in the air.
Flame detectors	Detect fires by sensing the ultraviolet (UV) or infrared (IR) radiation emitted by flames.
Gas detectors	Detects fires by sensing the presence of hydrogen gas.

- **Fire Suppression Systems**: These systems use water or foam to extinguish fires, reducing the risk of damage to equipment and protecting workers.
- **Fire Detectors**: Effective fire detection systems are important to quickly detect and alert individuals to the presence of a fire. Several types of detectors can be used to detect fires, each with a unique way of sensing the presence of a fire. Table 8.6 presents the commonly used types of fire detectors.

Gas detectors are devices designed to detect the presence of various gases in the air. They are essential in various industries, including manufacturing, mining, and oil and gas, where certain gases pose a significant safety risk to workers. Several types of gas detectors are available, each with a unique detection method. Table 8.7 presents the commonly used types of gas detectors.

8.13 CONSIDERATIONS FOR MATERIALS SELECTION TO BE USED IN HYDROGEN APPLICATIONS

The small molecule size of hydrogen can cause damage to storage, piping, and appurtenance materials. This can lead to corrosion, embrittlement, and cracking of materials used in hydrogen-fueled and stationary fuel cell storage systems. Several resources are available to assist in selecting appropriate materials for hydrogen technologies [11].

Table 8.8 summarizes the resources that provided specifications of materials commonly used in hydrogen storage, transportation, and applications.

TABLE 8.7
Types of Gas Detectors

Type of Detector	Description
Catalytic	A heated wire coil is used to detect the presence of combustible gases by measuring the temperature change caused by combustion reactions.
Infrared	Measures the absorption of specific wavelengths of infrared light by gases in the air to detect their presence. Different gases absorb different wavelengths of light, allowing for selective detection of specific gases.
Electrochemical	Measures the concentration of gases based on their electrochemical reaction with a sensing electrode. These detectors often detect toxic gases such as carbon monoxide or hydrogen sulfide.
Photoionization	Uses ultraviolet light to ionize gas molecules, creating a detectable current. These detectors are often used to detect volatile organic compounds.
Semiconductor	It uses a metal oxide semiconductor to detect changes in resistance caused by the presence of gases. These detectors often detect flammable gases such as methane or propane. An example is the MQ2 Gas sensor.

TABLE 8.8
Resources That Provide Specifications of Materials Commonly Used in Hydrogen

Resource	Information Provided
American Society of Materials (ASM) handbook series, Volume 13a [18]	A chapter on hydrogen's effects on materials.
ASME B31.12 Hydrogen Pipelines and Piping code [19]	Specifies materials for hydrogen piping.
ASME BPV code [20]	Extensive information on material selection and testing for hydrogen usage, including metals and composite materials for container construction.
Canadian Standards Association (CSA) document CSA CHMC-1 [21]	Material testing for hydrogen applications, including uniform test methods for evaluating material compatibility with compressed Hydrogen.
American Society of Testing Materials (ASTM) [22]	"Hydrogen Embrittlement: Prevention and Control ASTM STP 692" is a book covering test methods to evaluate metal susceptibility to hydrogen attack.
Sandia National Laboratories guide [23]	Detailed material selection criteria for various materials that could be used in hydrogen applications.

8.14 HYDROGEN-SPECIFIC CODES

Precise protocols must be adhered to during hydrogen handling and storage to avoid safety hazards. For this reason, numerous organizations worldwide have formulated hydrogen safety codes, rules, and standards relevant to its infrastructure. These regulations lay down the dos and don'ts regarding the safety of handling, storing, using, and transporting hydrogen, as well as designing and operating hydrogen infrastructure.

The National Fire Protection Association (NFPA) is one of the leading organizations formulating hydrogen safety codes. NFPA has also developed a series of standards for the storage and use of hydrogen, which are included in the NFPA 2 Hydrogen Technologies Code and NFPA 55 Compressed Gases and Cryogenic Fluids Code. The former incorporates provisions for safe siting, installation, operation, and maintenance of hydrogen systems. At the same time, the latter sets forth protective measures for the storage and handling of compressed gases, including hydrogen [24].

In addition to those mentioned above, the International Organization for Standardization (ISO) and the International Electrotechnical Commission (IEC) also have hydrogen infrastructure standards and codes. For instance, ISO 19880-1 issues stipulations concerning the operation and design of hydrogen refueling stations. The EU has also concurrently formulated a set of regulations known as Fuel Cells and Hydrogen Joint Undertaking (FCH JU), which governs the safe manipulation and application of Hydrogen [19,20].

For instance, ISO has already developed several hydrogen standards that cover almost every segment of the hydrogen supply chain, from production, storage, transportation, and utilization. A few examples of ISO hydrogen standards are included in Table 8.9.

Moreover, regulatory bodies in different countries adopt different codes and regulations related to hydrogen safety. Table 8.10 summarizes some major regulations, codes, and standards related to hydrogen infrastructure. The safety standards of H_2 worldwide are summarized in Table 8.11.

TABLE 8.9
Some Examples of ISO Hydrogen Standards [24–26]

Standard Number	Description
ISO 14687	Hydrogen fuel quality—Product specification
ISO 16111	Transportable gas storage devices
ISO 23273:2013	Fuel cell road vehicles—Safety specifications—Protection against hydrogen hazards for vehicles fueled with compressed hydrogen
ISO 22734-1:2008	Hydrogen pipelines—Design, construction, operation, and maintenance
ISO 23828	Fuel cell road vehicle-energy consumption measurement part 1: Vehicles fueled with compressed hydrogen
ISO 13985	Liquid hydrogen—land vehicle fuel tanks
ISO 19882	Gaseous hydrogen—Thermally activated pressure relief devices for compressed hydrogen vehicle fuel containers
ISO 13984	Liquid hydrogen—land vehicle fueling system interface
ISO 17268	Gaseous hydrogen Land Vehicle Refueling Connection Devices
ISO 19880-3	Gaseous hydrogen—Fueling stations—Part 3: Valves
ISO 19880-5	Gaseous hydrogen—Fueling Station—Part 5: Hoses
ISO 19880-8	Gaseous hydrogen Fueling Station—Part 8: Hydrogen Quality Control
ISO 19881:2018	Gaseous hydrogen—Land vehicle fuel containers
ISO 22734:2019	Hydrogen generators using water electrolysis—Industrial, commercial, and residential applications

TABLE 8.10
Main Regulations, Codes, and Standards Related to Hydrogen Infrastructure [24–26]

Organization	Description
National Fire Protection Association (NFPA2)	Guidelines for safe installation, operation, and maintenance of hydrogen systems.
National Fire Protection Association (NFPA 55)	Guidelines for safe handling and storage of compressed gases, including hydrogen.
International Organization for Standardization (ISO)	Guidelines for the design and operation of several aspects of the hydrogen supply chain.
UL 2075	Guidelines for hydrogen sensors.
Department of Transportation (DOT)	Guidelines for the safe transport of hydrogen.
Occupational Safety and Health Administration (OSHA), 1910.103—Hydrogen.	Guidelines for safe handling and storage of Hydrogen in the workplace.
ASME B31.3 and B31.12	Guidelines for piping design, installation, and material selection.
International Mechanical Code (IMC)	Guidelines for ventilation for hydrogen usage in buildings.
The International Code Council (ICC)	Provide requirements for safely using and storing Hydrogen in buildings and structures.

TABLE 8.11
Hydrogen Safety Standards Worldwide

Standard Number	Description
ISO/TR 15916	Basic considerations for the safety of hydrogen systems
ISO 26142	Hydrogen detection apparatus—stationary applications
IEC 60079-29-1	Explosive atmospheres—Part 29-1: Gas detectors- Performance requirements of detectors for flammable gases
IEC 60079-29-2	Explosive atmospheres—Part 29-2: Gas detectors—Selection, installation, use and maintenance of detectors for flammable gases and oxygen
ISO 16110-1	Hydrogen generators using fuel processing technologies—Safety
IEC 62282-6-100	Micro fuel cell power systems—Safety—General requirements
IEC 62282-6-107	Micro fuel cell power systems—Safety—Indirect water-reactive (Division 4.3) compounds
IEC 62282-5-100	Portable fuel cell appliances—Safety
Plan No. 20130687-T-604	Micro fuel cell power systems—Safety—General requirements
IEC 62282-2-100	Fuel cell modules—Safety
IEC 62282-3-100	Stationary fuel cell power systems—Safety
JIS C 8822	General safety code for small polymer electrolyte fuel cell power systems
JIS C 8831	Safety evaluation test for stationary polymer electrolyte fuel cell stack
JIS C 8841–2	Small solid oxide fuel cell systems—Part 2: General safety codes and safety testing methods fuel cell system installation criteria fuel cell system installation criteria

(Continued)

TABLE 8.11 (*Continued*)
Hydrogen Safety Standards Worldwide

Standard Number	Description
CNS 16013	Fuel cell technologies—Fuel cell power systems for telecommunication backup power—Safety
SAE AIR 6464 EUROCAE Working Group 80 / SAE AE-7AFC	Hydrogen fuel cells aircraft fuel cell safety guidelines
IEC 62282-4-101	Fuel cell technologies—Part 4–101: Fuel cell power systems for electrically powered industrial trucks—Safety
ISO 6469-3	Electrically propelled road vehicles—Safety specifications—Part 3: Protection of persons against electric shock
ISO 23273:2013	Fuel cell road vehicles—Safety specifications—Protection against hydrogen hazards for vehicles fueled with compressed hydrogen
ISO 6469-1	Electrically propelled road vehicles—Safety specifications—Part 1: On-board rechargeable energy storage systems (RESS)
ISO 6469-2	Electrically propelled road vehicles—Safety specifications—Part 2: Vehicle operational safety
GB/T 36288-2018	Fuel cell electric vehicle fuel cell stack safety requirements
GB/T 31498-2021	Post-crash safety requirements for electric vehicle
GB/T 24549-2020	Safety and manual requirements for fuel cell electric vehicles
GB/T 34544-2017	Safety test methods for onboard low-pressure hydrogen storage devices for small fuel cell vehicles
GB/T 34866-2017	Safety requirements for all vanadium flow batteries
GB/T 31036-2014	Proton exchange membrane fuel cell backup power system: Safety
GB/T 27748.1-2017	Stationary fuel cell generation power system—Part 1 Safety
GB/T 33983.1-2017	Direct methanol fuel cell system—Part 1: Safety
GB/T 30084-2013	Portable fuel cell power system—Safety
GB/Z 34541-2017	Safety operation management regulation for hydrogen fueling facility of hydrogen vehicles
GB/T 37563-2019	Safety requirements for pressurized water electrolysis system for hydrogen production

8.15 RISK ANALYSIS THROUGH DEDICATED SOFTWARE

Different software tools could be dedicated to modeling hydrogen safety. For example, CFD is a powerful tool for predicting how fluids (gases and liquids) will behave in different scenarios. CFD has many applications in the field of hydrogen safety, including consequence analysis, safety design, safety approvals, licensing permitting processes, and safety demonstration.

In consequence analysis, CFD can be used to model and predict the effects of a hydrogen release, such as the dispersion and concentration of the gas in the surrounding environment. In safety design, CFD can help engineers design and optimize safety systems, such as pressure relief devices, to ensure they function correctly and effectively.

CFD can be used to demonstrate the safety of hydrogen systems and identify and mitigate potential risks, ensuring that they are designed, built, and operated safely.

CFD tools are utilized in various case studies for different purposes. For instance, the SUSANA project uses CFD for safety engineering design and hydrogen and fuel cell technology assessment. The project aims to develop a model evaluation protocol covering all aspects of safety assessment modeling, from release to combustion, to inform stakeholders about the state-of-the-art and best practices [27].

CFD tools play a vital role in modeling unique properties such as reactivity and flammability in quantitative risk assessments of hydrogen applications, providing more accurate predictions [28]. CFD software is also crucial in predicting dynamic pressure loads from hydrogen combustion during severe accidents in pressurized water reactors [29].

8.16 CHAPTER SUMMARY

This chapter addressed the biological, mechanical, and chemical threats associated with hydrogen, including suffocation, fire, and explosions. It provided a historical overview of hydrogen-related accidents, underscoring the inherent risks. The chapter delved into hydrogen safety, explaining risk sources and causes, and clarified explosive limits through numerical examples. Physiological, physical, and chemical risks were categorized and examined. Safety considerations for hydrogen storage and transport were emphasized, including relevant codes and standards. Finally, the chapter explored risk minimization strategies, focusing on the use of dedicated software for materials selection in hydrogen applications.

SELF-ASSESSMENT QUESTIONS

1. What are the three main safety concerns related to hydrogen?
2. What is the main hazard associated with compressed hydrogen gas storage?
3. What are the potential risks of using hydrogen fuel cells in vehicles?
4. What are the primary causes of industrial hydrogen accidents?
5. What are the LEL and UEL of hydrogen in the air?
6. How does mixing hydrogen with methane affect the explosion risk?
7. What is the purpose of Le Chatelier's principle in hydrogen safety?
8. What are the physiological hazards associated with hydrogen?
9. What are the main challenges in storing and handling liquid hydrogen?
10. What safety systems are typically employed at hydrogen fueling stations?
11. What are the functions of pressure relief valves and leak detection systems in hydrogen applications?
12. What are some examples of emergency shutdown systems used in hydrogen applications?
13. What types of fire detectors are commonly used in hydrogen environments?
14. What resources can assist in selecting appropriate materials for hydrogen technologies?
15. Why do organizations such as NFPA and ISO develop some key hydrogen-specific codes and standards?

16. What software tools can be used for modeling hydrogen safety, and what are their applications?
17. What is the minimum ignition energy (MIE) of hydrogen?
18. What is hydrogen embrittlement?
19. What are the different ways hydrogen can be transported?
20. How are fuel cell electric vehicles (FCEVs) designed to mitigate safety risks in the event of a collision?
21. Calculate the LEL of a gas mixture containing 60% methane (LEL ≈ 5%) and 40% propane (LEL ≈ 2.1%) by volume using the Burgess–Wheeler equation.
22. Analyze the potential hazards associated with a hydrogen refueling station and propose safety measures to mitigate those risks.
23. Design a ventilation system for a hydrogen storage facility, considering factors such as ventilation type, inlet and outlet positioning, and detector locations.
24. Evaluate the material compatibility of various metals in a liquid hydrogen storage tank, considering factors such as embrittlement and thermal stability.
25. Develop an emergency response plan for a hydrogen leak at an industrial facility, outlining the steps to be taken to ensure the safety of workers and the surrounding environment

REFERENCES

1. Rigas, Fotis and Spyros Sklavounos. 2008. "Hydrogen safety." In *Hydrogen Fuel: Production, Transport, and Storage*, pp. 547–580. CRC Press. https://doi.org/10.1201/9781420045772.
2. Keçebaş, Ali and Muhammet Kayfeci 2019. "Hydrogen properties." In *Solar Hydrogen Production*, pp. 3–29. Academic Press. https://doi.org/10.1016/B978-0-12-814853-2.00001-1
3. Al-Dabbas, Mohammed Awwad and Ammar Ahmed Al-Rousan. 2009. "The simulation of using hydrogen fuel in gasoline internal combustion engines." *The Pacific Journal of Science and Technology* 10 (1). Available at: https://citeseerx.ist.psu.edu/document?repid=rep1&type=pdf&doi=e500cb4558ecbefea3758a7ee7b4cac2417ed16c.
4. Loving, Christopher, Garrett Mastantuono, Anthony C. Terracciano, et al. 2023. "Auto-ignition test results of hydrogen and natural gas fuels at atmospheric and elevated pressures for gas turbine safety." In *Turbo Expo: Power for Land, Sea, and Air*, vol. 86960, p. V03BT04A002. American Society of Mechanical Engineers. https://doi.org/10.1115/GT2023-102674.
5. Saffers, Jean Bernard and Vladimir Molkov. 2014. "Hydrogen safety engineering framework and elementary design safety tools." *International Journal of Hydrogen Energy* 39 (11): 6268–6285. https://doi.org/10.1016/j.ijhydene.2013.06.060.
6. Meda, Ujwal Shreenag, Nidhi Bhat, Aditi Pandey, et al. 2023. "Challenges associated with hydrogen storage systems due to the hydrogen embrittlement of high strength steels." *International Journal of Hydrogen Energy* 48 (47): 17894–17913. https://doi.org/10.1016/j.ijhydene.2023.01.292.

7. Roy, Brandon, Erin LaPointe, Andrew Holmes, et al. 2022. "Effect of hydrogen exposure temperature on hydrogen embrittlement in the palladium–copper alloy system (copper content 5–25 wt.%)." *Materials* 16 (1): 291. https://doi.org/10.3390/ma16010291.
8. Fu, Zhenghong, Bangjian Yang, Kefu Gan, et al. 2021. "Improving the hydrogen embrittlement resistance of a selective laser melted high-entropy alloy via modifying the cellular structures." *Corrosion Science* 190: 109695. https://doi.org/10.1016/j.corsci.2021.109695.
9. Li, Xinfeng, Yan Cui, and Jin Zhang. 2024. "Enhancement of hydrogen embrittlement resistance in CoCrFeNi high-entropy alloy through the addition of MoB elements." *International Journal of Hydrogen Energy* 92: 1306–1319. https://doi.org/10.1016/j.ijhydene.2024.10.092.
10. Compressed Gas Association. 2008. "Safe Handling of Compressed Gases in Containers." CGA P-1. Available at: https://law.resource.org/pub/us/cfr/ibr/003/cga.p-1.1965.pdf.
11. Abohamzeh, Elham, Fatemeh Salehi, Mohsen Sheikholeslami, et al. 2021. "Review of hydrogen safety during storage, transmission, and applications processes." *Journal of Loss Prevention in the Process Industries* 72: 104569. https://doi.org/10.1016/j.jlp.2021.104569.
12. Buttner, William J., Matthew B. Post, Robert Burgess, et al. 2011. "An overview of hydrogen safety sensors and requirements." *International Journal of Hydrogen Energy* 36 (3): 2462–2470. https://doi.org/10.1016/j.ijhydene.2010.04.176.
13. Aziz, Muhammed 2021. "Liquid hydrogen: A review on liquefaction, storage, transportation, and safety." *Energies* 14 (18): 5917. https://doi.org/10.3390/en14185917.
14. Heydorn, Edward C. 2013. "California Hydrogen Infrastructure Project." No. Final Report. Air Products and Chemicals, Inc. https://doi.org/10.2172/1068156.
15. Alternative Fuels Data Center. 2025. "Hydrogen Fueling Station Locations." Available at: https://afdc.energy.gov/fuels/hydrogen_locations.html#/station/61990.
16. Zhang, Caizhi, Xiujuan Cao, Piotr Bujlo, et al. 2022. "Review on the safety analysis and protection strategies of fast filling hydrogen storage system for fuel cell vehicle application." *Journal of Energy Storage* 45: 103451. https://doi.org/10.1016/j.est.2021.103451.
17. Söderholm, Dick. 2023. "Fire Safety in Hydrogen Processing Facilities-Design Considerations." Available at: https://www.theseus.fi/bitstream/handle/10024/815753/S%F6derholm_Dick.pdf;jsessionid=BDDED318D8A457B4BBEE626BE276F5DB?sequence=2.
18. Covino, Bernard S. and Stephen D. Cramer. 2003. *Corrosion: Fundamentals, Testing, and Protection*. ASM International. ISBN-10: 0871707055.
19. ASME. 2024. *B31.12—Hydrogen Pipelines and Piping Code*. ASME. ISBN: 9780791876176.
20. Burgess, Robert M., Mark McDougall, Norman Newhouse, et al. 2011. "Validation Testing In Support of Hydrogen Codes and Standards Developments." Available at: https://h2tools.org/sites/default/files/2019-08/paper_79.pdf.
21. CSA Group. 2014. "Test Methods For Evaluating Material Compatibility In Compressed Hydrogen Applications Phase 1—Metals." Available at: https://www.csagroup.org/store/product/ANSI-CSA%20CHMC%201-2014/?srsltid=AfmBOooi8_7hef2cmdgusKmiBhQrlAcFWKKQDx7eP0DwO9sjcv_HAJYu.
22. American Society for Testing and Materials. 1988. "Hydrogen Embrittlement: Prevention and Control; ASTM Special Technical Publication, No. 962." Available at: https://www.osti.gov/biblio/6715649.
23. Sandia National Laboratories. 2012. "Technical Reference for Hydrogen Compatibility of Materials." Available at: https://www.sandia.gov/matlstechref/.
24. NFPA. 2023. *NFPA 55, Compressed Gases and Cryogenic Fluids Code*. NEPA. ISBN: 1455929220.

25. NASA—Office of Safety and Mission Assurance. 1997. "Safety Standard for Hydrogen and Hydrogen Systems Guidelines for Hydrogen System Design, Materials Selection, Operations, Storage and Transportation." Available at: https://ntrs.nasa.gov/api/citations/19970033338/downloads/19970033338.pdf.
26. Rivkin, Christina, Richard Burgess, and William Buttner. 2015. "Hydrogen Technologies Safety Guide." No. NREL/TP-5400–60948. National Renewable Energy Lab. (NREL), Golden, CO. Available at: https://www.nrel.gov/docs/fy15osti/60948.pdf.
27. Baraldi, Daniele, Melideo, Daniele, Kotchourko, Alexei, et al. 2017. "Development of a model evaluation protocol for CFD analysis of hydrogen safety issues the SUSANA project." *International Journal of Hydrogen Energy* 42 (11): 7633–7643. https://doi.org/10.1016/j.ijhydene.2016.05.212.
28. Middha, Prankul and Olav R. Hansen. 2009. "Using computational fluid dynamics as a tool for hydrogen safety studies." *Journal of Loss Prevention in the Process Industries* 22 (3): 295–302. https://doi.org/10.1016/j.jlp.2008.10.006.
29. Sathiah, Pratap, Steven van Haren, Ed Komen, et al. 2012. "The role of CFD combustion modeling in hydrogen safety management—II: Validation based on homogeneous hydrogen–air experiments." *Nuclear Engineering and Design* 252: 289–302. https://doi.org/10.1016/j.nucengdes.2012.06.023.

9 Hydrogen Supply Chain Integrated with Artificial Intelligence

9.1 INTRODUCTION

The principles of artificial intelligence (AI) facilitate the activities AI systems undertake, including thinking, learning, problem-solving, and perception. The act of performing tasks that attempt to achieve human-like behavior for a particular machine is called AI. One of AI's enabling technologies is the rule-based system, or the simplest form of AI, which operates on a set of instructions for storing, organizing, and manipulating data given by a human. They can function exceptionally when the decision-making process can be put into precise, logical statements. Because systems tend to produce set outcomes by performing an input of tasks, it brings about the ability of a system to function. However, systems that depend too much on strict logic and lack reasoning when defining rules tend to lack adaptability to complex or vague environments. In the past 20 years, improved computer power and data access have enabled machine learning (ML) in applications such as data mining systems. Furthermore, the internet and the growth of available data in digitized form contributed significantly to these changes.

An important branch of AI is ML, which enables computers to execute tasks and achieve objectives autonomously without any specific directions provided. For example, AI algorithms have been programmed to enhance efficiency in large data sets. Data sets are analyzed using multiple algorithms that identify patterns and categorize data. After sufficient data is arranged in the required structure, ML algorithms can locate associations and predictive factors, enabling accurate projections.

Deep learning (DL), which focuses on artificial neural networks, has contributed to advances in ML systems. These multilayered systems process large amounts of information in intricate manners, resulting in numerous advancements in areas such as image and speech recognition. Lastly, natural language processing (NLP) is the final frontier of technology, aiming to enable machines to comprehend human phrases or words in their written or spoken form.

The advancement of AI has had a dramatic impact on several industries. AI technologies help solve some old and modern problems within the energy sector, such as improving the efficiency and effectiveness of systems used. Furthermore, AI has a purpose in smart grid systems where algorithms are implemented to analyze the vast amounts of data attained from the sensors and meters to optimize energy distribution, minimize losses, and improve the grid's stability. In addition, the remarkable speed enabled by AI facilitates monitoring and controlling in real time,

DOI: 10.1201/9781003639190-9

allowing supply to adjust to changes in demand quickly. Furthermore, enhanced fault prediction and detection technologies make power systems more reliable with AI implementations. For instance, AI can be used to predict renewable energy (RE) to decrease the effect of intermittency from sources such as solar or wind energy. This can be done by analyzing historical data and other relevant data; ML models provide intricate and accurate weather forecasts that aid in the seamless integration of grids. Another important AI use case is predictive maintenance, where the AI performs health monitoring to predict failures, reduce maintenance costs and unplanned outages, and extend asset life. AI is commonly implemented in demand response management to analyze real-time energy data for peak consumption periods, thereby managing and optimizing the supply–demand ratio, peak loads, and general energy consumption. Alternatively, AI is integrated into energy trading for market analysis, where real-time pricing data is constantly monitored to identify changes in supply and demand for optimal trading opportunities.

In energy storage systems, AI can help optimize the charging and discharging techniques of batteries. This will support the efficient and prolonged use of the systems and the integration of renewables. In the case of nuclear power plants, AI systems monitor and supervise day-to-day activities in the plant to enhance safety by preventing accidents using real-time anomaly detection and issue identification.

As hydrogen grows in importance, it earns its title as the main component of the change toward a cleaner and more sustainable energy future. This is due to its potential utilization as an energy carrier and the opportunity for zero-emission energy production taken advantage of in the decarbonization of transport, industry, and power generation. However, transitioning to green hydrogen is not easy, as there are dominant adoption issues throughout the supply chain. Concerns surrounding these problems include high production costs, efficiency, safety, and the current infrastructure. AI is anticipated to provide powerful solutions to better the hydrogen supply chain. Integrating AI methods such as ML, DL, and computer vision can enhance efficiency, safety, innovation, and integration at various supply chain stages.

9.2 THE ROLE OF AI IN HYDROGEN PRODUCTION

One of the key challenges in scaling green hydrogen production (GHP) technologies is the inherent weather dependence of RE sources. This variability necessitates precise prediction and optimization of RE energy. AI, particularly ML, emerges as a crucial tool alongside RE sources to address this challenge [1–4]. ML's capability to consistently resolve GHP problems has been demonstrated.

For example, AI algorithms can optimize energy transfer within the grid by analyzing real-time data on RE production, hydrogen generation, and consumer demand. This optimization maximizes hydrogen production efficiency and conserves clean energy. Furthermore, AI can proactively address potential grid issues arising from unstable renewable sources and fluctuating hydrogen demand, preventing disruptive blackouts. Figure 9.1 illustrates the multifaceted impact of AI on hydrogen production efficiency. This efficiency improvement is achieved through AI's ability to optimize electrolyzer operations by analyzing sensor data for ideal conditions,

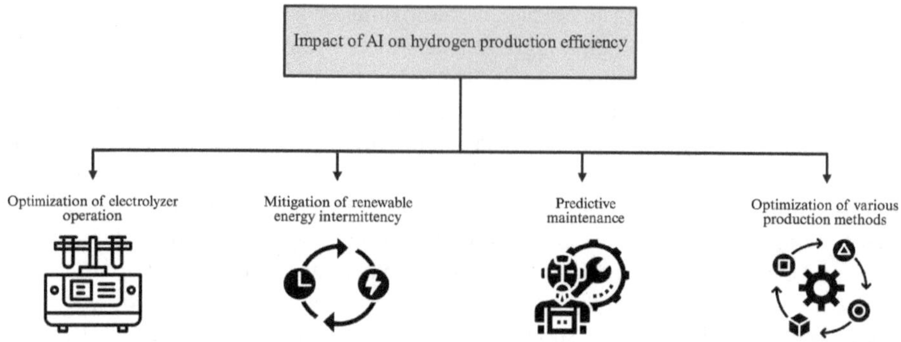

FIGURE 9.1 The main components in which AI can contribute to increase hydrogen production efficiency.

mitigate RE intermittency by predicting generation and optimizing grid transfer, enable predictive maintenance to prevent downtime and optimize various production methods such as electrolysis, photoelectrochemical (PEC) water splitting, and steam methane reforming (SMR) by fine-tuning parameters and maximizing yield.

Moreover, advanced optimization techniques are employed to lessen the effects of RE intermittency on hydrogen production. The Al-Biruni Earth Radius (BER) [5] and particle swarm optimization [6] algorithms have been used to optimize DL models, specifically recurrent neural networks (RNNs) [7]. These algorithms are effective in optimizing the dynamic hyperparameters of RNNs during the epochs when training algorithms process and forecast time-series data for solar hydrogen generation.

As previously mentioned, GHP relies on electrolyzers that split water into hydrogen and oxygen. AI can enhance their performance, efficiency, and cost-effectiveness. ML algorithms can determine the optimal operating conditions by analyzing data from various sensors within the electrolyzer, such as temperature, pressure, and current. This real-time optimization improves the electrolyzer's energy efficiency, reduces degradation, and lowers the cost of GHP.

Figure 9.2 depicts a system where RE input, such as solar or wind, powers the electrolyzer. Numerous sensors monitor parameters such as temperature, pressure, current, voltage, and hydrogen concentration, providing real-time operational data. A data acquisition system collects and digitizes this sensor data for processing by the AI control system. The AI control system, acting as the system's "brain," uses AI algorithms, often ML models, to analyze the data, identify patterns, predict optimal operating conditions, and detect potential issues. Based on this analysis, the AI control system sends control signals to the electrolyzer to adjust parameters such as voltage, current, electrolyte flow rate, temperature, and pressure. These AI-driven adjustments optimize the electrolyzer's performance, resulting in improved energy efficiency, increased hydrogen production rate, and enhanced durability. This optimized electrolysis process yields hydrogen gas (H_2), the desired output, and oxygen gas (O_2) as a byproduct [8,9].

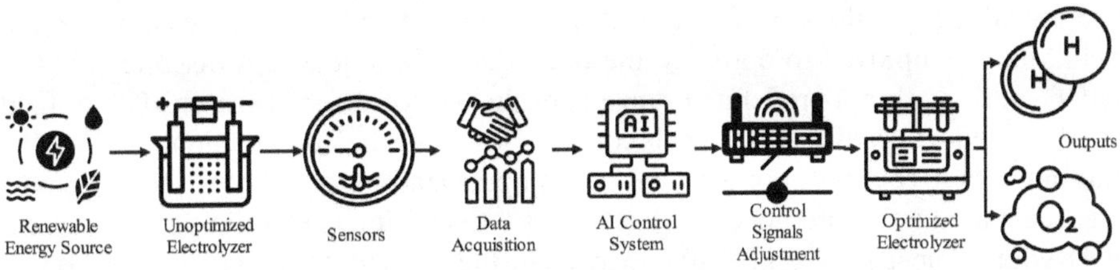

FIGURE 9.2 The principle of operation of electrolyzer with AI-driven controls.

The electrolytic production of hydrogen is a complex electrochemical process characterized by non-linear relationships between operational parameters and output metrics. Specifically, the hydrogen production rate (H), which is measured in Nm^3/h and is a function of applied voltage (V) in Volts and operating temperature (T) in °C. This relationship can be mathematically represented as

$$H = f(V,T) + \varepsilon \tag{9.1}$$

Where f(V, T) denotes the underlying, potentially nonlinear, functional relationship, and ε represents a residual error, encompassing unmodeled process variations and measurement noise. The objective of this analysis is to develop a data-driven model capable of accurately approximating f(V, T) and subsequently optimizing V and T to maximize H. A multilayer perceptron (MLP), a class of artificial neural networks, is employed to approximate the unknown function f(V, T).

The MLP is a universal function approximator, capable of modeling complex, nonlinear relationships. The mathematical representation of the MLP can be expressed as

$$y = f(x;y) = \sigma\left(w_2^T \sigma\left(w_1^T x + b_1\right) + b_2\right) \tag{9.2}$$

where $x = [V, T]$ is the input vector, w_1 and w_2 are weight matrices representing synaptic strengths, b_1 and b_2 are bias vectors, σ is a non-linear activation function, such as rectified linear unit (ReLU) or sigmoid, introducing non-linearity to the model, and y is the output, representing the predicted hydrogen production rate ($\hat{H}$).

The MLP is trained using a supervised learning approach. A dataset comprising experimentally obtained triplets (V_i, T_i, H_i) is utilized. The training process involves minimizing a loss function, typically the mean squared error (MSE), defined as

$$MSE = \frac{1}{N}\sum_{i=1}^{N}\left(H_i - \hat{H}_i\right)^2 \tag{9.3}$$

where N is the number of data points, gradient-based optimization algorithms, such as Adam (adaptive moment estimation) as in [10] or stochastic gradient descent (SGD) as in [11], can employed to iteratively adjust the weights and biases of the MLP to minimize the MSE. Once the MLP is trained, it can predict the hydrogen production rate for arbitrary values of V and T. A grid search or gradient-based optimization technique is employed to optimize the operational parameters. A discrete set of V and T values within a predefined range is evaluated for the grid search. The trained MLP is used to predict $\hat{H}_i$ for each combination, and the combination yielding the maximum $\hat{H}$ is selected. For gradient-based optimization, algorithms such as gradient ascent are used to iteratively adjust V and T to maximize $\hat{H}$ utilizing the gradient of $\hat{H}$ with respect to V and T obtained from the trained MLP. The application of neural network regression, specifically the MLP, provides a robust and data-driven approach for modeling and optimizing the electrolytic hydrogen production process. By accurately approximating the complex relationship between operational parameters and output metrics, this methodology facilitates the identification of optimal operating conditions, leading to enhanced efficiency and productivity.

Example 9.1: Electrolyzer Optimization with Neural Network Regression

An alkaline electrolyzer's efficiency, characterized by the hydrogen production rate (H, in Nm^3/h), is a function of operating voltage (V, in Volts) and temperature (T, in °C). Experimental data reveals the following discrete relationships: (V, T, H): (1.8 V, 60°C, 4.5 Nm^3/h), (2.0 V, 70°C, 5.5 Nm^3/h), (2.2 V, 80°C, 6.8 Nm^3/h).

Employ a neural network regression model to predict H at V = 2.1 V and T = 75°C and subsequently optimize V and T to maximize H.

Solution:

The first step in solving this example is data preprocessing. The data must be structured as input vectors x = [V, T] and output scalars y = H. To improve training stability, data normalization may be applied to scale V and T to a common range (e.g., [0, 1]).

The second step is to make a neural network model by choosing a MLP with a single hidden layer. Therefore, the model can be expressed using equation 9.2. After that, the model is trained using a gradient-based optimization algorithm (e.g., Adam) to minimize the MSE between predicted and actual H values.

The input vector x = [2.1 V, 75°C] is fed into the trained MLP. Accordingly, the model outputs predict the H value, e.g., H = 6.25 Nm^3/h.

The final step is to optimize V and T to obtain the maximum value of H. This can be fulfilled using a grid search or gradient-based optimization method (e.g., gradient ascent), which was applied in this example to find the optimal V and T values.

So, the grid search might test V values from 1.7 to 2.3 V in 0.05 V increments and T values from 55°C to 85°C in 2.5°C increments. Then, the trained neural network predicts the hydrogen output for each combination, and the combination that produces the highest value is retained. Therefore, the optimization may yield V = 2.3 V and T = 85°C, resulting in a predicted H = 6.27 Nm^3/h, as shown in Figure 9.3.

AI prediction tools extend beyond merely forecasting hydrogen production. User-friendly environments such as LabVIEW can analyze variations in hydrogen production, ensuring a stable and reliable supply of green hydrogen [12]. To enhance

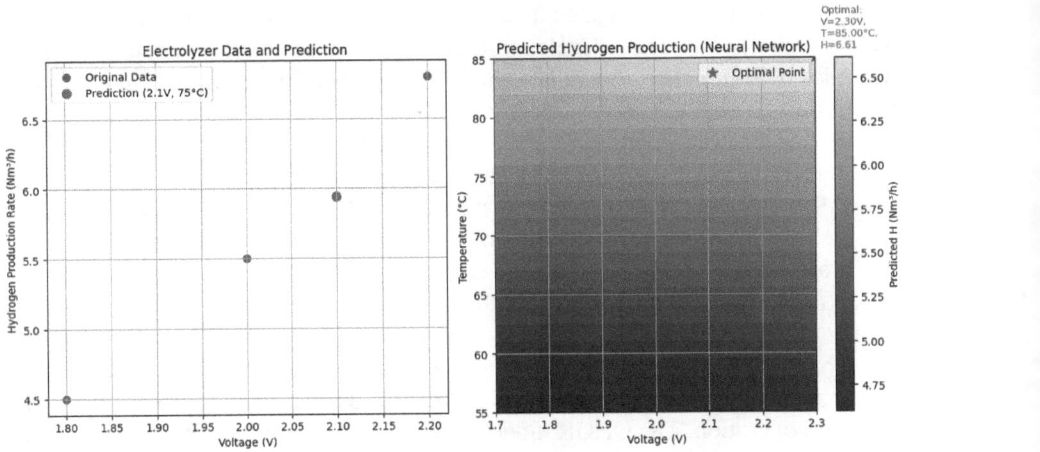

FIGURE 9.3 predicted hydrogen production versus optimal input values.

electrolyzer performance, adaptive neuro-fuzzy inference systems (ANFIS) and modern pelican optimization algorithms (POA) are combined. ANFIS models are used to develop predictive models of the water electrolysis process using experimental data [13]. POA determines the optimal operating conditions to maximize hydrogen production.

In PEC water splitting, AI can contribute to material optimization by analyzing material properties and predicting their suitability for PEC applications, accelerating the discovery of efficient and cost-effective photoelectrodes. AI can also optimize PEC system operating conditions, such as pH, temperature, and light intensity, to maximize hydrogen production efficiency. Furthermore, AI can aid in designing efficient and scalable PEC systems, considering factors such as light absorption, charge separation, and hydrogen evolution.

AI can also improve efficiency and reduce the environmental impact of SMR. AI-powered control systems can optimize SMR plant operating parameters, such as temperature, pressure, and steam-to-carbon ratio, maximizing hydrogen yield and minimizing energy consumption. AI can also reduce greenhouse gas emissions from SMR by optimizing carbon capture and utilization processes. Additionally, AI can predict equipment failures, enabling proactive maintenance and reducing downtime.

Abdelkareem et al. [14] explored the use of DL models to predict the performance of a solar-driven membrane reactor for hydrogen production. They employed a Bayesian-optimized long and short-term memory (BO-LSTM) network to predict methane conversion and hydrogen recovery under varying weather conditions. The results demonstrated the BO-LSTM model's high accuracy in both short-term (1-day) and long-term (40-day) predictions, outperforming other ML models such as random forest (RF), support vector machine (SVM), back propagation neural network (BPNN), and convolutional neural network (CNN) in terms of accuracy and stability.

Sirat et al. [15] investigated hydrogen production using alkaline water electrolysis (AWE) [15]. They developed a two-dimensional (2D) computational fluid dynamics (CFD) model to simulate the AWE process, validating it against experimental data. The CFD model accurately predicted key parameters, including gas flow and concentration, bubble behavior, and turbulence within the AWE cell. The model also analyzed the effects of current density and electrode–diaphragm distance. The researchers integrated the model's predictive capabilities with a neural network, enhancing its understanding of AWE cell behavior under diverse operating conditions.

The authors in [16] proposed an optimized hydrogen generation-based regression (OHGR) model to simulate hydrogen generation in fuel cells. The OHGR model, based on a DL architecture with stacked neural networks and crossover connections, predicts hydrogen generation at 1 mA/cm^2 and can exceed 10–20 mA/cm^2 after long-term operation. The OHGR model uses stochastic gradient optimization (SGD) to recommend optimal temperature, pressure, and humidity values for maximum hydrogen generation.

AI is expected to significantly reduce hydrogen production costs by enhancing efficiency and reliability. Integrating AI with other emerging technologies, such as the Internet of Things (IoT), will create new opportunities for developing interconnected energy systems [17], facilitating the integration of green hydrogen across various sectors and accelerating the transition to a sustainable energy future.

Figure 9.4 illustrates AI's role in optimizing RE for hydrogen production, while Figure 9.5 clarifies how AI can reduce hydrogen's high production cost. The process depicted in Figure 9.5 begins with gathering energy from renewable sources, followed by data collection from sensors and records. Using ML algorithms, AI-assisted energy forecasting analyzes this data to predict future energy production, enabling real-time estimation of renewable generation, a critical step in optimizing hydrogen production. Electrolyzer control, a key component of the hydrogen production process, benefits significantly from AI-driven automation. The operation of electrolyzers

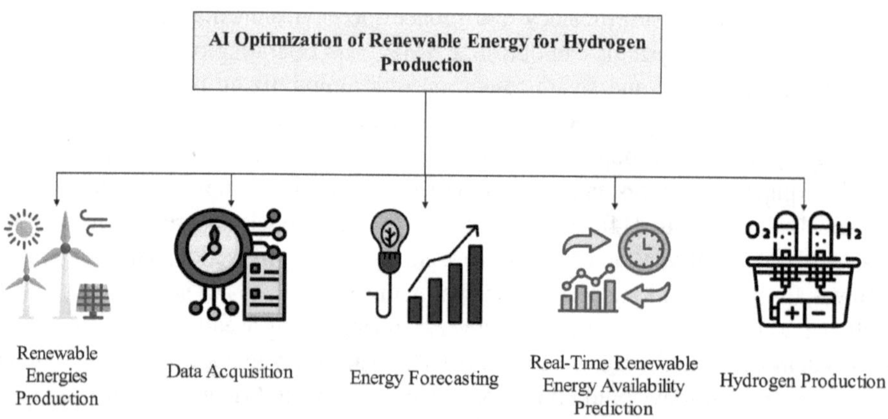

FIGURE 9.4 The use of AI in optimizing renewable energy for hydrogen production.

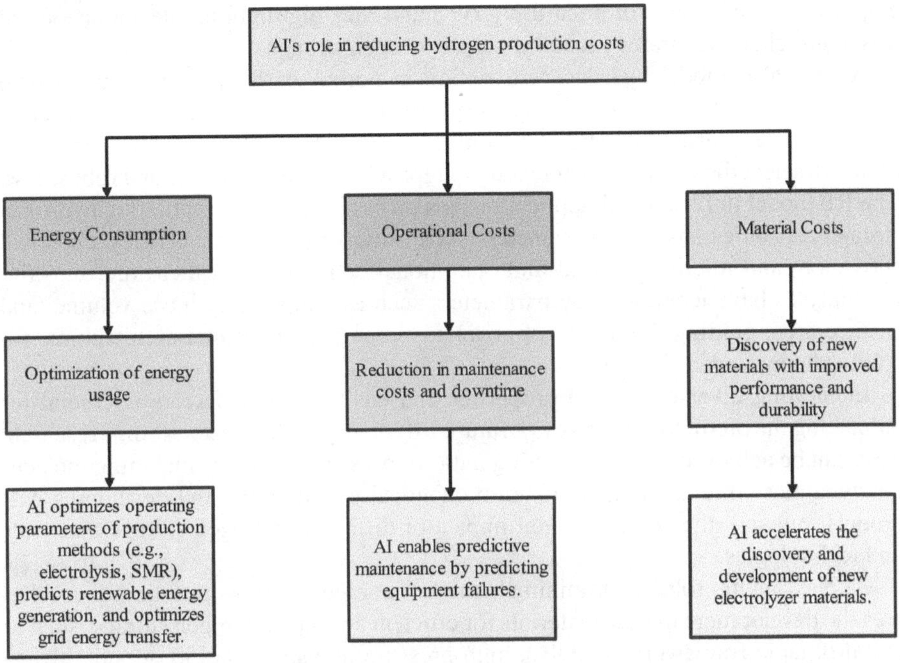

FIGURE 9.5 Predicted hydrogen production cost reduction with AI.

is dynamically adjusted based on predicted RE input, minimizing energy waste and maximizing hydrogen optimization.

9.3 AI IN HYDROGEN STORAGE AND TRANSPORTATION

Efficient and safe hydrogen storage is essential to transition to a sustainable energy future. However, challenges related to material selection, storage system design, and performance optimization have hindered widespread adoption. The emergence of AI, particularly ML and DL, provides a promising avenue to address these challenges and accelerate the development of advanced hydrogen storage solutions.

The search for ideal hydrogen storage materials is complex. Traditional experimental methods for material selection can be resource-intensive. AI offers data-driven approaches to expedite this process, demonstrating its efficiency in handling complex tasks [18]. ML techniques are expected to enhance decision tree regression, showing superior performance in analyzing datasets of metal hydrides, for example, as hydrogen storage materials [19]. Additionally, ML identifies the critical parameters that significantly influence storage material selection.

In Wang et al. [20] reviewed the use of AI in developing solid-state hydrogen storage materials. They discussed the advantages of high-throughput screening (HTS) and ML in this field. HTS allows for rapid screening of numerous materials, while ML can predict material performance and optimize material design. The authors emphasized the benefits of integrating HTS and ML, highlighting the application of this comprehensive strategy in solid-state hydrogen storage.

AI can also model hydrogen adsorption in porous carbon, as demonstrated in [21]. This work utilized various nature-inspired algorithms, such as Dragonfly, Grey Wolf, particle swarm, and genetic optimization, in conjunction with RF. The model aimed to select the most relevant features, improving model accuracy and robustness. The RF model in [21], an AI application, has proven effective in predicting hydrogen storage capacities based on parameters such as adsorbent characteristics, chemical activation, and micro-structural and operational conditions. Furthermore, sensitivity analyses have identified key parameters such as temperature, pore volume, and surface area, guiding the design of novel adsorbents with optimal architectures for solid-state hydrogen storage.

Integrating diverse material properties and environmental factors is crucial for enhancing the predictive accuracy of future ML models in hydrogen storage research. This can be achieved by incorporating data from various sources, including molecular dynamics simulations and quantum chemical calculations, and developing ML models to assess the environmental impact of different hydrogen storage materials and technologies.

AI plays a vital role in optimizing hydrogen storage and transportation. It accelerates the development of new materials for efficient and safe hydrogen storage, such as metal-organic frameworks (MOFs), high-pressure tanks, and liquid organic hydrogen carriers. AI also optimizes storage conditions, ensuring safety and maximizing capacity. This includes managing pressure and temperature, detecting leaks, and optimizing storage capacity.

The study in [22] demonstrated a novel integrated framework combining analytical techniques, numerical simulations, and DL to design lightweight, high-pressure hydrogen storage tanks. The use of deep neural networks to accurately predict stress components and potential failure points exemplifies AI's innovative solutions, enabling optimized tank design and safer solutions.

Future research should expand the application of deep reinforcement learning (DRL) in areas such as real-time control of hydrogen storage and release to optimize operational parameters, grid integration, and energy management to optimize hydrogen storage within smart grids, and predictive maintenance to analyze sensor data and forecast equipment failures.

AI is also instrumental in material science, crucial for hydrogen storage and production advancements. ML algorithms can analyze vast datasets of material properties to identify promising candidates for hydrogen storage, catalysis, and other applications, accelerating the discovery process and leading to the development of more efficient and cost-effective materials. This is key to addressing current limitations in hydrogen storage and transportation.

AI can monitor the condition of hydrogen storage tanks. By analyzing pressure, temperature, and hydrogen concentration sensor data, AI algorithms can identify potential leaks or safety hazards, allowing operators to take corrective action before major incidents occur.

Chi et al. [23] reviewed structural health monitoring (SHM) methods for hydrogen storage tanks based on flexible sensing technology. They summarized damage modes, reviewed flexible sensing technologies and their applications in SHM, and discussed potential advancements in SHM systems for hydrogen storage tanks.

A comprehensive overview of composite materials and AI for designing hydrogen storage vessels was conducted in [24]. The review examined different types of hydrogen storage vessels, including compressed hydrogen gas, cryogenic and liquid hydrogen, sorbents, metal hydrides, and chemical hydrides. The authors also discussed the design of hydrogen storage vessels, including composite materials and nanocomposites, and the use of AI and digital twin technologies for predicting the hygro-thermomechanical behavior of hydrogen storage vessels.

Neural networks can predict hydrogen storage capacity and temperature distribution in hydrogen absorption reactions. The neural network model in [25] uses a lumped parameter system, simplifying analysis by assuming uniformly distributed properties. This allows for a computationally efficient model without sacrificing accuracy. This approach enables a more accurate prediction of hydrogen absorption behavior in alloys, which is crucial for improving hydrogen storage system efficiency.

The application of AI in Mg-based hydrogen energy storage has been discussed in [26]. The review focused on ML techniques to improve the prediction of Mg-based composites' properties and hydrogen storage performance, highlighting the potential of ML in guiding experimental design and accelerating materials research.

Finally, Figure 9.6 summarizes the efforts in using AI in hydrogen storage development.

As discussed in Chapter 5, salt caverns are expected to be an efficient and cheap storage option for hydrogen. However, the prediction of pressure dynamics within an underground hydrogen storage cavern necessitates the development of a robust model that accounts for the complex interplay of several influencing factors. To this end, a data-driven approach employing RNNs can be adopted as in [53–55]. The underlying mathematical principles and the application of RNNs are meticulously detailed below. The pressure (P), measured in bar, within a salt cavern used for hydrogen storage, is conceptualized as a function of multiple governing parameters: time (t), denoting the duration of hydrogen injection measured in days; injection rate (Q), representing the rate of hydrogen influx measured in tons per day; initial pressure P_0, indicating the pressure within the cavern at the commencement of injection, measured in bar; cavern volume (V_c), specifying the total volumetric capacity of the cavern, measured in cubic meters (m^3); and salt rock permeability (k), quantifying the ease with which fluids permeate through the salt rock, measured in square meters (m^2). This intricate relationship is mathematically expressed as

$$P = f(t, Q, P_0, V_c, k) + \varepsilon \tag{9.4}$$

where f represents a complex, potentially non-linear, functional relationship encapsulates the system's dynamic behavior and ε accounts for residual errors arising from unmodeled factors and measurement inaccuracies. Given the inherent temporal nature of pressure variations, RNNs, specifically long short-term memory (LSTM) or gated recurrent unit (GRU) networks, are utilized to model the time series data. These networks are specifically designed to process sequential data by

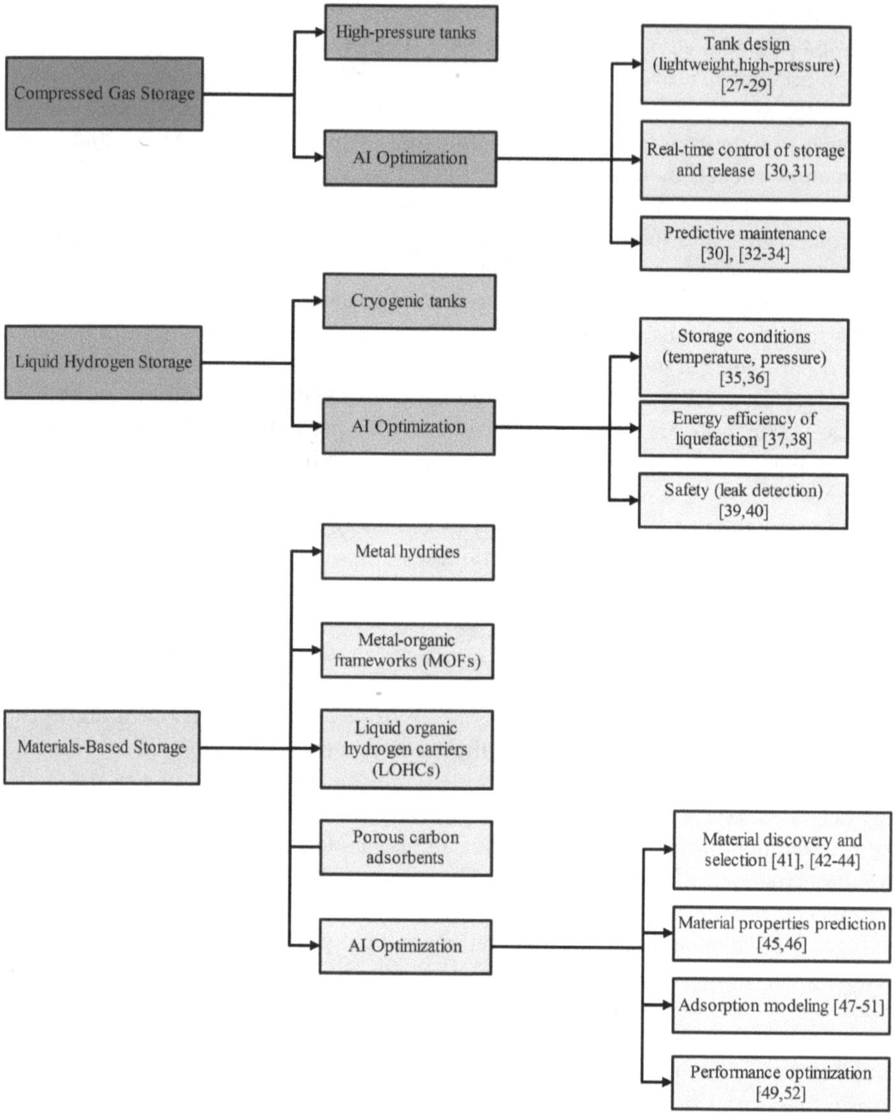

FIGURE 9.6 Hydrogen storage methods and AI optimization.

maintaining an internal memory that effectively captures temporal dependencies. At each time step, the input to the RNN is a vector comprising the governing parameters [t, Q, P_0, V_c, k].

The RNN maintains a hidden state that stores information from preceding time steps, enabling it to learn the temporal patterns inherent in the pressure data. The output of the RNN at each time step is the predicted pressure (P). Mathematically, the RNN's operational framework is summarized as follows:

$$h_t = \sigma(W_{xh}x_t + W_{hh}h_{t-1} + b_h) \tag{9.5}$$

$$y_t = W_{hy}h_t + b_y \tag{9.6}$$

where x_t is the input vector at time t, h_t is the hidden state at time t, y_t is the output (predicted pressure) at time t, W_{xh}, W_{hh}, and W_{hy} are weight matrices, b_h and b_y are bias vectors, and σ is an activation function, such as sigmoid as in [56] or ReLU as in [57]. Model training uses a time series of pressure measurements and corresponding governing parameters. The model's parameters are adjusted to minimize the MSE between the predicted and actual pressure values. Once trained, the RNN can predict the pressure at future time steps by feeding it the input sequence. Furthermore, the trained RNN can be employed to extrapolate pressure predictions beyond the available data, facilitating risk assessment of prolonged injection. By simulating various injection rates and cavern conditions, the RNN aids in optimizing operational parameters and ensuring safe storage practices. The input data is normalized before training using techniques such as Min–Max scaling. This normalization ensures that features with disparate scales do not disproportionately influence the model's learning process, thereby enhancing the model's stability and performance.

Example 9.2: Predicting Pressure Dynamics in Underground Hydrogen Storage Using RNNs

Predict the pressure (P, in bar) within a salt cavern used for large-scale hydrogen storage as a function of injection time (t, in days), hydrogen injection rate (Q, in tons/day), initial cavern pressure (P_0, in bar), cavern volume (V_c, in m^3), and the geological permeability of the salt rock (k, in m^2).

Given:

$P_0 = 100$ bar (initial pressure).

$Q = 20$ tons/day (constant injection rate).

$V_c = 500{,}000\ m^3$ (cavern volume).

$k = 1 \times 10^{-19}\ m^2$ (salt rock permeability).

Historical pressure data: at $t = 5$ days, $P = 120$ bar and at $t = 10$ days, $P = 145$ bar.

Predict the cavern pressure (P) at $t = 15$ days and assess potential risks associated with prolonged injection.

Solution:

The first step is to structure the data as a time series, where the input at each time step consists of [t, Q, P_0, V_c, k], and the output is the corresponding pressure (P). Then, the input and output data are normalized to improve the RNN's training stability. This is important when dealing with data with different units and widely different numerical ranges.

In this example, a LSTM or GRU network is chosen to model the temporal dependencies in the pressure data. These types of RNNs are well-suited for time series forecasting. RNN is trained using historical pressure data to learn the relationship between the input variables and the pressure dynamics. The loss function used for training is the MSE. The trained RNN is fed with the input sequence up to $t = 10$ days.

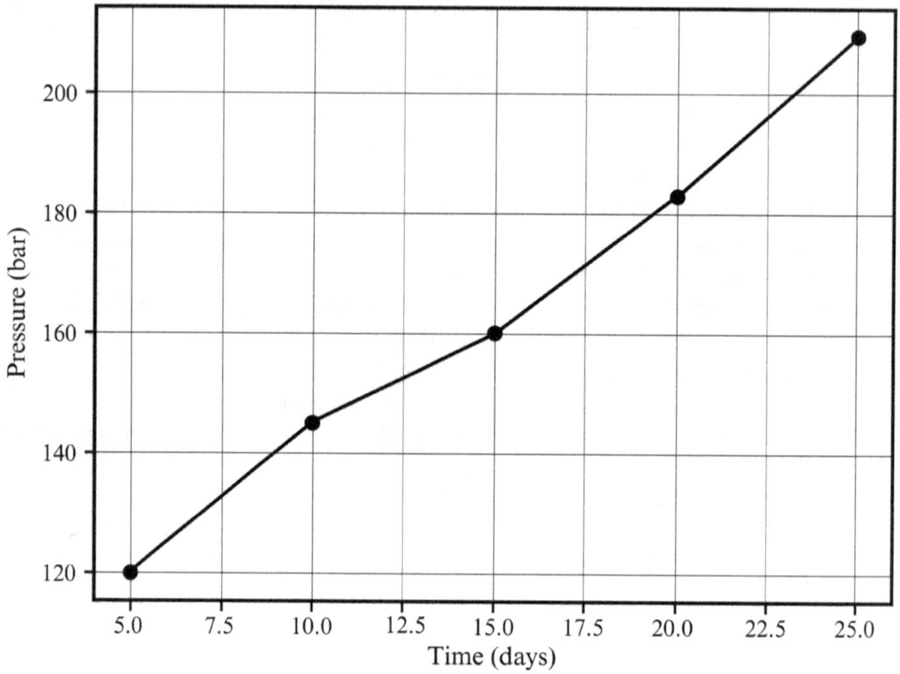

FIGURE 9.7 Prediction of salt cavern pressure versus storage days.

Accordingly, the RNN can predict the pressure at t=15 days, for example, P=165 bar, and the model can also predict values for any other time value.

The trained RNN is used to extrapolate pressure predictions for longer injection times. For example, the RNN might predict that at t=20 days, P=188 bar, and at t=25 days, P=210 bar.

If the cavern's maximum safe operating pressure is 200 bar, predicting 210 bar at 25 days triggers an alarm, indicating a high risk of exceeding the safe pressure limit. Additionally, simulations with varying injection rates (Q) can be performed to determine the maximum injection rate that maintains safe pressure levels within the cavern. Figure 9.7 depicts the pressure in the salt cavern per day versus days using RNN.

9.4 AI IN HYDROGEN UTILIZATION

AI can optimize the use of hydrogen in several applications, such as fuel cells, vehicles, and even related infrastructure. For instance, AI can optimize the operational parameters of fuel cells, improving their performance, durability, and fault diagnosis. Further, hydrogen-powered infrastructure and vehicles can tremendously benefit from AI's predictive maintenance capabilities, which monitor sensor data, predict maintenance, and optimize the maintenance schedule. Figure 9.8 depicts the role of AI in fuel cell performance improvement.

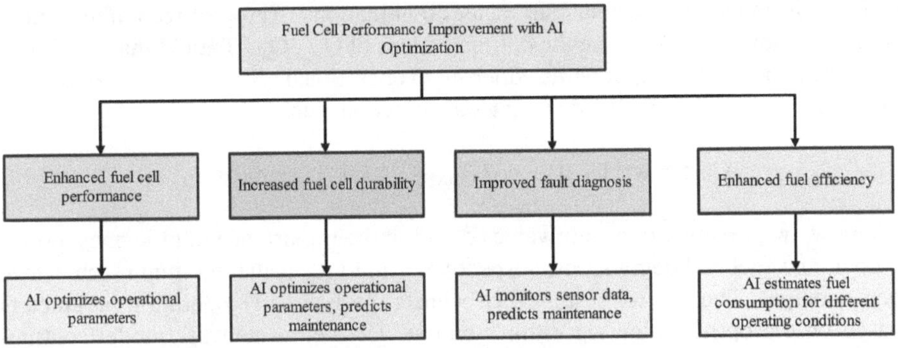

FIGURE 9.8 Fuel cell performance improvement with AI optimization.

The optimization of hydrogen fuel cell performance necessitates the development of a mathematical framework that accurately represents the complex relationships between operational parameters and performance metrics. The fuel cell's behavior is influenced by several key parameters, including hydrogen flow rate (Q_{H_2}), air flow rate (Q_{air}), operating temperature (T), and load current (I). These parameters directly impact the fuel cell's efficiency (η) and power output (P) [58,59]. Mathematically, the relationships between these parameters and performance metrics can be expressed as

$$P = f_{\eta}\left(Q_{H_2}, Q_{air}, T, I\right) + \varepsilon_{\eta} \tag{9.7}$$

$$P = f_{P}\left(Q_{H_2}, Q_{air}, T, I\right) + \varepsilon_{P} \tag{9.8}$$

where f_{η} and f_P represent complex, non-linear functions encapsulating the underlying electrochemical and thermodynamic processes, and ε_{η} ε_P account for residual errors due to unmodeled factors and measurement noise.

The objective is to approximate these functions using a data-driven approach and subsequently optimize the operational parameters to maximize efficiency and power output. A neural network model can be employed to achieve this. The model must be trained using a dataset comprising real-time sensor measurements of Q_{H_2}, Q_{air}, T, I, η, and P. The training process will involve minimizing the MSE between the predicted and actual performance metrics, defined as

$$MSE = \frac{1}{N}\sum_{i=1}^{N} \left(\eta_i - \widehat{\eta_i}\right)^2 + \left(P_i - \widehat{P_i}\right)^2 \tag{9.9}$$

where N represents the number of data points, η_i and P_i are the actual values, and $\hat{\eta}$ $\hat{P}$ are the predicted values. Once trained, the neural network can predict the efficiency and power output for arbitrary combinations of operational parameters. In real-time operation, the trained model receives continuous sensor data and predicts the fuel cell's performance for various parameter combinations. An optimization algorithm is then employed to determine the optimal values of Q_{H_2}, Q_{air}, T, and I that maximize efficiency and power output while adhering to operational constraints. The optimization process can be formalized using a cost function, such as

$$\text{Cost} = w_1(1-\eta) + w_2(P_{max} - P) + w_3(\text{Degradation}) \tag{9.10}$$

where w_1, w_2, and w_3 are weight values defining the importance of efficiency, power output, and fuel cell degradation, respectively, and P_{max} is the maximum achievable power output. The control system then adjusts the fuel cell's operating parameters based on the optimization algorithm's output. The neural network model continuously learns and adapts based on new data, refining its predictions and optimization capabilities over time. This approach enables real-time optimization and adaptive control of hydrogen fuel cell performance, improving efficiency, power output, and durability.

Example 9.3: Real-Time Optimization of Hydrogen Fuel Cell Performance

A vehicle equipped with a hydrogen fuel cell requires real-time optimization of its operating parameters to maximize efficiency and power output. Use AI to predict and optimize fuel cell efficiency (η) and power output (P).

Given that:

The current state of the fuel cell is

Load current (I): 50 A
Operating temperature (T): 70°C
Hydrogen flow rate (Q_{H_2}): 20 L/min
Air flow rate (Q_{air}): 100 L/min
Fuel cell efficiency (η): 55%
Power output (P): 3 kW

The AI prediction shows that if Q_{H_2} is increased to 25 L/min and Q_{air} is increased to 120 L/min, while I and T remain constant, the predicted fuel cell efficiency ($\hat{\eta}$) will be 60%, and the predicted power output ($\hat{P}$) will be 3.5 kW.

The AI control system must decide if it will change the flow rates.

Solution:

The first step in solving this example is to collect and preprocess the real-time sensor data. Then, the neural network must be trained to approximate f_η and f_P by minimizing the MSE using equation 9.9. Based on the AI's prediction, the control system evaluates the potential increase in efficiency and power output. The decision is made using a cost function in equation 9.10. The control system uses an optimization algorithm to minimize the cost function. If, for example, the cost function is weighted so that efficiency and power output are highly valued, and the degradation prediction is low, the optimization algorithm will suggest increasing the flow rates.

Therefore, the control system sends control signals to increase the hydrogen flow rate to 25 L/min and the air flow rate to 120 L/min. After the changes are

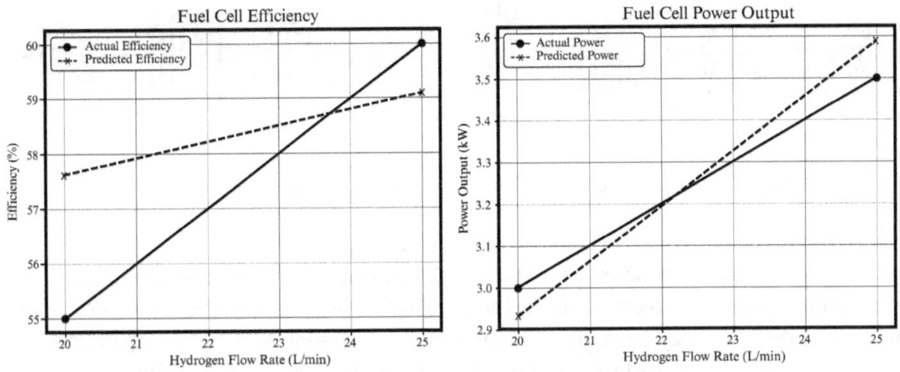

FIGURE 9.9 The efficiency of the fuel cell and output power versus the input hydrogen flow rate.

implemented, the sensors will provide new data confirming that the efficiency has increased to 60%, and the power output has increased to 3.5 kW. If the degradation prediction was high, the control system may decide to keep the flow rates as they were or increase them by a smaller amount. Figure 9.9 depicts the output of the optimization process provided in this example.

In addition, AI assistance can be provided to the control systems in charge of hydrogen refueling stations to manage the refueling technique, monitor the station's performance, and manage the hydrogen availability. AI can also support the development of energy management systems based on hydrogen by optimizing energy use with the grid and enabling demand response. Hydrogen refueling stations can receive hydrogen in multiple ways—transported via trucks, pipelines, or produced onsite via electrolysis. Hydrogen is most commonly stored in tanks at high pressure, and if need be, hydrogen supplied or stored for fueling is compressed to the necessary high levels using compressors. Various sensors, such as pressure, temperature, flow, and leak detection, help keep track of the fueling process and are essential for ensuring safety. All the information captured by the sensors is collected by a data acquisition system, which digitizes it and then sends it to the AI control system, which takes charge of everything.

Managing the fueling process using specific constraints, such as avoiding overfilling while optimizing efficiency, requires the use of optimization algorithms, which are programmed into the AI control system. The AI control system further guarantees safety by monitoring sensor data for leaks, sudden changes in pressure, or temperature rises in real time. Modern AI incorporated into systems can bypass human command and autonomously initiate default safety measures such as component isolation or system shutdowns in case set thresholds for safety are breached. The dispensing unit where hydrogen is injected into the vehicle's tank is called the refueling dispenser, consisting of the nozzle and interface. Ultimately, the hydrogen vehicle is the vehicle that is filled with hydrogen fuel. Figure 9.10 depicts the components of a hydrogen refueling station with AI-based controls.

AI can support demand-side management strategies where hydrogen-fueled devices and systems can autonomously modulate their use of energy in real time relative to the

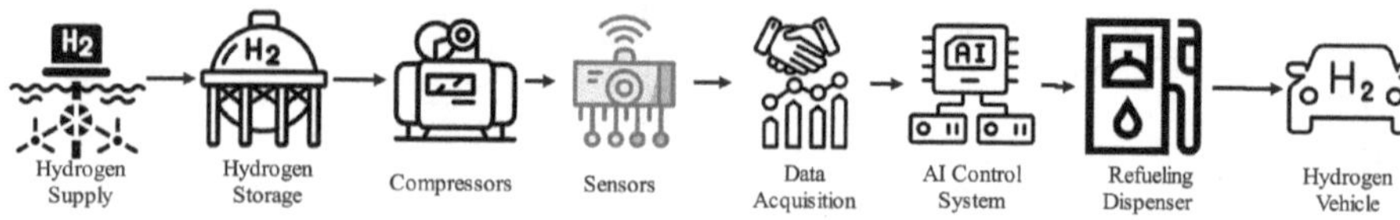

FIGURE 9.10 Hydrogen refueling station with AI-based controls.

status of the power grid. This principle assists with equilibrating supply with demand, alleviating excessive investments in the grid's infrastructure while yielding a more agile and robust energy system. AI also helps in the adoption of green hydrogen into the smart grid, which fosters a comprehensive sustainable energy solution at a low cost and with low emissions. In the end, AI can strengthen the reliability of the grid by preventing and mitigating malicious cyber intrusions and other risks that utilize the hydrogen infrastructure as an operational target while distorting the energy system's functions. The potential of AI can be fully harnessed to optimally control the hydrogen systems in the smart grids, which enables the transition to the hydrogen economy. AI grants it the capability to function as an environmentally benign and RE carrier.

The diagram in Figure 9.11 illustrates the integration of hydrogen systems into smart grids through the central role of AI. The AI-powered smart grid management system is positioned as the core element, emphasizing AI's integrative function by connecting and coordinating all aspects of the system. This centralized placement highlights AI's crucial role in overseeing the flow of both electricity and hydrogen within the smart grid system. The diagram also ensures a clearer flow of energy and hydrogen. Continuous arrows directly depict the movement of electricity from sources such as RE to consumption points and hydrogen flow from production to utilization. The AI system's central position implies the flow of data to and from it, indicating that AI receives information for analysis and sends control signals to optimize the system's operation, which is indicated by dashed arrows.

The diagram ensures no ambiguity in the flow of energy and hydrogen. Electricity from renewable sources to consumption and the transfer of hydrogen from production to use is illustrated with continuous arrows showing its direction. The position of AI implies the movement of data within the system. Therefore, AI takes in information for processing and sends commands to make the system work optimally.

Ramasubramanian et al. [60] studied the possibilities of developing sustainable fuel cells and batteries using AI. The research showed the role of AI in material development and device performance optimization for energy storage systems. The authors analyzed many sustainable electrochemical energy storage methodologies, including green materials and processes, renewables, circular economy approaches, regulatory policies, and even AI-based methodologies [60]. The study also illustrated a case of very high throughput property optimization of sustainable carbon materials for electrochemical energy system devices with ML.

In Peksen [61], the authors focused on using AI to enhance the fuel efficiency of hydrogen fuel cell electric vehicles. The study performed a multiphysics model-based simulation of the vehicle operating in different driving scenarios. The simulation results were used to train the ML model to estimate fuel consumption for different

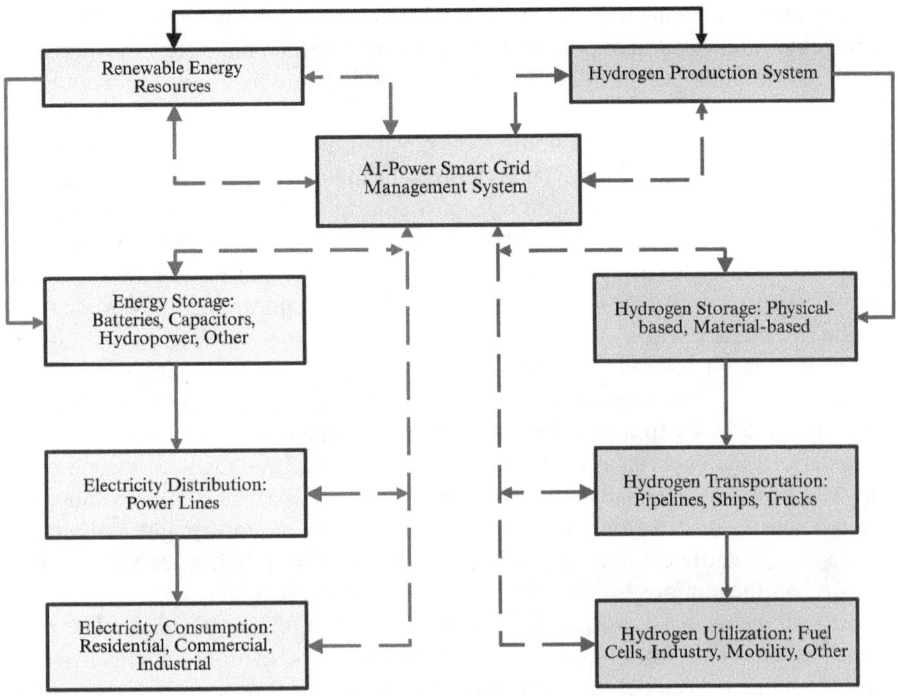

FIGURE 9.11 AI-enhanced integration of hydrogen systems in smart grids.

operating conditions. The refined ML model predicted when fuel consumption would peak with an error of less than 1%.

The idea of having digital counterparts of physical systems known as digital twins is becoming popular in the hydrogen industry. Digital twins are an excellent tool to visualize, synthesize, understand, and improve all the relevant steps in the hydrogen supply chain, from production, storage, and transportation to its end use. By devising a virtual model of real hydrogen infrastructure, stakeholders can observe, estimate, and influence the system's performance within safe limits. This effect is vital for boosting the hydrogen sector's safety, efficiency, and innovation.

One notable example of using digital twins along the hydrogen supply chain is improving hydrogen production processes. Digital twins assist in recognizing the most efficient and economically favorable processes by simulating different production scenarios and evaluating their productivity, cost, emission, and overall impact. This capability is essential for cost-effective hydrogen production while protecting the environment.

9.5 AI IN HYDROGEN ECONOMY

Developing a robust hydrogen economy requires sound policy and economic frameworks. AI can contribute to these areas by analyzing complex datasets related to energy markets, infrastructure costs, and environmental impacts. This analysis can inform policy decisions, helping to create incentives for the adoption of hydrogen technologies and to address potential economic and social challenges. Furthermore, AI can aid in modeling the economic viability of different hydrogen production and distribution scenarios, providing valuable insights for investors and policymakers.

One of the most significant hurdles in transitioning to a hydrogen economy lies in effectively integrating hydrogen systems with the existing energy infrastructure. This is where AI can play a crucial role, particularly within the framework of smart grids. Smart grids, essentially modernized electrical grids, leverage digital technology to enhance electricity delivery efficiency, reliability, and sustainability. AI has the potential to amplify these capabilities significantly, optimizing the way we manage and distribute energy.

Hou et al. [62] explored using AI to advance global economic policies and hydrogen production in the transition to a low-carbon economy. The study developed a comprehensive index to assess hydrogen export competitiveness globally, incorporating various factors such as economic viability, political and regulatory conditions, industrial experience, and resource availability. AI played a crucial role in data analysis and index construction, enabling accurate evaluation and weight assignment. The findings identified countries with competitive hydrogen industries and provided policy recommendations for enhancing hydrogen production and export.

The integration of DL, ML, and AI with the Internet of Things (IoT) holds transformative potential for both supply chain management and hydrogen energy systems, offering enhanced model development, interoperability, accelerated innovation, increased trust and transparency, and efficient deployment. Standardized datasets, protocols, and data formats are crucial for training accurate AI models, enabling seamless data exchange among stakeholders, fostering collaboration, and accelerating innovation [63]. While challenges such as computational resource demands, system integration, data quality issues, and the need for specialized talent exist, the benefits include improved operational efficiency, optimized energy use, and enhanced decision-making capabilities. Moreover, standardization promotes transparency and trust in AI models, which are essential for building public confidence in these technologies [64–66]. It facilitates smooth integration into existing infrastructures, driving the growth of the hydrogen economy [67].

The effective integration of hydrogen into the energy system relies heavily on accurate demand forecasting and comprehensive market analysis, both of which can be significantly enhanced by AI. AI's ability to analyze diverse data sources, including energy consumption patterns, economic trends, and policy developments, enables precise predictions of future hydrogen demand. For instance, AI systems can estimate hydrogen consumption by considering historical energy data, economic activities, and policy changes, facilitating optimized planning for production, storage, and distribution infrastructure, as demonstrated by Nikseresht's [68] hybrid time series model. Moreover, AI can identify untapped market potentials for hydrogen-based technologies by processing market data, aiding informed investments, and fostering the hydrogen economy, as evidenced by the fuzzy decision-making model used in [69] to assess green hydrogen generation. AI also plays a crucial role in optimizing hydrogen supply pricing by leveraging market data to formulate effective pricing strategies, influencing economic and environmental outcomes, as highlighted in [70] regarding the impact of carbon credits and varying green hydrogen supply rates. Furthermore, AI can evaluate the influence of policies on the hydrogen market, inform policy development, and provide real-time awareness of market developments, enabling strategic decision-making and investment planning to drive the transition towards a hydrogen economy.

9.6 AI IN THE SAFETY SUPERVISION OF HYDROGEN

Safety stands as the paramount concern within the hydrogen industry, necessitating robust monitoring and control systems. In this critical area, AI-supported technologies play a pivotal role in risk management. Modern AI sensors shift the paradigm from reactive to preventive measures, enabling the prediction of issues such as gas leaks, pressure fluctuations, and temperature anomalies. This proactive approach is essential for building public trust in hydrogen technologies.

Like any critical infrastructure, hydrogen facilities, encompassing production plants, storage, and transportation networks, are susceptible to safety and reliability concerns. Therefore, predictive maintenance becomes indispensable. AI technologies facilitate this by analyzing failure trends derived from sensor data. These patterns enable precise maintenance scheduling, prevent system outages, and ensure the continuous operation of the hydrogen infrastructure. Implementing AI in hydrogen supply chain predictive maintenance streamlines processes, eliminates costly downtimes, and ensures the proper functioning of infrastructure, accelerating the transition to a hydrogen economy.

Safety remains non-negotiable within the hydrogen industry, particularly given the highly flammable nature of hydrogen gas. AI significantly bolsters safety by integrating data from diverse sources, including sensors, historical records, and weather data, to identify and mitigate potential hazards. AI's capacity to process vast datasets enables accurate risk assessment, preventing accidents and safeguarding hydrogen transport and processing networks. These preventive measures are crucial for fostering public confidence and trust in hydrogen technologies.

To further enhance safety, AI enables effective monitoring, maintenance, and risk assessment of hydrogen storage and transport. AI-based sensors and monitoring systems actively supervise infrastructure for leaks, pressure changes, and other anomalies, providing early warnings. Predictive maintenance, powered by AI, anticipates system failures, while risk assessment identifies and mitigates critical threats.

AI contributes to hydrogen safety through several key functions:

- **Leak Detection:** AI analyzes sensor data to detect potential leaks in storage tanks, pipelines, and other infrastructure, preventing incidents.
- **Pressure and Temperature Supervision:** AI monitors real-time pressure and temperature, identifying risks such as overpressure and overheating, enabling timely corrective actions [71].
- **Predictive Maintenance:** AI forecasts maintenance needs by analyzing sensor data, preventing accidents, and ensuring safe operation [72].
- **Risk Evaluation:** AI assesses risks across hydrogen manufacturing, storage, transportation, and utilization, enabling the development of safety

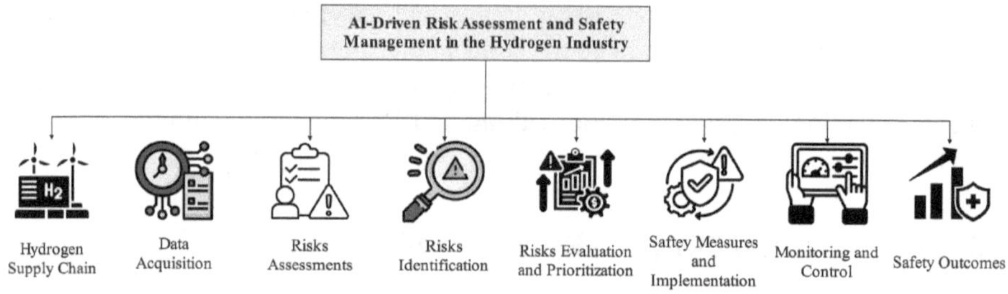

FIGURE 9.12 A proactive and data-driven approach to risk assessment and safety management in the hydrogen industry.

measures [73]. A risk assessment approach was developed using the Tuna Swarm Optimization algorithm and least squares SVM model.

Finally, AI facilitates strategically coordinated control during critical situations, minimizing damage and casualties. AI enables immediate information sharing and real-time action recommendations. Lyu et al. [74] developed the Intelligent Technology Empowered Active Emergency Command Strategy (ITAECS) for smart cities, focusing on hazardous chemical disasters, including hydrogen, through intelligent perception, IoT systems, disaster posture deduction, virtual reality emergency rescue, and immersive command platforms.

As a general overview, risk evaluation, and safety management in the hydrogen sector can be improved based on the efficiency of AI technologies. Different aspects of hydrogen safety to which AI technology contributes are illustrated in Figure 9.12. AI-driven systems can detect risks and evaluate them in terms of magnitude and severity by integrating and analyzing data from a variety of sensors as well as historical data and meteorological information AI identifies anomalies such as gas leaks, pressure changes, temperature changes, and even equipment failure and as a result, enables pro-active safety measures such as maintenance and emergency protocols AI monitored actions of this type markedly improves safety results and reduces accidents while increasing the integrity of hydrogen infrastructure.

9.7 AI AND HYDROGEN SUSTAINABILITY

The development of a hydrogen economy necessitates a strong focus on environmental sustainability. AI plays a crucial role in sustainability and lifecycle assessment by analyzing the ecological impacts of hydrogen production, storage, transportation, and utilization. AI-powered devices can assess energy consumption, emissions, and resource extraction by implementing AI, enabling comprehensive evaluations of the effectiveness of hydrogen technology. Ensuring sustainability across all aspects is paramount to preventing significant environmental harm from the hydrogen economy.

When considering AI's application in the hydrogen economy, socio-cultural and ethical issues must be addressed. The transition raises concerns about data privacy, algorithmic bias, and job displacement, which require careful resolution for a successful transition. Building trust within communities and ensuring equitable resource allocation can be achieved through bottom-up approaches encouraging participation from stakeholders, including communities, workers, and consumers.

Public perception poses a significant hurdle to the implementation of hydrogen technologies. AI can transform this through educational gaming, simulations, and tailored information campaigns that highlight the benefits and safety of hydrogen energy. Intelligent chatbots and virtual assistants can facilitate the promotion, questioning, and discussion of hydrogen technologies, fostering understanding and building trust in hydrogen energy within society. This approach encourages proactive engagement with hydrogen energy.

For instance, AI-powered interactive simulations can educate the public about the diverse applications of hydrogen energy and its safety features. AI-powered chatbots can be integrated into websites and social media platforms to provide instant, 24/7 responses and address concerns or misconceptions about hydrogen technologies. Interest- and concern-based messaging can be distributed through lower-tier AI systems on social media, simplifying the implementation of information campaigns.

Utilizing AI technologies for public engagement and education significantly benefits the hydrogen industry. Public trust in hydrogen technologies directly influences the pace of the transition to a hydrogen economy.

9.8 CHAPTER SUMMARY

This chapter has looked at both the novel and existing intuitive applications of AI within the hydrogen industry. The potential of AI in infusing innovation, enhancing safety, and optimizing processes throughout the hydrogen supply chain has been discussed comprehensively. AI has been demonstrated to optimize the utilization of RE, electrolyzer performance and efficiency, and new materials discovery in AI-supported hydrogen production. AI also significantly improves hydrogen storage and transportation by optimizing suitable storage material, conditions, and tank safety. In the scope of hydrogen utilization, AI has also been shown to improve fuel cell performance and durability, automate hydrogen refueling stations, and assist in developing hydrogen-based energy management systems. The chapter also analyzed the role of AI in assisting the formulation of policies, modeling economic feasibility, and safety aspects to aid in the development of a robust hydrogen economy. Further, the chapter analyzed the impact of AI on hydrogen sustainability and lifecycle assessment from a sociocultural and ethical standpoint. In overcoming challenges and fostering innovation, AI has the potential to impact the shift toward a hydrogen-powered energy future profoundly.

SELF-ASSESSMENT QUESTIONS

1. How can AI be used to optimize the use of renewable energy sources in hydrogen production?
2. What are some examples of AI algorithms that can be used to improve the performance of electrolyzers?

3. How can AI be used to accelerate the discovery of new materials for hydrogen storage?
4. What are some challenges in storing hydrogen efficiently and safely?
5. How can AI be used to optimize the conditions under which hydrogen is stored?
6. What are some benefits of using AI in hydrogen transportation?
7. How can AI be used to improve the performance and durability of fuel cells?
8. What are some of the ways AI can be used to optimize hydrogen refueling stations?
9. How can AI support the development of hydrogen-based energy management systems?
10. What are some of the economic and social challenges that need to be addressed in the transition to a hydrogen economy?
11. How can AI be used to inform policy decisions and ensure safety in the hydrogen industry?
12. What are some of the sociocultural and ethical considerations that need to be addressed in the development of a hydrogen economy?
13. How can AI be used to educate the public about the benefits and safety of hydrogen energy?
14. What are some key challenges and opportunities in integrating AI into the hydrogen sector?
15. How can AI contribute to developing a sustainable and equitable hydrogen economy?
16. An alkaline electrolyzer's efficiency, characterized by the hydrogen production rate (H, in Nm^3/h), is a function of operating voltage (V, in Volts) and temperature (T, in °C). Experimental data reveals the following discrete relationships: (V, T, H): (1.9 V, 65°C, 4.8 Nm^3/h), (2.1 V, 75°C, 5.9 Nm^3/h), (2.3 V, 85°C, 7.0 Nm^3/h). Employ a neural network regression model to predict H at V = 2.0 V and T = 70°C and subsequently optimize V and T to maximize H.
17. Predict the pressure (P, in bar) within a salt cavern used for large-scale hydrogen storage as a function of injection time (t, in days), hydrogen injection rate (Q, in tons/day), initial cavern pressure (P_0, in bar), cavern volume (V_c, in m^3), and the geological permeability of the salt rock (k, in m^2).

    Given:

    $P_0 = 110$ bar (initial pressure).

    $Q = 15$ tons/day (constant injection rate).

    $V_c = 450{,}000\,m^3$ (cavern volume).

    $k = 1.2 \times 10^{-19}\,m^2$ (salt rock permeability).

    Historical pressure data: at t = 6 days, P = 130 bar, and at t = 12 days, P = 155 bar. Predict the cavern pressure (P) at t = 18 days and assess potential risks associated with prolonged injection.
18. A vehicle with a hydrogen fuel cell requires real-time optimization of its operating parameters to maximize efficiency and power output. Use AI to predict and optimize fuel cell efficiency (η) and power output (P).

Given that:
The current state of the fuel cell is:
Load current (I): 45 A
Operating temperature (T): 65°C
Hydrogen flow rate (Q_{H_2}): 18 L/min
Air flow rate (Qair): 95 L/min
Fuel cell efficiency (η): 50%
Power output (P): 2.8 kW

The AI prediction shows that if Q_{H_2} is increased to 22 L/min and Qair is increased to 110 L/min, while I and T remain constant, the predicted fuel cell efficiency (η) will be 58%, and the predicted power output (P) will be 3.3 kW. The AI control system must decide whether to change the flow rates.

REFERENCES

1. Ju, Hyung Kuk, Donghyun Yoon, Sungyool Bong, et al. 2024. "Challenge and opportunity in scaling-up hydrogen production via electrochemical ammonia electrolysis process." *Current Opinion in Electrochemistry* 49: 101609. https://doi.org/10.1016/j.coelec.2024.101609.
2. Boretti, Alberto. 2024. "Artificial intelligence driven evolution of perovskite-based solar thermochemical systems for hydrogen production: A narrative review." *Emergent Materials* 1–15. https://doi.org/10.1007/s42247-024-00823-w.
3. Mani, Preeyanghaa, Sulakshana Shenoy, Prince J. J. Sagayaraj, et al. 2025. "Scaling up of photocatalytic systems for large-scale hydrogen generation." *Applied Physics Reviews* 12 (1). https://doi.org/10.1063/5.0223598.
4. Yao, Zhenpeng, Yanwei Lum, Andrew Johnston, et al. 2023. "Machine learning for a sustainable energy future." *Nature Reviews Materials* 8: 202–215. https://doi.org/10.1038/s41578-022-00490-5.
5. Alhussan, Amel Ali, El-Sayed M. El-Kenawy, Mohammed A. Saeed, et al. 2023. "Green hydrogen production ensemble forecasting based on hybrid dynamic optimization algorithm." *Frontiers in Energy Research* 11: 1221006. https://doi.org/10.3389/fenrg.2023.1221006.
6. Kilic, Heybet. 2024. "Improving the performance of microgrid-based Power-to-X systems through optimization of renewable hydrogen generation." *International Journal of Hydrogen Energy* 75: 106–120. https://doi.org/10.1016/j.ijhydene.2024.01.066.
7. Aghamolaei, Reihaneh. 2024. "Recent advancements in applying machine learning in power-to-X processes: A literature eeview." *Sustainability* 16: 9555. https://doi.org/10.3390/su16219555.
8. Mahmoudi, Mehrdad, Ahad Ghaemi, Ahamd Rahbar Kelishami, et al. 2024. "Predictive modeling of membrane reactor efficiency using advanced artificial neural networks for green hydrogen production." *Scientific Reports* 14: 24211. https://doi.org/10.1038/s41598-024-75068-y.
9. Mohamed, Amira, Hatem Ibrahem, Rui Yang, et al. 2022. "Optimization of proton exchange membrane electrolyzer cell design using machine learning" *Energies* 15 (18): 6657. https://doi.org/10.3390/en15186657.
10. Mavrogiorgos, Konstantinos, Athanasios Kiourtis, Argyro Mavrogiorgou, et al. 2024. "Bias in machine learning: A literature review" *Applied Sciences* 14 (19): 8860. https://doi.org/10.3390/app14198860.
11. Mert, İlker. 2021. "Agnostic deep neural network approach to the estimation of hydrogen production for solar-powered systems." *International Journal of Hydrogen Energy* 46 (9): 6272–6285. https://doi.org/10.1016/j.ijhydene.2020.11.161.

12. Egeland-Eriksen, Torbjørn, Amin Hajizadeh, and Sabrina Sartori. 2021. "Hydrogen-based systems for integrating renewable energy in power systems: Achievements and perspectives." *International Journal of Hydrogen Energy* 46 (63): 31963–31983. https://doi.org/10.1016/j.ijhydene.2021.06.218.
13. Rezk, Hegazy, Abdul Ghani Olabi, Mohammad Ali Abdelkareem, et al. 2023. "Maximizing green hydrogen production from water electrocatalysis: Modeling and optimization." *Journal of Marine Science and Engineering* 11 (3): 617. https://doi.org/10.3390/jmse11030617.
14. Abdelkareem, Mohammad A., Bassel Soudan, Mohamed S. Mahmoud, et al. 2022. "Progress of artificial neural networks applications in hydrogen production." *Chemical Engineering Research and Design* 182: 66–86. https://doi.org/10.1016/j.cherd.2022.03.030.
15. Sirat, Abdullah, Sher Ahmad, Iftikhar Ahmad, et al. 2024. "Integrative CFD and AI/ML-based modeling for enhanced alkaline water electrolysis cell performance for hydrogen production." *International Journal of Hydrogen Energy* 83: 1120–1131. https://doi.org/10.1016/j.ijhydene.2024.08.184.
16. Ahmed, Rania, Sara A. Shehab, Omar M. Elzeki, et al. 2024. "An explainable AI for green hydrogen production: A deep learning regression model." *International Journal of Hydrogen Energy* 83: 1226–1242. https://doi.org/10.1016/j.ijhydene.2024.08.064.
17. Nwankwo, Samuel, Omolayo M. Ikumapayi, Imhade P. Okokpujie, et al. 2024. "Application and optimization of IoT in storage devices for hydrogen energy." In *International Conference on Science, Engineering and Business for Driving Sustainable Development Goals (SEB4SDG)*, Omu-Aran, Nigeria, pp. 1–7. IEEE. https://doi.org/10.1109/SEB4SDG60871.2024.10629891.
18. Wang, Guoqing, Zongmin Luo, Halefom G. Desta, et al. 2024. "AI-driven development of high-performance solid-state hydrogen storage." *Energy Reviews* 4 (1): 100106. https://doi.org/10.1016/j.enrev.2024.100106.
19. Rahnama, Alireza, Guilherme Zepon, and Seetharaman Sridhar. 2019. "Machine learning based prediction of metal hydrides for hydrogen storage, part I: Prediction of hydrogen weight percent." *International Journal of Hydrogen Energy* 44 (14): 7337–7344. https://doi.org/10.1016/j.ijhydene.2019.01.261.
20. Nachtane, Mourad, M. El Fallaki Idrissi, Mostapha Tarfaoui, et al. 2024. "Deep learning-driven predictive tools for damage prediction and optimization in composite hydrogen storage tanks." *Composites Communications* 51: 102079. https://doi.org/10.1016/j.coco.2024.102079.
21. Thanh, Hung Vo, Sajad Ebrahimnia Taremsari, Benyamin Ranjbar, et al. 2023. "Hydrogen storage on porous carbon adsorbents: Rediscovery by nature-derived algorithms in random forest machine learning model." *Energies* 16 (5): 2348. https://doi.org/10.3390/en16052348.
22. Lan, Penghang, She Chen, Qihang Li, et al. 2024. "Intelligent hydrogen-ammonia combined energy storage system with deep reinforcement learning." *Renewable Energy* 237: 121725. https://doi.org/10.1016/j.renene.2024.121725.
23. Chi, Guidong, Shuang Xu, Dehai Yu, et al. 2024. "A brief review of structural health monitoring based on flexible sensing technology for hydrogen storage tank." *International Journal of Hydrogen Energy* 80: 980–998. https://doi.org/10.1016/j.ijhydene.2024.07.215.
24. Nachtane, Mourad, Mostapha Tarfaoui, Mohamed Amine Abichou, et al. 2023. "An overview of the recent advances in composite materials and artificial intelligence for hydrogen storage vessels design." *Journal of Composites Science* 7 (3): 119. https://doi.org/10.3390/jcs7030119.
25. Ma, Kun, Lingyu Xu, Azher M. Abed, et al. 2023. "An artificial intelligence approach study for assessing hydrogen energy materials for energy saving in building." *Sustainable Energy Technologies and Assessments* 56: 103052. https://doi.org/10.1016/j.seta.2023.103052.

26. Huang, Song-Jeng, Matoke Peter Mose, and Sathiyalingam Kannaiyan. 2021. "Artificial intelligence application in solid state Mg-based hydrogen energy storage." *Journal of Composites Science* 5 (6): 145. https://doi.org/10.3390/jcs5060145.
27. Qarssis, Youssef, Mourad Nachtane, Ayoub Karine, et al. 2024. "Machine learning-based analytical approach for mechanical analysis of composite hydrogen storage tanks under internal pressure." *International Journal of Hydrogen Energy* 89: 1440–1453. https://doi.org/10.1016/j.ijhydene.2024.09.404.
28. Baldwin, Donald. 2017. "Development of High Pressure Hydrogen Storage Tank for Storage and Gaseous Truck Delivery." No. DOE-HEXAGON-GO18062. Hexagon Lincoln LLC, Lincoln, NE. https://doi.org/10.2172/1373926.
29. Mikroni, Maria, Grigorios Koutsoukis, Dimitrios Vlachos, et al. 2024. "Design, analysis, and testing of a type V composite pressure vessel for hydrogen storage." *Polymers* 16 (24): 3576. https://doi.org/10.3390/polym16243576.
30. An, Na Yeon, Jung Hyun Yang, Eunyong Song, et al. 2024. "Digital twin-based hydrogen refueling station (HRS) safety model: CNN-based decision-making and 3D simulation." *Sustainability* 16 (21): 9482. https://doi.org/10.3390/su16219482.
31. Ramadan, Montaser, Mohammed Ali, Shin Yee Khoo, et al. 2024. "Real-time IoT-powered AI system for monitoring and forecasting of air pollution in industrial environment." *Ecotoxicology and Environmental Safety* 283: 116856. https://doi.org/10.1016/j.ecoenv.2024.116856.
32. Maleki, Akbar and Alireza Askarzadeh. 2014. "Comparative study of artificial intelligence techniques for sizing of a hydrogen-based stand-alone photovoltaic/wind hybrid system." *international Journal of Hydrogen Energy* 39 (19): 9973–9984. https://doi.org/10.1016/j.ijhydene.2014.04.147.
33. Nwulu, Emmanuella Onyinye, Tari Yvonne Elete, Ovie Vincent Erhueh, et al. 2023. "Machine learning applications in predictive maintenance: Enhancing efficiency across the oil and gas industry." *International Journal of Engineering Research Updates* 5 (1): 17–30. https://doi.org/10.53430/ijeru.2023.5.1.0017.
34. Stoicescu, Alexandru-Adrian, Razvan George Ripeanu, Maria Tănase, et al. 2025. "Current methods and technologies for storage tank condition assessment: A comprehensive review." *Materials* 18 (5): 1074. https://doi.org/10.3390/ma18051074.
35. Rahman, Anas A., Haoren Wang, Bo Wang, et al. 2024. "Prediction of pressure evolution in non-venting self-pressurized liquid hydrogen tanks using artificial neural network approach." *International Journal of Hydrogen Energy* 68: 428–439. https://doi.org/10.1016/j.ijhydene.2024.04.261.
36. Klepp, Georg. 2024. "Modelling activated carbon hydrogen storage tanks using machine learning models." *Energy* 306: 132318. https://doi.org/10.1016/j.energy.2024.132318.
37. Rehman, Ali, Bo Zhang, Amjad Riaz, et al. 2024. "Artificial intelligence-based surrogate modeling for computational cost-effective optimization of hydrogen liquefaction process." *International Journal of Hydrogen Energy*. https://doi.org/10.1016/j.ijhydene.2024.04.331.
38. Ghorbani, Bahram, Sohrab Zendehboudi, Zahra Alizadeh Afrouzi, et al. 2024. "Efficient design of the hydrogen liquefaction system: Thermodynamic, economic, environmental, and uncertainty perspectives." *Industrial & Engineering Chemistry Research* 63 (33): 14668–14699. https://doi.org/10.1021/acs.iecr.3c04286.
39. Correa-Jullian, Camila and Katrina M. Groth. 2022. "Opportunities and data requirements for data-driven prognostics and health management in liquid hydrogen storage systems." *International Journal of Hydrogen Energy* 47 (43): 18748–18762. https://doi.org/10.1016/j.ijhydene.2022.04.048.
40. Wu, Yujian, Gang Yang, Jiangang Sun, et al. 2024. "Leakage monitoring and diagnosis of LNG storage tanks with temperature sensing network integration and artificial intelligence algorithm." *Measurement Science and Technology* 35 (5): 055113. https://dx.doi.org/10.1088/1361-6501/ad25e2.

41. Osman, Ahmed I., Mahmoud Nasr, Abdelazeem S. Eltaweil, et al. 2024. "Advances in hydrogen storage materials: harnessing innovative technology, from machine learning to computational chemistry, for energy storage solutions." *International Journal of Hydrogen Energy*. https://doi.org/10.1016/j.ijhydene.2024.03.223.
42. Yang, Xi, Yuting Li, Yitao Liu, et al. 2024. "Crystal structure prediction and performance assessment of hydrogen storage materials: Insights from computational materials science." *Energies* 17 (14): 3591. https://doi.org/10.3390/en17143591.
43. Huang, Guochang, Fuqiang Huang, and Wujie Dong. 2024. "Machine learning in energy storage material discovery and performance prediction." *Chemical Engineering Journal* 492: 152294. https://doi.org/10.1016/j.cej.2024.152294.
44. Delpisheh, Mostafa, Benyamin Ebrahimpour, Abolfazl Fattahi, et al. 2024. "Leveraging machine learning in porous media." *Journal of Materials Chemistry A*. https://doi.org/10.1039/D4TA00251B.
45. Banerjee, Tanumoy, Kevin Ji, Weiyi Xia, et al. 2025. "Machine Learning Driven Search of Hydrogen Storage Materials." arXiv preprint arXiv:2503.04027. https://doi.org/10.48550/arXiv.2503.04027.
46. Rahnama, Alireza, Guilherme Zepon, and Seetharaman Sridhar. 2019. "Machine learning based prediction of metal hydrides for hydrogen storage, part II: Prediction of material class." *International Journal of Hydrogen Energy* 44 (14): 7345–7353. https://doi.org/10.1016/j.ijhydene.2019.01.264.
47. Hai, Tao, Farhan A. Alenizi, Adil Hussein Mohammed, et al. 2023. "Machine learning-aided modeling of the hydrogen storage in zeolite-based porous media." *International Communications in Heat and Mass Transfer* 145: 106848. https://doi.org/10.1016/j.icheatmasstransfer.2023.106848.
48. Davoodi, Shadfar, Hung Vo Thanh, David A. Wood, et al. 2023. "Machine-learning models to predict hydrogen uptake of porous carbon materials from influential variables." *Separation and Purification Technology* 316: 123807. https://doi.org/10.1016/j.seppur.2023.123807.
49. Peng, Chuchuan, Xianyang Liu, Zhichun Liu, et al. 2024. "Performance optimization of adsorption hydrogen storage system via computation fluid dynamics and machine learning." *Chemical Engineering Research and Design* 207: 100–109. https://doi.org/10.1016/j.cherd.2024.05.022.
50. Zhang, Xuan, Qing-Rong Zheng, and Hong-Zhou He. 2022. "Machine-learning-based prediction of hydrogen adsorption capacity at varied temperatures and pressures for MOFs adsorbents." *Journal of the Taiwan Institute of Chemical Engineers* 138: 104479. https://doi.org/10.1016/j.jtice.2022.104479.
51. Heravi, Mohammad Javad Tavakkoli, Elham Yasari, et al. 2022. "Data-driven modelling and optimization of hydrogen adsorption on carbon nanostructures." *International Journal of Hydrogen Energy* 47 (61): 25704–25723. https://doi.org/10.1016/j.ijhydene.2022.05.301.
52. Kumar, Abhijit, Saurabh Tiwari, Nandlal Gupta, et al. 2024. "Machine learning modelling and optimization for metal hydride hydrogen storage systems." *Sustainable Energy & Fuels* 8 (9): 2073–2086. https://doi.org/10.1039/D4SE00031E.
53. Nassabeh, Mehdi, Zhenjiang You, Alireza Keshavarz, et al. 2025. "Data-driven strategy for contact angle prediction in underground hydrogen storage using machine learning." *Journal of Energy Storage* 114: 115908. https://doi.org/10.1016/j.est.2025.115908.
54. Han, Zhilei, Zeeshan Tariq, and Bicheng Yan. 2024. "A Novel Robust Optimization Framework Based on Surrogate Modeling for Underground Hydrogen Storage in Depleted Natural Gas Reservoirs." Available at SSRN 5065309. https://dx.doi.org/10.2139/ssrn.5065309.
55. Saeed, Motaz, and Prashant Jadhawar. 2024. "Modelling underground hydrogen storage: A state-of-the-art review of fundamental approaches and findings." *Gas Science and Engineering* 121: 205196. https://doi.org/10.1016/j.jgsce.2023.205196.

56. Longe, Promise O., Shadfar Davoodi, Mohammad Mehrad, et al. 2024. "Combined deep learning and optimization for hydrogen-solubility prediction in aqueous systems appropriate for underground hydrogen storage reservoirs." *Energy & Fuels* 38 (22): 22031–22049. https://doi.org/10.1021/acs.energyfuels.4c03376.
57. Youn, Ye Rim and Jinkeun Hong. 2024. "Optimization of model based on relu activation function in MLP neural network model." *International Journal of Advanced Smart Convergence* 13 (2): 80–87. https://doi.org/10.7236/IJASC.2024.13.2.80.
58. Saco, Arun, P. Shanmuga Sundari, and Anand Paul. 2022. "An optimized data analysis on a real-time application of PEM fuel cell design by using machine learning algorithms." *Algorithms* 15 (10): 346. https://doi.org/10.3390/a15100346.
59. Bicer, Yusuf, Ibrahim Dincer, and M. U. R. A. T. Aydin. 2016. "Maximizing performance of fuel cell using artificial neural network approach for smart grid applications." *Energy* 116: 1205–1217. https://doi.org/10.1016/j.energy.2016.10.050.
60. Ramasubramanian, Brindha, Rayavarapu Prasada Rao, Vijila Chellappan, et al. 2022. "Towards sustainable fuel cells and batteries with an AI perspective." *Sustainability* 14 (23): 16001. https://doi.org/10.3390/su142316001.
61. Peksen, Murphy M. 2022. "Artificial intelligence-based machine learning toward the solution of climate-friendly hydrogen fuel cell electric vehicles." *Vehicles* 4 (3): 663–680. https://doi.org/10.3390/vehicles4030038.
62. Hou, Wenjia, Xingchen Pan, Yang Pu, Rui Ma, et al. 2024. "Harnessing artificial intelligence for energy strategies: Advancing global economic policies and hydrogen production in the transition to a low-carbon economy." *Energy Strategy Reviews* 56: 101568. https://doi.org/10.1016/j.esr.2024.101568.
63. Khedr, Ahmed M. 2024. "Enhancing supply chain management with deep learning and machine learning techniques: A review." *Journal of Open Innovation: Technology, Market, and Complexity* 10: 100379. https://doi.org/10.1016/j.joitmc.2024.100379.
64. Joshua, Salaki Reynaldo, Sanguk Park, and Kihyeon Kwon. 2024. "Knowledge-based modeling approach: A schematic design of artificial intelligence of things (AIoT) for hydrogen energy system." In *2024 IEEE 14th Annual Computing and Communication Workshop and Conference (CCWC)*, Las Vegas, NV, pp. 0235–0241. IEEE. https://doi.org/10.1109/CCWC60891.2024.10427681.
65. Olawale, Mokanmiyo Adedeji, Ayomide Abdulbaqi Ayeh, Faruk Obasanjo Adekola, et al. 2023. "A review on the intersection of artificial intelligence on building resilient infrastructure, promoting inclusive and sustainable industrialization and fostering innovation." *International Journal of Engineering and Modern Technology* 9 (3): 1–31. https://doi.org/10.56201/ijemt.v9.no3.2023.pg44.74.
66. Arunthavanathana, Rajeevan, Zaman Sajida, and Md Tanjin Amina. 2024. "Ethics in AI for energy systems safety." In *Method of Process Systems in Energy Systems: Current System Part* I, vol. 8, p. 81. Elsevier. https://doi.org/10.1016/bs.mcps.2024.08.002
67. Jayabal, Ravikumar. 2025. "Hydrogen energy storage with artificial intelligent-powered strategies for a sustainable future: A review." *Journal of Mechanical Science and Technology* 39: 1–8. https://doi.org/10.1007/s12206-025-0240-3.
68. Nikseresht, Ali. 2025. "Dynamic hydrogen demand forecasting using hybrid time series models: Insights for renewable energy systems." *Renewable Energy* 244: 122737. https://doi.org/10.1016/j.renene.2025.122737.
69. Yüksel, Serhat, Hasan Dinçer, Merve Acar, et al. 2024. "Financial multidimensional assessment of a green hydrogen generation process via an integrated artificial intelligence-based four-stage fuzzy decision-making model." *International Journal of Hydrogen Energy* 83: 577–588. https://doi.org/10.1016/j.ijhydene.2024.08.140.
70. Song, Geunseo, Pouya Ifaei, Jiwoo Ha, et al. 2024. "The AI circular hydrogen economist: Hydrogen supply chain design via hierarchical deep multi-agent reinforcement learning." *Chemical Engineering Journal* 497: 154464. https://doi.org/10.1016/j.cej.2024.154464.

71. Alarcon, Claudio, Sofia Alarcon, Alvaro Hoffer, et al. 2025. "Hydrogen 5.0: Interdisciplinary development of a proof-of-concept smart system for green hydrogen leak detection." *Processes* 13 (3): 639. https://doi.org/10.3390/pr13030639.
72. García, Jesús Enrique Sierra, Mario Peñacoba Yagüe, and Pedro Jesús Cabrera Santana. 2025 "XIX Simposio CEA de Control Inteligente: libro de actas." In *XIX Simposio CEA de Control Inteligente: libro de actas*. Servicio de Publicaciones e Imagen Institucional. https://doi.org/10.36443/9788418465970.
73. Liu, Jicheng, Yanan Song, and Xue Yu. 2024. "Risk assessment study of hydrogen energy storage system based on KPCA-TSO-LSSVM." *International Journal of Hydrogen Energy* 79: 931–942. https://doi.org/10.1016/j.ijhydene.2024.07.070.
74. Lyu, Jieyin, Shouqin Zhou, Jingang Liu, et al. 2023. "Intelligent-technology-empowered active emergency command strategy for urban hazardous chemical disaster management." *Sustainability* 15 (19): 14369. https://doi.org/10.3390/su151914369.

Index

Note: **Bold** page numbers refer to tables and *italic* page numbers refer to figures.

FSC
www.fsc.org
MIX
Papier | Fördert
gute Waldnutzung
FSC® C083411

Kolibri 360 GmbH
Ferdinand-Jühlke-Straße 7
99095 Erfurt, Deutschland
produktsicherheit@kolibri360.de